BUSINESS AND FINANCIAL

PLANNING TABLES

DESK BOOK

by

The IBP Research and Editorial Staff

Institute for Business Planning, Inc.

IBP PLAZA, ENGLEWOOD CLIFFS, N.J. 07632

This publication is designed to provide accurate and authoritative information in regard to the subject matter covered. It is sold with the understanding that the publisher is not engaged in rendering legal, accounting or other professional service. If legal advice or other expert assistance is required, the services of a competent professional person should be sought.

—From a Declaration of Principles jointly adopted by a Committee of the American Bar Association and a Committee of Publishers and Associations.

How to Use This Desk Book

Here, in one handy reference volume, you will find over 100 up-to-date tables for use in your business or profession. These tables are arranged under three general headings: *Business Planning Aids, Tax Planning Aids,* and *Estate Planning Aids.* The section under the first heading, *Business Planning Aids,* begins with the seven basic interest tables. These are followed by tax tables that show the tax effect on interest and other accumulations. Then come several useful real estate tables. Finally, there are numerous tables covering the current state laws on various financial transactions, as well as an industry-by-industry review of key business ratios. Also included under this heading are tables dealing with stocks, covering commission rates and price-earning ratio projections.

Under the second heading, *Tax Planning Aids,* are current tables for state and federal taxes, as well as tables showing the tax effect of gifts, capital gains, dividends, and transfers of income-producing property. You will find apportionment tables for state income taxes in addition to other state tax tables. Finally, there are tables covering all aspects of tax depreciation and tables for the minimum and maximum taxes, as well as for investment credit.

Under the third heading, *Estate Planning Aids,* are mortality and expectancy tables, estate and gift tax tables, and valuation tables for estate tax purposes. You will find up-to-date tables on Social Security, Medicare, and insurance costs, as well as tables covering the expected returns on numerous annuity contracts. You will also find tables and information on all the computations under the estate tax, including state-by-state coverage of executors' fees and commissions.

Of course, many of the tables listed under one heading are useful in other areas as well. However, your book will be most helpful if you familiarize yourself with the tables under each general heading and look under the proper heading first. You should be able to find everything you need in this way. And the more familiar you become with the various tables, the more useful they will be to you.

Finally, you will find the Index a valuable aid in locating a table for a particular purpose, such as discounting or compounding depreciation, if you cannot quickly find it under one of the general headings. All the tables are cross-referenced in the Index under their various uses.

It is hoped that this book will become one of your most useful and handy references. To this end, IBP welcomes suggestions on these and any other tables that you would consider useful in later editions. Please direct any suggestions or inquiries to:

IBP Research and Editorial Staff
IBP Plaza
Englewood Cliffs, N.J. 07632

iii

TABLE OF CONTENTS

Section A—Business Planning Aids

Section B—Tax Planning Aids

Section C—Estate Planning Aids

SECTION A
BUSINESS PLANNING AIDS

[¶101]

Simple Interest Table

Example of use of this table:

Find amount of $500 in 8 years at 6% simple interest.

From table at 8 years and 6% for $1	1.48
Value in 8 years for $500 (500 × 1.48)	$740

Interest Rate

Number of Years	3%	3½%	4%	4½%	5%	6%	7%	8%
1	1.03	1.035	1.04	1.045	1.05	1.06	1.07	1.08
2	1.06	1.070	1.08	1.090	1.10	1.12	1.14	1.16
3	1.09	1.105	1.12	1.135	1.15	1.18	1.21	1.24
4	1.12	1.140	1.16	1.180	1.20	1.24	1.28	1.32
5	1.15	1.175	1.20	1.225	1.25	1.30	1.35	1.40
6	1.18	1.210	1.24	1.270	1.30	1.36	1.42	1.48
7	1.21	1.245	1.28	1.315	1.35	1.42	1.49	1.56
8	1.24	1.280	1.32	1.360	1.40	1.48	1.56	1.64
9	1.27	1.315	1.36	1.405	1.45	1.54	1.63	1.72
10	1.30	1.350	1.40	1.450	1.50	1.60	1.70	1.80
11	1.33	1.385	1.44	1.495	1.55	1.66	1.77	1.88
12	1.36	1.420	1.48	1.540	1.60	1.72	1.84	1.96
13	1.39	1.455	1.52	1.585	1.65	1.78	1.91	2.04
14	1.42	1.490	1.56	1.630	1.70	1.84	1.98	2.12
15	1.45	1.525	1.60	1.675	1.75	1.90	2.05	2.20
16	1.48	1.560	1.64	1.720	1.80	1.96	2.12	2.28
17	1.51	1.595	1.68	1.765	1.85	2.02	2.19	2.36
18	1.54	1.630	1.72	1.810	1.90	2.08	2.26	2.44
19	1.57	1.665	1.76	1.855	1.95	2.14	2.33	2.52
20	1.60	1.700	1.80	1.900	2.00	2.20	2.40	2.60
21	1.63	1.735	1.84	1.945	2.05	2.26	2.47	2.68
22	1.66	1.770	1.88	1.990	2.10	2.32	2.54	2.76
23	1.69	1.805	1.92	2.035	2.15	2.38	2.61	2.84
24	1.72	1.840	1.96	2.080	2.20	2.44	2.68	2.92
25	1.75	1.875	2.00	2.125	2.25	2.50	2.75	3.00
26	1.78	1.910	2.04	2.170	2.30	2.56	2.82	3.08
27	1.81	1.945	2.08	2.215	2.35	2.62	2.89	3.16
28	1.84	1.980	2.12	2.260	2.40	2.68	2.96	3.24
29	1.87	2.015	2.16	2.305	2.45	2.74	3.03	3.32
30	1.90	2.050	2.20	2.350	2.50	2.80	3.10	3.40
31	1.93	2.085	2.24	2.395	2.55	2.86	3.17	3.48
32	1.96	2.120	2.28	2.440	2.60	2.92	3.24	3.56
33	1.99	2.155	2.32	2.485	2.65	2.98	3.31	3.64
34	2.02	2.190	2.36	2.530	2.70	3.04	3.38	3.72
35	2.05	2.225	2.40	2.575	2.75	3.10	3.45	3.80
36	2.08	2.260	2.44	2.620	2.80	3.16	3.52	3.88
37	2.11	2.295	2.48	2.665	2.85	3.22	3.59	3.96
38	2.14	2.330	2.52	2.710	2.90	3.28	3.66	4.04
39	2.17	2.365	2.56	2.755	2.95	3.34	3.73	4.12
40	2.21	2.400	2.60	2.800	3.00	3.40	3.80	4.20

Simple Interest Table *(continued)*

Interest Rate

Number of Years	9%	10%	11%	12%	13%	14%	15%	20%
1	1.09	1.10	1.11	1.12	1.13	1.14	1.15	1.20
2	1.18	1.20	1.22	1.24	1.26	1.28	1.30	1.40
3	1.27	1.30	1.33	1.36	1.39	1.42	1.45	1.60
4	1.36	1.40	1.44	1.48	1.52	1.56	1.60	1.80
5	1.45	1.50	1.55	1.60	1.65	1.70	1.75	2.00
6	1.54	1.60	1.66	1.72	1.78	1.84	1.90	2.20
7	1.63	1.70	1.77	1.84	1.91	1.98	2.05	2.40
8	1.72	1.80	1.88	1.96	2.04	2.12	2.20	2.60
9	1.81	1.90	1.99	2.08	2.17	2.26	2.35	2.80
10	1.90	2.00	2.10	2.20	2.30	2.40	2.50	3.00
11	1.99	2.10	2.21	2.32	2.43	2.54	2.65	3.20
12	2.08	2.20	2.32	2.44	2.56	2.68	2.80	3.40
13	2.17	2.30	2.43	2.56	2.69	2.82	2.95	3.60
14	2.26	2.40	2.54	2.68	2.82	2.96	3.10	3.80
15	2.35	2.50	2.65	2.80	2.95	3.10	3.25	4.00
16	2.44	2.60	2.76	2.92	3.08	3.24	3.40	4.20
17	2.53	2.70	2.87	3.04	3.21	3.38	3.55	4.40
18	2.62	2.80	2.98	3.16	3.34	3.52	3.70	4.60
19	2.71	2.90	3.09	3.28	3.47	3.66	3.85	4.80
20	2.80	3.00	3.20	3.40	3.60	3.80	4.00	5.00
21	2.89	3.10	3.31	3.52	3.73	3.94	4.15	5.20
22	2.98	3.20	3.42	3.64	3.86	4.08	4.30	5.40
23	3.07	3.30	3.53	3.76	3.99	4.22	4.45	5.60
24	3.16	3.40	3.64	3.88	4.12	4.36	4.60	5.80
25	3.25	3.50	3.75	4.00	4.25	4.50	4.75	6.00
26	3.34	3.60	3.86	4.12	4.38	4.64	4.90	6.20
27	3.43	3.70	3.97	4.24	4.51	4.78	5.05	6.40
28	3.52	3.80	4.08	4.36	4.64	4.92	5.20	6.60
29	3.61	3.90	4.19	4.48	4.77	5.06	5.35	6.80
30	3.70	4.00	4.30	4.60	4.90	5.20	5.50	7.00
31	3.79	4.10	4.41	4.72	5.03	5.34	5.65	7.20
32	3.88	4.20	4.52	4.84	5.16	5.48	5.80	7.40
33	3.97	4.30	4.63	4.96	5.29	5.62	5.95	7.60
34	4.06	4.40	4.74	5.08	5.42	5.76	6.10	7.80
35	4.15	4.50	4.85	5.20	5.55	5.90	6.25	8.00
36	4.24	4.60	4.96	5.32	5.68	6.04	6.40	8.20
37	4.33	4.70	5.07	5.44	5.81	6.18	6.55	8.40
38	4.42	4.80	5.18	5.56	5.94	6.32	6.70	8.60
39	4.51	4.90	5.29	5.68	6.07	6.46	6.85	8.80
40	4.60	5.00	5.40	5.80	6.20	6.60	7.00	9.00

¶101

[¶102]

Compound Interest Table

Example of use of this table:

Find how much $1,000 now in bank will grow to in 14 years at 4% interest.

From table 14 years at 4%   1.7317

Value in 14 years of $1,000 $1,731.70

Interest Rate

Number of Years	3%	3½%	4%	4½%	5%	6%	7%	8%
1	1.0300	1.0350	1.0400	1.0450	1.0500	1.0600	1.0700	1.0800
2	1.0609	1.0712	1.0816	1.0920	1.1025	1.1236	1.1449	1.1664
3	1.0927	1.1087	1.1249	1.1412	1.1576	1.1910	1.2250	1.2597
4	1.1255	1.1475	1.1699	1.1925	1.2155	1.2624	1.3107	1.3604
5	1.1593	1.1877	1.2167	1.2462	1.2763	1.3332	1.4025	1.4693
6	1.1941	1.2293	1.2653	1.3023	1.3401	1.4135	1.5007	1.5868
7	1.2299	1.2723	1.3159	1.3609	1.4071	1.5030	1.6057	1.7138
8	1.2668	1.3168	1.3686	1.4221	1.4775	1.5938	1.7181	1.8509
9	1.3048	1.3629	1.4233	1.4861	1.5513	1.6894	1.8384	1.9990
10	1.3439	1.4106	1.4802	1.5530	1.6289	1.7908	1.9671	2.1589
11	1.3842	1.4600	1.5395	1.6229	1.7103	1.8982	2.1048	2.3316
12	1.4258	1.5111	1.6010	1.6959	1.7959	2.0121	2.2521	2.5181
13	1.4685	1.5640	1.6651	1.7722	1.8856	2.1329	2.4098	2.7196
14	1.5126	1.6187	1.7317	1.8519	1.9799	2.2609	2.5785	2.9371
15	1.5580	1.6753	1.8009	1.9353	2.0789	2.3965	2.7590	3.1721
16	1.6047	1.7340	1.8730	2.0224	2.1829	2.5403	2.9521	3.4259
17	1.6528	1.7947	1.9479	2.1134	2.2920	2.6927	3.1588	3.7000
18	1.7024	1.8575	2.0258	2.2085	2.4066	2.8543	3.3799	3.9960
19	1.7535	1.9225	2.1068	2.3079	2.5270	3.0255	3.6165	4.3157
20	1.8061	1.9898	2.1911	2.4117	2.6533	3.2075	3.8696	4.6609
21	1.8603	2.0594	2.2788	2.5202	2.7860	3.3995	4.1405	5.0338
22	1.9161	2.1315	2.3699	2.6337	2.9253	3.6035	4.4304	5.4365
23	1.9736	2.2061	2.4647	2.7522	3.0715	3.8197	4.7405	5.8714
24	2.0328	2.2833	2.5633	2.8760	3.2251	4.0489	5.0723	6.3411
25	2.0938	2.3632	2.6658	3.0054	3.3864	4.2918	5.4274	6.8484
26	2.1566	2.4460	2.7725	3.1407	3.5557	4.5493	5.8073	7.3963
27	2.2213	2.5316	2.8834	3.2820	3.7335	4.8223	6.2138	7.9880
28	2.2879	2.6202	2.9987	3.4297	3.9201	5.1116	6.6488	8.6271
29	2.3566	2.7119	3.1187	3.5840	4.1161	5.4183	7.1142	9.3172
30	2.4273	2.8068	3.2434	3.7453	4.3219	5.7434	7.6122	10.5582
31	2.5001	2.9050	3.3731	3.9139	4.5380	6.0881	8.1451	10.8676
32	2.5751	3.0067	3.5081	4.0900	4.7649	6.4533	8.7152	11.7370
33	2.6523	3.1119	3.6484	4.2740	5.0032	6.8408	9.3253	12.6760
34	2.7319	3.2209	3.7943	4.4664	5.2533	7.2510	9.9781	13.6901
35	2.8139	3.3336	3.9461	4.6673	5.5160	7.6860	10.6765	14.7853
36	2.8983	3.4503	4.1039	4.8774	5.7918	8.1479	11.4239	15.9681
37	2.9852	3.5710	4.2681	5.0969	6.0814	8.6360	12.2236	17.2456
38	3.0748	3.6960	4.4388	5.3262	6.3855	9.1542	13.0792	18.6252
39	3.1670	3.8254	4.6164	5.5659	6.7048	9.7035	13.9948	20.1152
40	3.2620	3.9593	4.8010	5.8164	7.0400	10.2857	14.9744	21.7245

Compound Interest Table *(continued)*

Interest Rate

Number of Years	9%	10%	11%	12%	13%	14%	15%	20%
1	1.0900	1.1000	1.1100	1.1200	1.1300	1.1400	1.1500	1.2000
2	1.1881	1.2100	1.2321	1.2544	1.2769	1.2996	1.3225	1.4400
3	1.2950	1.3310	1.3576	1.4049	1.4428	1.4815	1.5208	1.7280
4	1.4115	1.4647	1.5180	1.5735	1.6304	1.6389	1.7490	2.0736
5	1.5386	1.6105	1.6350	1.7623	1.8424	1.9254	2.0113	2.4883
6	1.6771	1.7715	1.8704	1.9738	2.0819	2.1949	2.3130	2.9859
7	1.8230	1.9487	2.0761	2.2106	2.3526	2.5022	2.6600	3.5831
8	1.9925	2.1435	2.3045	2.4759	2.6584	2.8525	3.0590	4.2998
9	2.1718	2.3579	2.5580	2.7730	3.0040	3.2519	3.5178	5.1597
10	2.3673	2.5937	2.8394	3.1058	3.3945	3.7072	4.0455	6.1917
11	2.5804	2.8531	3.1517	3.4785	3.8358	4.2262	4.6523	7.4300
12	2.8126	3.1384	3.4984	3.8959	4.3345	4.8179	5.3502	8.9161
13	3.0658	3.4522	3.8832	4.3634	4.8980	5.4924	6.1527	10.6993
14	3.3417	3.7974	4.3104	4.8871	5.5347	6.2613	7.0757	12.8391
15	3.6424	4.1772	4.7845	5.4735	6.2542	7.1379	8.1370	15.4070
16	3.9703	4.5949	5.3108	6.1303	7.0673	8.1372	9.3576	18.4884
17	4.3276	5.0544	5.8950	6.8660	7.9860	9.2764	10.7612	22.1861
18	4.7171	5.5599	6.5435	7.6899	9.0242	10.5751	12.3754	26.6233
19	5.1416	6.1159	7.2633	8.6127	10.1974	12.0556	14.2317	31.9479
20	5.6044	6.7274	8.0623	9.6462	11.5230	13.7434	16.3665	38.3375
21	6.1088	7.4002	8.9491	10.8038	13.0210	15.6675	18.8215	46.0051
22	6.6586	8.1402	9.9335	12.1003	14.7138	17.8610	21.6447	55.2061
23	7.2578	8.9543	11.0262	13.5523	16.6266	20.3615	24.8914	66.2473
24	7.9110	9.8497	12.2391	15.1786	18.7880	23.2122	28.6251	79.4968
25	8.6230	10.8347	13.5854	17.0000	21.2305	26.4619	32.9189	95.3962
26	9.3991	11.9181	15.0793	19.0400	23.9905	30.1665	37.8567	114.4754
27	10.2450	13.1099	16.7386	21.3248	27.1092	34.3899	43.5353	137.3705
28	11.1671	14.4209	18.5799	23.8838	30.6334	39.2044	50.0656	164.8446
29	12.1721	15.8630	20.6236	26.7499	34.6158	44.6931	57.5754	197.8135
30	13.2676	17.4494	22.8922	29.9599	39.1158	50.9501	66.2117	237.3763
31	14.4617	19.1943	25.4104	33.5551	44.2009	58.0831	76.1435	284.8515
32	15.7633	21.1137	28.2055	37.5817	49.9470	66.2148	87.5650	341.8218
33	17.1820	23.2251	31.3082	42.0915	56.4402	75.4849	100.6998	410.1862
34	18.7284	25.5476	34.7521	47.1425	63.7774	86.0527	115.8048	492.2235
35	20.4139	28.1024	38.5748	52.7996	72.0685	98.1001	133.1755	590.6682
36	22.2512	30.9128	42.8180	59.1355	81.4374	111.8342	153.1518	708.8018
37	24.2538	34.0039	47.5280	66.2318	92.0242	127.4909	176.1246	850.5622
38	26.4366	37.4048	52.7561	74.1796	103.9874	145.3397	202.5433	1020.6746
39	28.8159	41.1447	58.5593	83.0812	117.5057	165.6872	232.9248	1224.8096
40	31.4094	45.2592	65.0008	93.0509	132.7815	188.8835	267.8635	1469.7715

¶102

[¶103]

Periodic Deposit Table

Example of use of this table:

How much is $1,000 a year invested at 5% worth in 20 years?

At 5% for 20 years, the figure is   34.719

For $1,000 a year, the amount is $34,719

Interest Rate

Number of Years	3%	3½%	4%	4½%	5%	6%	7%	8%
1	1.030	1.035	1.040	1.045	1.050	1.060	1.070	1.080
2	2.091	2.106	2.122	2.137	2.153	2.183	2.215	2.246
3	3.184	3.215	3.246	3.278	3.310	3.375	3.440	3.506
4	4.309	4.362	4.416	4.471	4.526	4.637	4.751	4.867
5	5.468	5.550	5.633	5.717	5.802	5.975	6.153	6.336
6	6.662	6.779	6.898	7.019	7.142	7.394	7.654	7.923
7	7.892	8.052	8.214	8.380	8.549	8.897	9.260	9.637
8	9.159	9.368	9.583	9.802	10.027	10.491	10.978	11.488
9	10.464	10.731	11.006	11.288	11.578	12.181	12.816	13.487
10	11.808	12.142	12.486	12.841	13.207	13.972	14.784	15.645
11	13.192	13.602	14.026	14.464	14.917	15.870	16.888	17.977
12	14.618	15.113	15.627	16.160	16.713	17.882	19.141	20.495
13	16.086	16.677	17.292	17.932	18.599	20.015	21.550	23.215
14	17.599	18.296	19.024	19.784	20.579	22.276	24.129	26.152
15	19.157	19.971	20.825	21.719	22.657	24.673	26.888	29.324
16	20.762	21.705	22.698	23.742	24.840	27.213	29.840	32.750
17	22.414	23.500	24.645	25.855	27.132	29.906	32.999	36.450
18	24.117	25.357	26.671	28.064	29.539	32.760	36.379	40.446
19	25.870	27.280	28.778	30.371	32.066	35.786	39.995	44.762
20	27.676	29.269	30.969	32.783	34.719	38.993	43.865	49.423
21	29.537	31.329	33.248	35.303	37.505	42.392	48.006	54.457
22	31.453	33.460	35.618	37.937	40.430	45.996	52.436	59.893
23	33.426	35.667	38.083	40.689	43.502	49.816	57.177	65.765
24	35.459	37.950	40.646	43.565	46.727	53.865	62.249	72.106
25	37.553	40.313	43.312	46.571	50.113	58.156	67.676	78.954
26	39.710	42.759	46.084	49.711	53.669	62.706	73.484	86.351
27	41.931	45.291	48.968	52.993	57.403	67.528	79.698	94.339
28	44.219	47.911	51.966	56.423	61.323	72.640	86.347	102.966
29	46.575	50.623	55.085	60.007	65.439	78.058	93.461	112.283
30	49.003	53.429	58.328	63.752	69.761	83.802	101.073	122.346
31	51.503	56.335	61.701	67.666	74.299	89.890	109.218	133.214
32	54.078	59.341	65.210	71.756	79.064	96.343	117.933	144.951
33	56.730	62.453	68.858	76.030	84.067	103.184	127.259	157.627
34	59.462	65.674	72.652	80.497	89.320	110.435	137.237	171.317
35	62.276	69.008	76.598	85.164	94.336	118.121	147.913	186.102
36	65.174	72.458	80.702	90.041	100.628	126.268	159.337	202.070
37	68.159	76.029	84.970	95.138	106.710	134.904	171.561	219.316
38	71.234	79.725	89.409	100.464	113.095	144.058	184.640	237.941
39	74.401	83.550	94.026	106.030	119.800	153.762	198.635	258.057
40	77.663	87.510	98.827	111.847	126.840	164.048	213.610	279.781

Periodic Deposit Table *(continued)*

Interest Rate

Number of Years	9%	10%	11%	12%	13%	14%	15%
1	1.090	1.100	1.110	1.120	1.130	1.140	1.150
2	2.278	2.310	2.342	2.374	2.407	2.440	2.473
3	3.573	3.641	3.710	3.779	3.850	3.921	3.993
4	4.985	5.105	5.228	5.353	5.480	5.610	5.742
5	6.523	6.716	6.913	7.115	7.323	7.536	7.754
6	8.200	8.487	8.783	9.089	9.405	9.730	10.067
7	10.028	10.436	10.859	11.300	11.757	12.233	12.727
8	12.021	12.579	13.164	13.776	14.416	15.085	15.768
9	14.193	14.937	15.722	16.549	17.420	18.337	19.304
10	16.560	17.531	18.561	19.655	20.814	22.045	23.349
11	19.141	20.384	21.713	23.133	24.650	26.271	28.002
12	21.953	23.523	25.212	27.029	28.985	31.089	33.352
13	25.019	26.975	29.095	31.393	33.883	36.581	39.505
14	28.361	30.772	33.405	36.280	39.417	42.842	46.580
15	32.003	34.950	38.190	41.753	45.672	49.980	54.717
16	35.974	39.545	43.501	47.884	52.739	58.118	64.075
17	40.301	44.599	49.396	54.750	60.725	67.394	74.836
18	45.018	50.159	55.939	62.440	69.749	77.969	87.212
19	50.160	56.275	63.203	71.052	79.947	90.025	101.444
20	55.765	63.002	71.265	80.699	91.470	103.768	117.810
21	61.873	70.403	80.214	91.503	104.491	119.436	136.632
22	68.532	78.543	90.148	103.603	119.205	137.297	158.276
23	75.790	87.497	101.174	117.155	135.831	157.659	183.168
24	83.701	97.347	113.413	132.334	154.620	180.871	211.793
25	92.324	108.182	126.999	149.334	175.850	207.333	244.712
26	101.723	120.100	142.079	168.374	199.841	237.499	282.569
27	111.968	133.210	158.817	189.699	226.950	271.889	326.104
28	123.135	147.631	177.397	213.583	257.583	311.094	376.170
29	135.308	163.494	198.021	240.333	292.199	355.787	433.745
30	148.575	180.943	220.913	270.293	331.315	406.737	499.957
31	163.037	200.138	246.324	303.848	375.516	464.820	576.100
32	178.800	221.252	274.529	341.429	425.463	531.035	663.666
33	195.982	244.477	305.837	383.521	481.903	606.520	764.365
34	214.711	270.024	340.590	430.663	545.681	692.573	880.170
35	235.125	298.127	379.164	483.463	617.749	790.673	1013.346
36	257.376	329.039	421.982	542.599	699.187	902.507	1166.498
37	281.630	363.043	469.511	608.831	791.211	1029.998	1342.622
38	308.066	400.448	522.267	683.010	895.198	1175.338	1545.165
39	336.882	441.593	580.826	766.091	1012.704	1341.025	1778.090
40	368.292	486.852	645.827	859.142	1145.486	1529.909	2045.954

¶103

[¶104]

Annuity Table

Example of use of this table:

If I put $1,000 in the bank, how much can I take out each year for 5 years to use up the entire sum if the interest rate is 3%?

From the table at 5 years	0.2184
For $1,000 (1000 x 0.2184)	$218.40

Note: Not necessarily representative of annuity payable under an annuity or insurance contract.

Interest Rate

Number of Years	3%	3½%	4%	4½%	5%	6%	7%	8%
1	1.0300	1.0350	1.0400	1.0450	1.0500	1.0600	1.0700	1.0800
2	0.5226	0.5264	0.5302	0.5340	0.5378	0.5454	0.5501	0.5608
3	0.3535	0.3569	0.3603	0.3638	0.3672	0.3741	0.3811	0.3880
4	0.2690	0.2723	0.2755	0.2787	0.2820	0.2886	0.2952	0.3019
5	0.2184	0.2215	0.2246	0.2278	0.2310	0.2374	0.2439	0.2505
6	0.1846	0.1877	0.1908	0.1939	0.1970	0.2034	0.2098	0.2163
7	0.1605	0.1635	0.1666	0.1697	0.1728	0.1791	0.1856	0.1921
8	0.1425	0.1455	0.1485	0.1516	0.1547	0.1610	0.1675	0.1740
9	0.1284	0.1314	0.1345	0.1376	0.1407	0.1470	0.1535	0.1601
10	0.1172	0.1202	0.1233	0.1264	0.1295	0.1359	0.1424	0.1490
11	0.1081	0.1111	0.1141	0.1172	0.1204	0.1268	0.1334	0.1401
12	0.1005	0.1035	0.1066	0.1097	0.1128	0.1193	0.1259	0.1327
13	0.0940	0.0971	0.1001	0.1033	0.1065	0.1129	0.1197	0.1265
14	0.0885	0.0916	0.0947	0.0978	0.1010	0.1076	0.1143	0.1213
15	0.0838	0.0868	0.0899	0.0931	0.0963	0.1029	0.1098	0.1168
16	0.0796	0.0827	0.0858	0.0890	0.0923	0.0989	0.1059	0.1129
17	0.0760	0.0790	0.0822	0.0854	0.0887	0.0954	0.1024	0.1096
18	0.0727	0.0758	0.0790	0.0822	0.0855	0.0924	0.0994	0.1067
19	0.0698	0.0729	0.0761	0.0794	0.0827	0.0896	0.0968	0.1041
20	0.0672	0.0704	0.0736	0.0769	0.0802	0.0872	0.0944	0.1019
21	0.0649	0.0680	0.0713	0.0746	0.0780	0.0850	0.0923	0.0998
22	0.0627	0.0659	0.0692	0.0725	0.0760	0.0837	0.0904	0.0980
23	0.0608	0.0640	0.0673	0.0707	0.0741	0.0813	0.0887	0.0964
24	0.0590	0.0623	0.0656	0.0690	0.0725	0.0797	0.0872	0.0949
25	0.0574	0.0607	0.0640	0.0674	0.0710	0.0782	0.0858	0.0937
26	0.0559	0.0592	0.0626	0.0660	0.0696	0.0769	0.0846	0.0925
27	0.0546	0.0579	0.0612	0.0647	0.0683	0.0757	0.0834	0.0914
28	0.0533	0.0566	0.0600	0.0635	0.0671	0.0746	0.0824	0.0905
29	0.0521	0.0554	0.0589	0.0624	0.0660	0.0736	0.0814	0.0896
30	0.0510	0.0544	0.0578	0.0614	0.0651	0.0726	0.0806	0.0888
31	0.0500	0.0534	0.0569	0.0604	0.0641	0.0718	0.0798	0.0881
32	0.0490	0.0524	0.0559	0.0596	0.0633	0.0710	0.0791	0.0875
33	0.0482	0.0516	0.0551	0.0587	0.0625	0.0703	0.0784	0.0869
34	0.0473	0.0508	0.0543	0.0580	0.0618	0.0696	0.0778	0.0863
35	0.0465	0.0500	0.0536	0.0573	0.0611	0.0686	0.0772	0.0858
36	0.0458	0.0493	0.0529	0.0566	0.0604	0.0684	0.0767	0.0853
37	0.0451	0.0486	0.0522	0.0560	0.0598	0.0679	0.0762	0.0849
38	0.0445	0.0480	0.0516	0.0554	0.0593	0.0674	0.0758	0.0845
39	0.0438	0.0474	0.0511	0.0549	0.0588	0.0669	0.0754	0.0842
40	0.0433	0.0468	0.0505	0.0543	0.0583	0.0665	0.0750	0.0839

Annuity Table (continued)

Interest Rate

Number of Years	9%	10%	11%	12%	13%	14%	15%
1	1.0900	1.1000	1.1100	1.1200	1.1300	1.1400	1.1500
2	0.5685	0.5762	0.5839	0.5917	0.5995	0.6073	0.6151
3	0.3951	0.4021	0.4092	0.4163	0.4235	0.4307	0.4379
4	0.3087	0.3155	0.3223	0.3292	0.3362	0.3432	0.3503
5	0.2571	0.2638	0.2706	0.2774	0.2843	0.2913	0.2983
6	0.2229	0.2296	0.2364	0.2432	0.2502	0.2572	0.2642
7	0.1987	0.2054	0.2121	0.2191	0.2261	0.2332	0.2404
8	0.1807	0.1874	0.1943	0.2013	0.2084	0.2156	0.2229
9	0.1668	0.1736	0.1806	0.1877	0.1949	0.2022	0.2096
10	0.1558	0.1627	0.1698	0.1769	0.1843	0.1917	0.1993
11	0.1469	0.1539	0.1611	0.1684	0.1758	0.1834	0.1911
12	0.1397	0.1468	0.1540	0.1614	0.1689	0.1767	0.1845
13	0.1336	0.1408	0.1482	0.1556	0.1634	0.1712	0.1791
14	0.1284	0.1357	0.1432	0.1509	0.1587	0.1666	0.1747
15	0.1241	0.1315	0.1391	0.1468	0.1547	0.1628	0.1710
16	0.1203	0.1278	0.1355	0.1434	0.1514	0.1596	0.1679
17	0.1170	0.1247	0.1325	0.1404	0.1486	0.1569	0.1654
18	0.1142	0.1219	0.1298	0.1379	0.1462	0.1546	0.1632
19	0.1117	0.1195	0.1276	0.1358	0.1441	0.1527	0.1613
20	0.1095	0.1175	0.1256	0.1339	0.1424	0.1509	0.1598
21	0.1076	0.1156	0.1238	0.1322	0.1408	0.1495	0.1584
22	0.1059	0.1140	0.1223	0.1308	0.1395	0.1483	0.1573
23	0.1044	0.1126	0.1209	0.1296	0.1383	0.1472	0.1563
24	0.1030	0.1113	0.1198	0.1285	0.1373	0.1463	0.1554
25	0.1018	0.1102	0.1187	0.1275	0.1364	0.1455	0.1547
26	0.1007	0.1092	0.1178	0.1267	0.1357	0.1448	0.1541
27	0.0997	0.1083	0.1169	0.1259	0.1349	0.1442	0.1535
28	0.0989	0.1075	0.1163	0.1252	0.1344	0.1437	0.1531
29	0.0981	0.1067	0.1156	0.1247	0.1338	0.1432	0.1527
30	0.0973	0.1061	0.1150	0.1241	0.1334	0.1428	0.1523
31	0.0967	0.1055	0.1145	0.1237	0.1330	0.1425	0.1519
32	0.0961	0.1049	0.1140	0.1233	0.1327	0.1421	0.1517
33	0.0956	0.1045	0.1136	0.1229	0.1323	0.1419	0.1515
34	0.0951	0.1041	0.1133	0.1226	0.1321	0.1416	0.1513
35	0.0946	0.1037	0.1129	0.1223	0.1318	0.1414	0.1511
36	0.0942	0.1033	0.1126	0.1221	0.1316	0.1413	0.1509
37	0.0939	0.1030	0.1124	0.1218	0.1314	0.1411	0.1508
38	0.0935	0.1027	0.1121	0.1216	0.1313	0.1409	0.1507
39	0.0932	0.1025	0.1119	0.1215	0.1311	0.1409	0.1506
40	0.0929	0.1023	0.1117	0.1213	0.1309	0.1407	0.1506

[¶105]

Present Worth Table—Single Future Payment

Example of use of this table:
Find how much $10,000 payable in 12 years is worth now at an interest rate of 4%.
From table for 12 years 4%   .6246
Present value of $10,000 in 12 years (10,000 x .6246) $6,246

Interest Rate

Number of Years	3%	3½%	4%	4½%	5%	6%	7%	8%
1	0.9709	0.9662	0.9615	0.9569	0.9524	0.9434	0.9346	0.9259
2	0.9426	0.9335	0.9246	0.9157	0.9070	0.8900	0.8734	0.8573
3	0.9151	0.9019	0.8890	0.8763	0.8638	0.8396	0.8163	0.7938
4	0.8885	0.8714	0.8548	0.8386	0.8227	0.7921	0.7629	0.7350
5	0.8626	0.8420	0.8219	0.8025	0.7835	0.7473	0.7130	0.6806
6	0.8375	0.8135	0.7903	0.7679	0.7462	0.7050	0.6663	0.6302
7	0.8131	0.7860	0.7599	0.7348	0.7107	0.6651	0.6227	0.5835
8	0.7894	0.7594	0.7307	0.7032	0.6768	0.6274	0.5820	0.5403
9	0.7664	0.7337	0.7026	0.6729	0.6446	0.5919	0.5439	0.5002
10	0.7441	0.7089	0.6756	0.6439	0.6139	0.5584	0.5083	0.4632
11	0.7224	0.6849	0.6496	0.6162	0.5847	0.5268	0.4751	0.4289
12	0.7014	0.6618	0.6246	0.5897	0.5568	0.4970	0.4440	0.3971
13	0.6810	0.6394	0.6006	0.5643	0.5303	0.4688	0.4150	0.3677
14	0.6611	0.6178	0.5775	0.5400	0.5051	0.4423	0.3878	0.3405
15	0.6419	0.5969	0.5553	0.5167	0.4810	0.4173	0.3624	0.3152
16	0.6232	0.5767	0.5339	0.4945	0.4581	0.3936	0.3387	0.2919
17	0.6050	0.5572	0.5134	0.4732	0.4363	0.3714	0.3166	0.2703
18	0.5874	0.5384	0.4936	0.4528	0.4155	0.3503	0.2959	0.2502
19	0.5703	0.5202	0.4746	0.4333	0.3957	0.3305	0.2765	0.2317
20	0.5537	0.5026	0.4564	0.4146	0.3769	0.3118	0.2584	0.2145
21	0.5375	0.4856	0.4388	0.3968	0.3589	0.2942	0.2415	0.1987
22	0.5219	0.4692	0.4220	0.3797	0.3418	0.2775	0.2257	0.1839
23	0.5067	0.4533	0.4057	0.3634	0.3256	0.2618	0.2109	0.1703
24	0.4919	0.4380	0.3901	0.3477	0.3101	0.2470	0.1971	0.1577
25	0.4776	0.4231	0.3751	0.3327	0.2953	0.2330	0.1842	0.1460
26	0.4637	0.4088	0.3607	0.3184	0.2812	0.2198	0.1722	0.1352
27	0.4502	0.3950	0.3468	0.3047	0.2678	0.2074	0.1609	0.1252
28	0.4371	0.3817	0.3335	0.2916	0.2551	0.1956	0.1504	0.1159
29	0.4243	0.3687	0.3207	0.2790	0.2429	0.1846	0.1406	0.1073
30	0.4120	0.3563	0.3083	0.2670	0.2314	0.1741	0.1314	0.0994
31	0.4000	0.3442	0.2965	0.2555	0.2204	0.1643	0.1228	0.0920
32	0.3883	0.3326	0.2851	0.2445	0.2099	0.1550	0.1147	0.0852
33	0.3770	0.3213	0.2741	0.2340	0.1999	0.1462	0.1072	0.0789
34	0.3660	0.3105	0.2636	0.2239	0.1904	0.1379	0.1002	0.0730
35	0.3554	0.3000	0.2534	0.2143	0.1813	0.1301	0.0937	0.0676
36	0.3450	0.2898	0.2437	0.2050	0.1727	0.1227	0.0875	0.0626
37	0.3350	0.2800	0.2343	0.1962	0.1644	0.1158	0.0818	0.0580
38	0.3252	0.2706	0.2253	0.1877	0.1566	0.1092	0.0765	0.0536
39	0.3158	0.2614	0.2166	0.1797	0.1491	0.1031	0.0715	0.0497
40	0.3066	0.2526	0.2083	0.1719	0.1420	0.0972	0.0668	0.0460

Present Worth Table—Single Future Payment *(continued)*

Interest Rate

Number of Years	9%	10%	11%	12%	13%	14%	15%
1	0.9174	0.9091	0.9009	0.8929	0.8850	0.8772	0.8696
2	0.8417	0.8264	0.8116	0.7972	0.7831	0.7695	0.7561
3	0.7722	0.7513	0.7312	0.7118	0.6931	0.6750	0.6575
4	0.7084	0.6830	0.6587	0.6355	0.6133	0.5921	0.5718
5	0.6499	0.6209	0.5935	0.5674	0.5428	0.5194	0.4972
6	0.5963	0.5645	0.5346	0.5066	0.4803	0.4556	0.4323
7	0.5470	0.5132	0.4816	0.4523	0.4251	0.3996	0.3759
8	0.5019	0.4665	0.4339	0.4039	0.3762	0.3506	0.3269
9	0.4604	0.4241	0.3909	0.3606	0.3329	0.3075	0.2843
10	0.4224	0.3855	0.3522	0.3220	0.2946	0.2697	0.2472
11	0.3875	0.3505	0.3173	0.2875	0.2607	0.2366	0.2149
12	0.3555	0.3186	0.2858	0.2567	0.2307	0.2076	0.1869
13	0.3262	0.2897	0.2575	0.2292	0.2042	0.1821	0.1625
14	0.2992	0.2633	0.2320	0.2046	0.1807	0.1597	0.1413
15	0.2745	0.2394	0.2090	0.1827	0.1599	0.1401	0.1229
16	0.2519	0.2176	0.1883	0.1631	0.1415	0.1229	0.1069
17	0.2311	0.1978	0.1696	0.1456	0.1252	0.1078	0.0929
18	0.2120	0.1799	0.1528	0.1300	0.1108	0.0946	0.0808
19	0.1945	0.1635	0.1377	0.1161	0.0981	0.0829	0.0703
20	0.1784	0.1486	0.1240	0.1037	0.0868	0.0728	0.0611
21	0.1637	0.1351	0.1117	0.0926	0.0768	0.0638	0.0531
22	0.1502	0.1228	0.1007	0.0826	0.0680	0.0560	0.0462
23	0.1378	0.1117	0.0907	0.0738	0.0601	0.0491	0.0402
24	0.1264	0.1015	0.0817	0.0660	0.0532	0.0431	0.0349
25	0.1160	0.0923	0.0736	0.0588	0.0471	0.0378	0.0304
26	0.1064	0.0829	0.0663	0.0525	0.0417	0.0331	0.0264
27	0.0976	0.0763	0.0597	0.0470	0.0369	0.0291	0.0230
28	0.0895	0.0693	0.0538	0.0420	0.0326	0.0255	0.0200
29	0.0822	0.0630	0.0485	0.0374	0.0289	0.0224	0.0174
30	0.0754	0.0573	0.0437	0.0334	0.0256	0.0196	0.0151
31	0.0691	0.0521	0.0394	0.0298	0.0226	0.0172	0.0131
32	0.0634	0.0474	0.0354	0.0266	0.0200	0.0151	0.0114
33	0.0582	0.0431	0.0319	0.0238	0.0177	0.0132	0.0099
34	0.0534	0.0391	0.0288	0.0212	0.0157	0.0116	0.0086
35	0.0490	0.0356	0.0259	0.0189	0.0139	0.0102	0.0075
36	0.0449	0.0323	0.0234	0.0169	0.0123	0.0089	0.0065
37	0.0412	0.0294	0.0210	0.0151	0.0109	0.0078	0.0057
38	0.0378	0.0267	0.0189	0.0135	0.0096	0.0069	0.0049
39	0.0347	0.0243	0.0171	0.0120	0.0085	0.0060	0.0043
40	0.0318	0.0221	0.0154	0.0107	0.0075	0.0053	0.0037

¶105

[¶106]

Present Worth Table—Periodic Future Payments

Example of use of this table:

To find the cost now of $1,000 of income per year for 20 years at 7%.

From table for 20 years at 7%   10.5940

Cost of $1,000 per year ($1,000 × 10.5940) $10,594

Interest Rate

Number of Years	3%	3½%	4%	4½%	5%	6%	7%	8%
1	0.9709	0.9662	0.9615	0.9569	0.9524	0.9434	0.9346	0.9259
2	1.9135	1.8997	1.8861	1.8727	1.8594	1.8334	1.8080	1.7833
3	2.8286	2.8016	2.7751	2.7490	2.7233	2.6730	2.6243	2.5771
4	3.7171	3.6731	3.6299	3.5875	3.5459	3.4651	3.3872	3.3121
5	4.5797	4.5151	4.4518	4.3900	4.3295	4.2124	4.1002	3.9927
6	5.4172	5.3286	5.2421	5.1579	5.0757	4.9173	4.7665	4.6229
7	6.2303	6.1145	6.0021	5.8927	5.7864	5.5824	5.3893	5.2064
8	7.0197	6.8740	6.7327	6.5959	6.4632	6.2098	5.9713	5.7466
9	7.7861	7.6077	7.4353	7.2688	7.1078	6.8017	6.5152	6.2469
10	8.5302	8.3166	8.1109	7.9127	7.7217	7.3601	7.0236	6.7101
11	9.2526	9.0016	8.7605	8.5289	8.3064	7.8869	7.4987	7.1390
12	9.9540	9.6633	9.3851	9.1186	8.8633	8.3838	7.9427	7.5361
13	10.6350	10.3027	9.9856	9.6829	9.3936	8.8527	8.3577	7.9038
14	11.2961	10.9205	10.5631	10.2228	9.8986	9.2950	8.7455	8.2442
15	11.9379	11.5174	11.1184	10.7395	10.3797	9.7122	9.1079	8.5595
16	12.5611	12.0941	11.6523	11.2340	10.8378	10.1059	9.4466	8.8514
17	13.1661	12.6513	12.1657	11.7072	11.2741	10.4773	9.7632	9.1216
18	13.7535	13.1897	12.6593	12.1600	11.6896	10.8276	10.0591	9.3719
19	14.3238	13.7098	13.1339	12.5933	12.0853	11.1581	10.3356	9.6036
20	14.8775	14.2124	13.5903	13.0079	12.4622	11.4699	10.5940	9.8181
21	15.4150	14.6980	14.0292	13.4047	12.8212	11.7641	10.8355	10.0168
22	15.9369	15.1671	14.4511	13.7844	13.1630	12.0416	11.0612	10.2007
23	16.4436	15.6204	14.8568	14.1478	13.4886	12.3034	11.2722	10.3711
24	16.9355	16.0584	15.2470	14.4955	13.7986	12.5504	11.4693	10.5288
25	17.4131	16.4815	15.6221	14.8282	14.0939	12.7834	11.6536	10.6748
26	17.8768	16.8904	15.9828	15.1466	14.3752	13.0032	11.8258	10.8100
27	18.3270	17.2854	16.3296	15.4513	14.6430	13.2105	11.9867	10.9352
28	18.7641	17.6670	16.6631	15.7429	14.8981	13.4062	12.1371	11.0511
29	19.1885	18.0358	16.9837	16.0219	15.1411	13.5907	12.2777	11.1584
30	19.6004	18.3920	17.2920	16.2889	15.3725	13.7648	12.4090	11.2578
31	20.0004	18.7363	17.5885	16.5444	15.5928	13.9291	12.5318	11.3498
32	20.3888	19.0689	17.8736	16.7889	15.8027	14.0840	12.6466	11.4350
33	20.7658	19.3902	18.1476	17.0229	16.0025	14.2302	12.7538	11.5139
34	21.1318	19.7007	18.4112	17.2468	16.1929	14.3681	12.8540	11.5869
35	21.4872	20.0007	18.6646	17.4610	16.3742	14.4982	12.9477	11.6546
36	21.8323	20.2905	18.9083	17.6660	16.5469	14.6210	13.0352	11.7172
37	22.1672	20.5705	19.1426	17.8622	16.7113	14.7368	13.1170	11.7752
38	22.4925	20.8411	19.3679	18.0500	16.8679	14.8460	13.1935	11.8289
39	22.8082	21.1025	19.5845	18.2297	17.0170	14.9491	13.2649	11.8786
40	23.1148	21.3551	19.7928	18.4016	17.1591	15.0463	13.3317	11.9246

¶106

Present Worth Table—Periodic Future Payments (continued)

Interest Rate

Number of Years	9%	10%	11%	12%	13%	14%	15%
1	0.9174	0.9091	0.9009	0.8929	0.8850	0.8772	0.8696
2	1.7591	1.7355	1.7125	1.6901	1.6681	1.6467	1.6257
3	2.5313	2.4869	2.4437	2.4018	2.3612	2.3216	2.2832
4	3.2397	3.1699	3.1024	3.0373	2.9745	2.9137	2.8550
5	3.8897	3.7908	3.6959	3.6048	3.5172	3.4331	3.3522
6	4.4859	4.3553	4.2305	4.1114	3.9975	3.8887	3.7845
7	5.0330	4.8684	4.7122	4.5638	4.4226	4.2883	4.1604
8	5.5348	5.3349	5.1461	4.9676	4.7988	4.6389	4.4873
9	5.9952	5.7590	5.5370	5.3282	5.1317	4.9464	4.7716
10	6.4177	6.1446	5.8892	5.6502	5.4262	5.2161	5.0188
11	6.8052	6.4951	6.2065	5.9377	5.6869	5.4527	5.2337
12	7.1607	6.8137	6.4924	6.1944	5.9176	5.6603	5.4206
13	7.4869	7.1034	6.7499	6.4235	6.1218	5.8424	5.5831
14	7.7862	7.3667	6.9819	6.6282	6.3025	6.0021	5.7245
15	8.0607	7.6061	7.1909	6.8109	6.4624	6.1422	5.8474
16	8.3126	7.8237	7.3792	6.9740	6.6039	6.2651	5.9542
17	8.5436	8.0216	7.5488	7.1196	6.7291	6.3729	6.0472
18	8.7556	8.2014	7.7016	7.2497	6.8399	6.4674	6.1279
19	8.9501	8.3649	7.8393	7.3658	6.9380	6.5504	6.1982
20	9.1285	8.5136	7.9633	7.4694	7.0248	6.6231	6.2593
21	9.2922	8.6487	8.0751	7.5620	7.1016	6.6870	6.3125
22	9.4424	8.7715	8.1757	7.6446	7.1695	6.7429	6.3587
23	9.5802	8.8832	8.2664	7.7184	7.2297	6.7921	6.3988
24	9.7066	8.9847	8.3481	7.7843	7.2829	6.8351	6.4338
25	9.8226	9.0770	8.4217	7.8431	7.3299	6.8729	6.4641
26	9.9290	9.1609	8.4881	7.8957	7.3717	6.9061	6.4906
27	10.0266	9.2372	8.5478	7.9426	7.4086	6.9352	6.5135
28	10.1161	9.3066	8.6016	7.9844	7.4412	6.9607	6.5335
29	10.1983	9.3696	8.6501	8.0218	7.4701	6.9830	6.5509
30	10.2737	9.4269	8.6938	8.0552	7.4957	7.0027	6.5660
31	10.3428	9.4790	8.7331	8.0850	7.5183	7.0199	6.5791
32	10.4062	9.5264	8.7686	8.1116	7.5383	7.0350	6.5905
33	10.4644	9.5694	8.8005	8.1354	7.5560	7.0482	6.6005
34	10.5178	9.6086	8.8293	8.1566	7.5717	7.0599	6.6091
35	10.5668	9.6442	8.8552	8.1755	7.5856	7.0701	6.6166
36	10.6118	9.6765	8.8786	8.1924	7.5979	7.0790	6.6231
37	10.6530	9.7059	8.8996	8.2075	7.6087	7.0868	6.6288
38	10.6908	9.7327	8.9186	8.2210	7.6183	7.0937	6.6338
39	10.7255	9.7569	8.9357	8.2330	7.6268	7.0997	6.6380
40	10.7574	9.7791	8.9511	8.2438	7.6344	7.1050	6.6418

¶106

[¶ 107]

Sinking Fund Requirements Table

Example of use of this table:

To find the amount of money which must be deposited at the end of each year to grow to $10,000 in 19 years at 8%.

From table for 19 years at 8%　　　　　　.02413
Amount of each deposit ($10,000 × .02413)　$241.30

Interest Rate

Number of Years	3%	3½%	4%	4½%	5%	6%	7%	8%
1	1.00000	1.00000	1.00000	1.00000	1.00000	1.00000	1.00000	1.00000
2	.49261	.49140	.49020	.48900	.48780	.48544	.48309	.48077
3	.32353	.32193	.32035	.31877	.31721	.31411	.31105	.30803
4	.23903	.23725	.23549	.23374	.23201	.22859	.22523	.22192
5	.18835	.18618	.18463	.18279	.18097	.17740	.17389	.17046
6	.15460	.15267	.15076	.14588	.14702	.14336	.13979	.13632
7	.13051	.12854	.12661	.12470	.12282	.11913	.11555	.11207
8	.11246	.11048	.10853	.10661	.10472	.10104	.09747	.09401
9	.09843	.09645	.09449	.09257	.09069	.08702	.08349	.08008
10	.08723	.08524	.08329	.08138	.07950	.07587	.07238	.06903
11	.07808	.07609	.07415	.07225	.07039	.06679	.06336	.06008
12	.07046	.06848	.06655	.06467	.06283	.05928	.05590	.05269
13	.06403	.06206	.06014	.05828	.05646	.05296	.04965	.04652
14	.05853	.05657	.05467	.05282	.05102	.04758	.04434	.04129
15	.05377	.05183	.04994	.04811	.04634	.04206	.03979	.03683
16	.04961	.04768	.04582	.04402	.04227	.03895	.03586	.03298
17	.04595	.04404	.04220	.04042	.03870	.03544	.03243	.02963
18	.04271	.04082	.03899	.03724	.03555	.03236	.02941	.02670
19	.03981	.03794	.03614	.03441	.03275	.02962	.02675	.02413
20	.03722	.03536	.03358	.03188	.03024	.02718	.02439	.02185
21	.03487	.03304	.03128	.02960	.02800	.02500	.02229	.01983
22	.03275	.03093	.02920	.02755	.02597	.02305	.02041	.01803
23	.03081	.02902	.02731	.02568	.02414	.02128	.01871	.01642
24	.02905	.02727	.02559	.02399	.02247	.01968	.01719	.01498
25	.02743	.02567	.02401	.02244	.02095	.01823	.01581	.01368
26	.02594	.02421	.02257	.02102	.01956	.01690	.01456	.01251
27	.02456	.02285	.02124	.01972	.01829	.01570	.01343	.01145
28	.02329	.02160	.02001	.01852	.01712	.01459	.01239	.01049
29	.02211	.02045	.01888	.01741	.01605	.01358	.01145	.00962
30	.02102	.01937	.01783	.01639	.01505	.01265	.01059	.00883
31	.02000	.01837	.01686	.01544	.01413	.01179	.00979	.00811
32	.01905	.01744	.01595	.01456	.01328	.01100	.00907	.00745
33	.01816	.01657	.01510	.01374	.01249	.01027	.00841	.00685
34	.01732	.01576	.01431	.01298	.01176	.00960	.00779	.00630
35	.01654	.01500	.01358	.01227	.01107	.00897	.00723	.00580
36	.01580	.01428	.01289	.01161	.01043	.00839	.00676	.00534
37	.01511	.01361	.01224	.01098	.00984	.00786	.00624	.00492
38	.01446	.01298	.01163	.01040	.00928	.00736	.00579	.00454
39	.01384	.01239	.01106	.00986	.00876	.00689	.00539	.00419
40	.01326	.01183	.01052	.00934	.00828	.00646	.00501	.00386

Sinking Fund Requirements Table *(continued)*

Interest Rate

Number of Years	9%	10%	11%	12%	13%	14%	15%
1	1.00000	1.00000	1.00000	1.00000	1.00000	1.00000	1.00000
2	.47847	.47619	.47393	.47169	.46948	.46729	.46512
3	.30505	.30211	.29921	.29635	.29352	.29073	.28798
4	.21867	.21547	.21233	.20923	.20619	.20320	.20027
5	.16709	.16379	.16057	.15741	.15431	.15128	.14832
6	.13292	.12961	.12638	.12323	.12015	.11716	.11424
7	.10869	.10541	.10222	.09912	.09611	.09319	.09036
8	.09067	.08744	.08432	.08130	.07839	.07557	.07285
9	.07679	.07364	.07060	.06768	.06487	.06217	.05957
10	.06582	.06275	.05980	.05698	.05429	.05171	.04925
11	.05695	.05396	.05112	.04846	.04584	.04339	.04107
12	.04965	.04676	.04403	.04144	.03899	.03667	.03448
13	.04357	.04078	.03815	.03568	.03335	.03116	.02911
14	.03843	.03575	.03323	.03087	.02867	.02661	.02469
15	.03406	.03147	.02907	.02682	.02474	.02281	.02102
16	.03030	.02782	.02552	.02339	.02143	.01962	.01795
17	.02705	.02466	.02247	.02046	.01861	.01692	.01537
18	.02421	.02193	.01984	.01794	.01620	.01462	.01319
19	.02173	.01955	.01756	.01576	.01413	.01266	.01134
20	.01955	.01746	.01558	.01388	.01235	.01099	.00976
21	.01762	.01562	.01384	.01224	.01081	.00954	.00842
22	.01590	.01401	.01231	.01081	.00948	.00830	.00727
23	.01438	.01257	.01097	.00956	.00832	.00723	.00628
24	.01302	.01129	.00979	.00846	.00731	.00630	.00543
25	.01181	.01017	.00874	.00749	.00643	.00550	.00469
26	.01072	.00916	.00781	.00665	.00565	.00480	.00407
27	.00973	.00826	.00699	.00590	.00498	.00419	.00353
28	.00885	.00745	.00626	.00524	.00439	.00366	.00306
29	.00806	.00673	.00561	.00466	.00387	.00320	.00265
30	.00734	.00608	.00502	.00414	.00341	.00280	.00230
31	.00669	.00549	.00451	.00369	.00301	.00245	.00199
32	.00609	.00497	.00404	.00328	.00266	.00215	.00173
33	.00556	.00449	.00363	.00292	.00234	.00188	.00150
34	.00508	.00407	.00326	.00260	.00207	.00165	.00131
35	.00464	.00369	.00293	.00232	.00183	.00144	.00113
36	.00424	.00334	.00263	.00206	.00162	.00126	.00099
37	.00387	.00303	.00236	.00184	.00143	.00111	.00086
38	.00354	.00275	.00213	.00164	.00126	.00097	.00074
39	.00324	.00249	.00191	.00146	.00112	.00085	.00065
40	.00296	.00226	.00172	.00130	.00099	.00075	.00056

¶107

After-Tax Cost of Interest Paid

The interest tables do not, of course, tell the whole story, since money you receive as interest is often taxable income to you. Likewise, in most cases, the money you pay as interest is deductible. Depending on your tax bracket, this fact can go a long way towards reducing your actual dollar costs of borrowing. For example, if you paid $2,000 in interest, and you are in the 36% tax bracket, your actual cost is only 64% of $2,000, or $1,280. So, if you borrowed the money at 10%, the real interest rate you pay is 6.4%, or 64% of 10%. You would find this in the table by finding the figure for your tax bracket and 10% interest, which is 6.4%.

You should remember, however, that if you pay enough interest to reduce your tax bracket, the amount of tax benefit decreases. For example, if your taxable income was $56,000, and you reduced it to $36,000 by paying $20,000 interest at 10%, the first $4,000 would cost you 4.7%; the next $8,000 would cost you 5%; the next $4,000, 5.2%; and the last $4,000, 5.5%. So, the actual after-tax cost of the $20,000 payment was $10,160, or 5.08%.

Also, individuals, partners, estates, and shareholders of Subchapter S corporations have to watch out for excess investment interest. This is defined as interest payments on investments which exceed $25,000 plus your investment income for the year plus your long-term capital gains for the year. The penalty for these excess interest payments is to disallow one-half of the interest deduction. Thus, if you paid $100,000 in investment interest, and you had investment income of $30,000 and no long-term capital gains, $45,000 of the interest payment would be excess, and you could only deduct one-half of it, or $22,500, for a total deduction of $77,500. However, you could carry over the $22,500 to your next tax year (see Internal Revenue Code § 163(d)(1) and (2)).

Interest Rate

Taxable Income (thousands of dollars) Married Taxpayer Separate Return	Joint Return	6%	6½%	7%	7½%	8%	8½%	9%	9½%	10%	10½%
$ 2-4	$ 4-8	4.86	5.26	5.67	6.07	6.48	6.89	7.29	7.70	8.10	8.51
4-6	8-12	4.68	5.07	5.46	5.85	6.24	6.63	7.02	7.41	7.80	8.19
6-8	12-16	4.50	4.88	5.25	5.63	6.00	6.38	6.75	7.13	7.50	7.88
8-10	16-20	4.32	4.68	5.04	5.40	5.76	6.12	6.48	6.84	7.20	7.56
10-12	20-24	4.08	4.42	4.76	5.10	5.44	5.88	6.12	6.46	6.80	7.14
12-14	24-28	3.84	4.16	4.48	4.80	5.12	5.44	5.76	6.08	6.40	6.72
14-16	28-32	3.66	3.97	4.27	4.58	4.88	5.19	5.49	5.80	6.10	6.41
16-18	32-36	3.48	3.77	4.06	4.35	4.64	4.93	5.22	5.51	5.80	6.09
18-20	36-40	3.30	3.58	3.85	4.13	4.40	4.68	4.95	5.23	5.50	5.78
20-22	40-44	3.12	3.38	3.64	3.90	4.16	4.42	4.68	4.94	5.20	5.46
22-26	44-52	3.00	3.25	3.50	3.75	4.00	4.25	4.50	4.75	5.00	5.25
26-32	52-64	2.82	3.06	3.29	3.53	3.76	4.00	4.23	4.47	4.70	4.94
32-38	64-76	2.70	2.93	3.15	3.38	3.60	3.83	4.05	4.28	4.50	4.73
38-44	76-88	2.52	2.73	2.94	3.15	3.36	3.57	3.78	3.99	4.20	4.41
44-50	88-100	2.40	2.60	2.80	3.00	3.20	3.40	3.60	3.80	4.00	4.20
50-60	100-120	2.28	2.47	2.66	2.85	3.04	3.23	3.42	3.61	3.80	3.99
60-70	120-140	2.16	2.34	2.52	2.70	2.88	3.06	3.24	3.42	3.60	3.78
70-80	140-160	2.04	2.21	2.38	2.55	2.72	2.89	3.06	3.23	3.40	3.57
80-90	160-180	1.92	2.08	2.24	2.40	2.56	2.72	2.88	3.04	3.20	3.36
90-100	180-200	1.86	2.02	2.17	2.33	2.48	2.64	2.79	2.95	3.10	3.26
100 and over	200 and over	1.80	1.95	2.10	2.25	2.40	2.55	2.70	2.85	3.00	3.15

After-Tax Cost of Interest Paid *(continued)*

Taxable Income (thousands of dollars) Married Taxpayer		Interest Rate								
Separate Return	Joint Return	11%	11½%	12%	12½%	13%	13½%	14%	14½%	15%
$ 2-4	$ 4-8	8.91	9.32	9.72	10.13	10.53	10.94	11.34	11.75	12.15
4-6	8-12	8.58	8.97	9.36	9.75	10.14	10.53	10.92	11.31	11.70
6-8	12-16	8.25	8.63	9.00	9.38	9.75	10.13	10.50	10.88	11.25
8-10	16-20	7.92	8.28	8.64	9.00	9.36	9.72	10.08	10.44	10.80
10-12	20-24	7.48	7.82	8.16	8.50	8.84	9.18	9.52	9.86	10.20
12-14	24-28	7.04	7.36	7.68	8.00	8.32	8.64	8.96	9.28	9.60
14-16	28-32	6.71	7.02	7.32	7.63	7.93	8.24	8.54	8.85	9.15
16-18	32-36	6.38	6.67	6.96	7.25	7.54	7.83	8.12	8.41	8.70
18-20	36-40	6.05	6.33	6.60	6.88	7.15	7.43	7.70	7.98	8.25
20-22	40-44	5.75	5.98	6.24	6.50	6.76	7.02	7.28	7.54	7.80
22-26	44-52	5.50	5.75	6.00	6.25	6.50	6.75	7.00	7.25	7.50
26-32	52-64	5.17	5.41	5.64	5.88	6.11	6.35	6.58	6.82	7.05
32-38	64-76	4.95	5.18	5.40	5.63	5.85	6.08	6.30	6.53	6.75
38-44	76-88	4.62	4.83	5.04	5.25	5.46	5.67	5.88	6.09	6.30
44-50	88-100	4.40	4.60	4.80	5.00	5.20	5.54	5.60	5.80	6.00
50-60	100-120	4.18	4.37	4.56	4.75	4.94	5.13	5.32	5.51	5.70
60-70	120-140	3.96	4.14	4.32	4.50	4.68	4.86	5.14	5.22	5.40
70-80	140-160	3.74	3.91	4.08	4.25	4.42	4.59	4.76	4.93	5.10
80-90	160-180	3.52	3.68	3.84	4.00	4.16	4.32	4.48	4.64	4.80
90-100	180-200	3.41	3.57	3.72	3.88	4.03	4.19	4.34	4.50	4.65
100 and over	200 and over	3.30	3.45	3.60	3.75	3.90	4.05	4.20	4.35	4.50

After-Tax Value of Income

Tax-free income—includes municipal bonds, life insurance return, special income earned abroad, etc.

Dividends and interest—includes interest on taxable bonds, dividends (not considering the $100 exclusion), compensation, net rental income, etc., which are taxable at ordinary income tax rates.

Long-term capital gains—includes sales and exchanges of capital items, dividends on stock where company has no earnings and profits, etc.

Taxable Income (thousands of dollars)		Tax-Free Income	Long-Term Capital Gains[1]	Dividends or Interest	How Much More Capital Gain Nets Over Dividend and Interest	How Much More Tax-Free Income Nets Over:	
Single Individual	Joint Return					Capital Gains	Dividends & Interest
$ 2-4	$ 4-8	$1.00	$.905	$.81	12%	10	23
4-6	8-12	1.00	.89	.78	14	12	28
6-8	12-16	1.00	.87½	.75	17	14	33
8-10	16-20	1.00	.86	.72	19	16	39
10-12	20-24	1.00	.84	.68	24	19	47
12-14	24-28	1.00	.82	.64	28	22	56
14-16	28-32	1.00	.805	.61	32	24	64
16-18	32-36	1.00	.79	.58	36	27	72
18-20	36-40	1.00	.78½	.55	41	28	78
20-22	40-44	1.00	.76	.52	.46	29	82
22-26	44-52	1.00	.75	.50	50	32	92
26-32	52-64	1.00	.75	.47	60	33	100
32-38	64-76	1.00	.75	.45	67	33	113
38-44	76-88	1.00	.75	.42	79	33	122
44-50	88-100	1.00	.75	.40	88	33	138
50-60	100-120	1.00	.75	.38	97	33	150
60-70	120-140	1.00	.75	.36	108	33	163
70-80	140-160	1.00	.75	.34	121	33	194
80-90	160-180	1.00	.75	.32	134	33	213
90-100	180-200	1.00	.75	.31	142	33	223
100-150	200-300	1.00	.75	.30	150	33	233
150-200	300-400	1.00	.75	.30	150		
200 and over	400 and over	1.00	.75	.30	150		

[1]The return indicated in this table is applicable only with respect to the first $50,000 of capital gains. The maximum tax on these gains is 25%. Where a taxpayer has net long-term capital gains above $50,000, half of the gain is included in ordinary income and is taxed at a maximum 35%. Nor does it take into account the 10% tax on tax-preference income (which includes half of an individual's capital gains) in excess of $30,000 plus the amount of regular taxes.

[¶110]

Wealth Accumulator Table

Regular investment in sound ventures can be a sure road to wealth. The table below dramatically shows the wealth-accumulating power of regular investment at different buildup rates. If the buildup is taxable, only an approximation is possible. Take the actual rate of buildup and reduce it by your estimated average tax bracket over the period covered. Thus, if the actual return is 4% and your estimated average tax rate is 25%, your rate of buildup would be 3%.

A regular investment of $1,000 per year with a buildup at rate of:	5 YRS	10 YRS	15 YRS	20 YRS	25 YRS	30 YRS	35 YRS	40 YRS
3%	$ 5,468	$11,807	$19,156	$27,676	$ 37,553	$ 49,002	$ 62,275	$ 77,663
3½	5,550	12,141	19,971	29,269	40,313	53,429	69,007	87,509
4	5,632	12,486	20,824	30,969	43,311	58,328	76,598	98,826
4½	5,716	12,841	21,719	32,783	46,570	63,752	85,163	111,846
5	5,801	13,206	22,657	34,719	50,113	69,760	94,836	126,839
5½	5,888	13,583	23,641	36,786	53,965	76,419	105,765	144,118
6	5,975	13,971	24,672	38,992	58,156	83,801	118,120	164,047
6½	6,063	14,371	25,754	41,348	62,715	91,989	132,096	187,047
7	6,153	14,783	26,888	43,865	67,676	101,073	147,913	213,609
7½	6,244	15,208	28,077	46,552	73,076	111,154	165,820	244,300
8	6,335	15,645	29,324	49,422	78,954	122,345	186,102	279,781
8½	6,429	16,096	30,632	52,489	85,354	134,772	209,081	320,815
9	6,523	16,560	32,003	55,764	92,323	148,575	235,124	368,291
9½	6,618	17,038	33,441	59,263	99,914	163,907	264,648	423,239
10	6,715	17,531	34,949	63,002	108,181	180,943	298,126	486,851
11	6,912	18,561	38,189	71,265	126,998	220,913	379,164	645,826
12	7,115	19,654	41,753	80,698	149,333	270,292	483,463	859,142
13	7,322	20,814	45,671	91,469	175,850	331,315	617,749	1,145,485
14	7,535	22,044	49,980	103,768	207,332	406,737	790,672	1,529,908
15	7,753	23,349	54,717	117,810	244,711	499,956	1,013,345	2,045,953
20	8,929	31,150	86,442	224,025	566,377	1,418,257	3,538,009	8,812,629
25	10,258	41,566	137,108	428,680	1,318,488	4,033,967	12,320,951	37,610,819

[¶111]
Self-Liquidating Mortgages—Constant Monthly Payment Table

The following table shows the constant monthly payment required to liquidate a mortgage loan of $1,000 running for any number of whole years between 5 and 40 years inclusive and at interest rates running from 4% through 12% at ¼% intervals. All fractions are rounded to the next higher cent, thus making the final payment slightly smaller than the others.

Interest Rate

Years of Loan	4%	4¼%	4½%	4¾%	5%	5¼%	5½%	5¾%	6%	6¼%	6½%
1	85.15	85.26	85.38	85.49	85.61	85.72	85.84	85.95	86.07	86.18	86.30
2	43.42	43.54	43.65	43.76	43.87	43.98	44.10	44.21	44.32	44.43	44.55
3	29.52	29.63	29.75	29.86	29.97	30.08	30.20	30.31	30.42	30.54	30.65
4	22.58	22.69	22.80	22.92	23.03	23.14	23.26	23.37	23.48	23.60	23.71
5	18.42	18.53	18.64	18.76	18.87	18.99	19.10	19.22	19.33	19.45	19.57
6	15.65	15.76	15.87	15.99	16.10	16.22	16.34	16.45	16.57	16.69	16.81
7	13.67	13.78	13.90	14.02	14.13	14.25	14.37	14.49	14.61	14.73	14.85
8	12.19	12.31	12.42	12.54	12.66	12.78	12.90	13.02	13.14	13.26	13.39
9	11.04	11.16	11.28	11.40	11.52	11.64	11.76	11.88	12.01	12.13	12.25
10	10.12	10.24	10.36	10.48	10.61	10.73	10.85	10.98	11.10	11.23	11.35
11	9.38	9.50	9.62	9.74	9.86	9.99	10.11	10.24	10.37	10.49	10.62
12	8.76	8.88	9.00	9.12	9.25	9.37	9.50	9.63	9.76	9.89	10.02
13	8.23	8.35	8.48	8.60	8.73	8.86	8.99	9.12	9.25	9.38	9.51
14	7.78	7.91	8.03	8.16	8.29	8.42	8.55	8.68	8.81	8.95	9.08
15	7.40	7.52	7.65	7.78	7.91	8.04	8.17	8.30	8.44	8.57	8.71
16	7.06	7.19	7.32	7.45	7.58	7.71	7.84	7.98	8.11	8.25	8.39
17	6.76	6.89	7.02	7.15	7.29	7.42	7.56	7.69	7.83	7.97	8.11
18	6.50	6.63	6.76	6.90	7.03	7.17	7.30	7.44	7.58	7.72	7.87
19	6.27	6.40	6.53	6.67	6.80	6.94	7.08	7.22	7.36	7.50	7.65
20	6.06	6.19	6.33	6.46	6.60	6.74	6.88	7.02	7.16	7.31	7.46
21	5.87	6.01	6.14	6.28	6.42	6.56	6.70	6.84	6.99	7.14	7.28
22	5.70	5.84	5.97	6.11	6.25	6.39	6.54	6.68	6.83	6.98	7.13
23	5.55	5.68	5.82	5.96	6.10	6.25	6.39	6.54	6.69	6.84	6.99
24	5.41	5.54	5.68	5.83	5.97	6.11	6.26	6.41	6.56	6.71	6.86
25	5.28	5.42	5.56	5.70	5.85	5.99	6.14	6.29	6.44	6.60	6.75
26	5.16	5.30	5.44	5.59	5.73	5.88	6.03	6.18	6.34	6.49	6.65
27	5.05	5.19	5.34	5.48	5.63	5.78	5.93	6.08	6.24	6.40	6.55
28	4.95	5.09	5.24	5.39	5.54	5.69	5.84	5.99	6.15	6.31	6.47
29	4.86	5.00	5.15	5.30	5.45	5.60	5.75	5.91	6.07	6.23	6.39
30	4.77	4.92	5.07	5.22	5.37	5.52	5.68	5.84	5.99	6.16	6.32
31	4.69	4.84	4.99	5.14	5.29	5.45	5.61	5.77	5.93	6.09	6.25
32	4.62	4.77	4.92	5.07	5.23	5.38	5.54	5.70	5.86	6.03	6.19
33	4.55	4.70	4.85	5.01	5.16	5.32	5.48	5.64	5.80	5.97	6.14
34	4.49	4.64	4.79	4.94	5.10	5.26	5.42	5.59	5.75	5.92	6.09
35	4.43	4.58	4.73	4.89	5.05	5.21	5.37	5.53	5.70	5.87	6.04
36	4.37	4.52	4.68	4.84	5.00	5.16	5.32	5.49	5.66	5.83	6.00
37	4.32	4.47	4.63	4.79	4.95	5.11	5.28	5.44	5.61	5.78	5.96
38	4.27	4.42	4.58	4.74	4.90	5.07	5.23	5.40	5.57	5.75	5.92
39	4.22	4.38	4.54	4.70	4.86	5.03	5.19	5.36	5.54	5.71	5.89
40	4.18	4.34	4.50	4.66	4.82	4.99	5.16	5.33	5.50	5.68	5.85

Self-Liquidating Mortgages—Constant Monthly
Payment Table (continued)

Interest Rate

Years of Loan	6¾%	7%	7¼%	7½%	7¾%	8%	8¼%	8½%	8¾%	9%	9¼%
1	86.41	86.53	86.64	86.76	86.87	86.99	87.10	87.22	87.33	87.45	87.57
2	44.66	44.77	44.89	45.00	45.11	45.23	45.34	45.46	45.57	45.68	45.80
3	30.76	30.88	30.99	31.11	31.22	31.34	31.45	31.57	31.68	31.80	31.92
4	23.83	23.95	24.06	24.18	24.30	24.41	24.53	24.65	24.77	24.88	25.00
5	19.68	19.80	19.92	20.04	20.16	20.28	20.40	20.52	20.64	20.76	20.88
6	16.93	17.05	17.17	17.29	17.41	17.53	17.65	17.78	17.90	18.03	18.15
7	14.97	15.09	15.22	15.34	15.46	15.59	15.71	15.84	15.96	16.09	16.22
8	13.51	13.63	13.76	13.88	14.01	14.14	14.26	14.39	14.52	14.65	14.78
9	12.38	12.51	12.63	13.76	12.89	13.02	13.15	13.28	13.41	13.54	13.68
10	11.48	11.61	11.74	11.87	12.00	12.13	12.26	12.40	12.53	12.67	12.80
11	10.75	10.88	11.02	11.15	11.28	11.42	11.55	11.69	11.82	11.96	12.10
12	10.15	10.28	10.42	10.55	10.69	10.82	10.96	11.10	11.24	11.38	11.52
13	9.65	9.78	9.92	10.05	10.19	10.33	10.47	10.61	10.75	10.90	11.04
14	9.22	9.35	9.49	9.63	9.77	9.91	10.06	10.20	10.34	10.49	10.64
15	8.85	8.99	9.13	9.27	9.41	9.56	9.70	9.85	9.99	10.14	10.29
16	8.53	8.67	8.81	8.96	9.10	9.25	9.40	9.54	9.69	9.84	10.00
17	8.25	8.40	8.54	8.69	8.83	8.98	9.13	9.28	9.43	9.59	9.74
18	8.01	8.15	8.30	8.45	8.60	8.75	8.90	9.05	9.21	9.36	9.52
19	7.79	7.94	8.09	8.24	8.39	8.54	8.70	8.85	9.01	9.17	9.33
20	7.60	7.75	7.90	8.06	8.21	8.36	8.52	8.68	8.84	9.00	9.16
21	7.43	7.58	7.74	7.89	8.05	8.20	8.36	8.52	8.68	8.85	9.01
22	7.28	7.43	7.59	7.74	7.90	8.06	8.22	8.38	8.55	8.71	8.88
23	7.14	7.30	7.46	7.61	7.77	7.93	8.10	8.26	8.43	8.59	8.76
24	7.02	7.18	7.34	7.50	7.66	7.82	7.98	8.15	8.32	8.49	8.66
25	6.91	7.07	7.23	7.39	7.55	7.72	7.88	8.05	8.22	8.39	8.56
26	6.81	6.97	7.13	7.29	7.46	7.63	7.79	7.96	8.13	8.31	8.48
27	6.72	6.88	7.04	7.21	7.37	7.54	7.71	7.88	8.06	8.23	8.41
28	6.63	6.80	6.96	7.13	7.30	7.47	7.64	7.81	7.99	8.16	8.34
29	6.56	6.72	6.89	7.06	7.23	7.40	7.57	7.75	7.92	8.10	8.28
30	6.49	6.65	6.82	6.99	7.16	7.34	7.51	7.69	7.89	8.05	8.23
31	6.42	6.59	6.76	6.93	7.11	7.28	7.46	7.64	7.81	8.00	8.18
32	6.36	6.53	6.71	6.88	7.05	7.23	7.41	7.59	7.77	7.95	8.13
33	6.31	6.48	6.65	6.83	7.01	7.18	7.36	7.54	7.73	7.91	8.09
34	6.26	6.43	6.61	6.78	6.96	7.14	7.32	7.50	7.69	7.87	8.06
35	6.21	6.39	6.56	6.74	6.92	7.10	7.28	7.47	7.65	7.84	8.03
36	6.17	6.35	6.53	6.70	6.88	7.07	7.25	7.44	7.62	7.81	8.00
37	6.13	6.31	6.49	6.67	6.85	7.03	7.22	7.41	7.59	7.78	7.97
38	6.10	6.28	6.46	6.64	6.82	7.00	7.19	7.38	7.57	7.76	7.95
39	6.06	6.24	6.42	6.61	6.79	6.98	7.16	7.35	7.54	7.73	7.93
40	6.03	6.21	6.40	6.58	6.77	6.95	7.14	7.33	7.52	7.71	7.91

¶111

Self-Liquidating Mortgages—Constant Monthly
Payment Table *(continued)*

Interest Rate

Years of Loan	9½%	9¾%	10%	10¼%	10½%	10¾%	11%	11¼%	11½%	11¾%	12%
1	87.68	87.80	87.92	88.03	88.15	88.27	88.38	88.50	88.62	88.73	88.85
2	45.91	46.03	46.14	46.26	46.38	46.49	46.61	46.72	46.84	46.96	47.07
3	32.03	32.15	32.27	32.38	32.50	32.62	32.74	32.86	32.98	33.10	33.21
4	25.12	25.24	25.36	25.48	25.60	25.72	25.85	25.97	26.09	26.21	26.33
5	21.00	21.12	21.25	21.37	21.49	21.62	21.74	21.87	21.99	22.12	22.24
6	18.27	18.40	18.53	18.65	18.78	18.91	19.03	19.16	19.29	19.42	19.55
7	16.34	16.47	16.60	16.73	16.86	16.99	17.12	17.25	17.39	17.52	17.65
8	14.91	15.04	15.17	15.31	15.44	15.57	15.71	15.84	15.98	16.12	16.25
9	13.81	13.94	14.08	14.21	14.35	14.49	14.63	14.76	14.90	15.04	15.18
10	12.94	13.08	13.21	13.35	13.49	13.63	13.78	13.92	14.06	14.20	14.35
11	12.24	12.38	12.52	12.66	12.80	12.95	13.09	13.24	13.38	13.53	13.68
12	11.66	11.81	11.95	12.10	12.24	12.39	12.54	12.68	12.83	12.98	13.13
13	11.19	11.33	11.48	11.63	11.78	11.92	12.08	12.23	12.38	12.53	12.69
14	10.78	10.93	11.08	11.23	11.38	11.54	11.69	11.85	12.00	12.16	12.31
15	10.44	10.59	10.75	10.90	11.05	11.21	11.37	11.52	11.68	11.84	12.00
16	10.15	10.30	10.46	10.62	10.77	10.93	11.09	11.25	11.41	11.57	11.74
17	9.90	10.05	10.21	10.37	10.53	10.69	10.85	11.02	11.18	11.35	11.51
18	9.68	9.84	10.00	10.16	10.32	10.49	10.65	10.82	10.98	11.15	11.32
19	9.49	9.65	9.81	9.98	10.14	10.31	10.47	10.64	10.81	10.98	11.15
20	9.32	9.48	9.65	9.82	9.98	10.15	10.32	10.49	10.66	10.84	11.01
21	9.17	9.34	9.51	9.68	9.85	10.02	10.19	10.36	10.54	10.71	10.89
22	9.04	9.21	9.38	9.55	9.73	9.90	10.07	10.25	10.42	10.60	10.78
23	8.93	9.10	9.27	9.44	9.62	9.79	9.97	10.15	10.33	10.51	10.69
24	8.83	9.00	9.17	9.35	9.52	9.70	9.88	10.06	10.24	10.42	10.60
25	8.74	8.91	9.09	9.26	9.44	9.62	9.80	9.98	10.16	10.35	10.53
26	8.66	8.83	9.01	9.19	9.37	9.55	9.73	9.91	10.10	10.28	10.47
27	8.58	8.76	8.94	9.12	9.30	9.49	9.67	9.85	10.04	10.23	10.41
28	8.52	8.70	8.88	9.06	9.25	9.43	9.61	9.80	9.99	10.18	10.37
29	8.46	8.64	8.82	9.01	9.19	9.38	9.57	9.75	9.94	10.13	10.32
30	8.41	8.59	8.78	8.96	9.15	9.33	9.52	9.71	9.90	10.09	10.29
31	8.36	8.55	8.73	8.92	9.11	9.30	9.48	9.68	9.87	10.06	10.25
32	8.32	8.50	8.69	8.88	9.07	9.26	9.45	9.64	9.84	10.03	10.22
33	8.28	8.47	8.66	8.85	9.04	9.23	9.42	9.61	9.81	10.00	10.20
34	8.25	8.44	8.63	8.82	9.01	9.20	9.39	9.59	9.78	9.98	10.18
35	8.22	8.41	8.60	8.79	8.98	9.18	9.37	9.56	9.76	9.96	10.16
36	8.19	8.38	8.57	8.76	8.96	9.15	9.35	9.54	9.74	9.94	10.14
37	8.16	8.35	8.55	8.74	8.94	9.13	9.33	9.53	9.72	9.92	10.12
38	8.14	8.33	8.53	8.72	8.92	9.11	9.31	9.51	9.71	9.91	10.11
39	8.12	8.31	8.51	8.70	8.90	9.10	9.30	9.50	9.70	9.90	10.10
40	8.10	8.30	8.49	8.69	8.89	9.08	9.28	9.48	9.68	9.88	10.08

¶111

Self-Liquidating Mortgages—Constant Monthly
Payment Table *(continued)*

Interest Rate

Years of Loan	12⅜%	12¾%	13⅛%	13½%	13⅞%	14¼%	14⅝%	15%	15⅜%	15¾%
1	89.02	89.20	89.38	89.55	89.73	89.90	90.08	90.26	90.44	90.61
2	47.25	47.42	47.60	47.78	47.95	48.13	48.31	48.49	48.67	48.84
3	33.39	33.57	33.75	33.94	34.12	34.30	34.48	34.67	34.85	35.03
4	26.52	26.70	26.89	27.08	27.26	27.45	27.64	27.83	28.02	28.21
5	22.43	22.63	22.82	23.01	23.20	23.40	23.59	23.79	23.99	24.19
6	19.75	19.94	20.14	20.34	20.54	20.74	20.94	21.15	21.35	21.55
7	17.85	18.06	18.26	18.46	18.67	18.88	19.09	19.30	19.51	19.72
8	16.46	16.67	16.88	17.09	17.30	17.51	17.73	17.95	18.16	18.38
9	15.40	15.61	15.83	16.04	16.26	16.48	16.70	16.92	17.15	17.37
10	14.56	14.78	15.00	15.23	15.45	15.68	15.90	16.13	16.36	16.60
11	13.90	14.13	14.35	14.58	14.81	15.04	15.27	15.51	15.75	15.98
12	13.36	13.59	13.82	14.06	14.29	14.53	14.77	15.01	15.25	15.49
13	12.92	13.15	13.39	13.63	13.87	14.11	14.36	14.60	14.85	15.10
14	12.55	12.79	13.03	13.28	13.52	13.77	14.02	14.27	14.52	14.78
15	12.24	12.49	12.73	12.98	13.23	13.49	13.74	14.00	14.25	14.51
16	11.98	12.23	12.48	12.74	12.99	13.25	13.51	13.77	14.03	14.29
17	11.76	12.02	12.27	12.53	12.79	13.05	13.31	13.58	13.84	14.11
18	11.57	11.83	12.09	12.35	12.62	12.88	13.15	13.42	13.69	13.96
19	11.41	11.67	11.94	12.20	12.47	12.74	13.01	13.28	13.56	13.83
20	11.27	11.54	11.80	12.07	12.34	12.62	12.89	13.17	13.45	13.73

¶111

[¶112]

Self-Liquidating Mortgages—Annual Interest, Amortization, And Remaining Balance

The following tables are useful planning tools, helping you to determine quickly the constant monthly payments, the annual interest, the annual amortization payment, and the remaining balance for self-liquidating mortgage loans at various interest rates and at payout terms of 5, 10, 15, 20, 25 and 30 years on each $1,000 of mortgage loan.

Note: The final payment shown in these tables usually is lower than the regular monthly payments, since, in the monthly payments, fractional sums are stated at the next higher cent.

5-Year Term

	4% interest—$18.42 monthly payment			4½% interest—$18.65 monthly payment			5% interest—$18.88 monthly payment		
Year	Interest	Amort.	Balance	Interest	Amort.	Balance	Interest	Amort.	Balance
1	36.63	184.41	815.59	41.26	182.54	817.46	45.91	180.65	819.35
2	29.14	191.90	623.69	32.88	190.92	626.54	36.64	189.92	629.43
3	21.31	199.73	423.96	24.11	199.69	426.85	26.92	199.64	429.79
4	13.19	207.85	216.11	14.94	208.86	217.99	16.74	209.82	219.97
5	4.67	216.11	0	5.34	217.99	0	5.99	219.97	0

	5¼% interest—$18.99 monthly payment			5½% interest—$19.11 monthly payment			5¾% interest—$19.22 monthly payment		
Year	Interest	Amort.	Balance	Interest	Amort.	Balance	Interest	Amort.	Balance
1	48.23	179.65	820.35	50.53	178.79	821.21	52.85	177.79	822.21
2	38.55	189.33	631.02	40.44	188.88	632.33	42.37	188.27	633.94
3	28.37	199.51	431.51	29.79	199.53	432.80	31.26	199.38	434.56
4	17.65	210.23	221.28	18.58	210.74	222.06	19.47	211.17	223.39
5	6.33	221.28	0	6.63	222.06	0	7.00	223.39	0

	6% interest—$19.34 monthly payment			6¼% interest—$19.45 monthly payment			6½% interest—$19.57 monthly payment		
Year	Interest	Amort.	Balance	Interest	Amort.	Balance	Interest	Amort.	Balance
1	55.19	176.89	823.11	57.51	175.89	824.11	59.85	174.99	825.01
2	44.29	187.79	635.32	46.18	187.22	636.89	48.13	186.71	638.30
3	32.70	199.38	435.94	34.16	199.24	437.65	35.63	199.21	439.09
4	20.39	211.68	224.26	21.35	212.05	225.60	22.30	212.54	226.55
5	7.33	224.26	0	7.72	225.60	0	8.04	226.55	0

	6¾% interest—$19.69 monthly payment			7% interest—$19.81 monthly payment			7¼% interest—$19.92 monthly payment		
Year	Interest	Amort.	Balance	Interest	Amort.	Balance	Interest	Amort.	Balance
1	62.20	174.08	825.92	64.51	173.21	826.79	66.85	172.19	827.81
2	50.07	186.21	639.71	52.01	185.71	641.08	53.95	185.09	642.72
3	37.10	199.18	440.53	38.57	199.15	441.93	40.06	198.98	443.74
4	23.23	213.05	227.48	24.18	213.54	228.39	25.17	213.87	229.87
5	8.41	227.48	0	8.75	228.39	0	9.13	229.87	0

Self-Liquidating Mortgages—Annual Interest, Amortization, And Remaining Balance (continued)

Five-Year Term (continued)

	7½% interest—$20.04 monthly payment			7¾% interest—$20.16 monthly payment			8% interest—$20.28 monthly payment		
Year	Interest	Amort.	Balance	Interest	Amort.	Balance	Interest	Amort.	Balance
1	69.18	171.30	828.70	71.54	170.38	829.62	73.87	169.49	830.51
2	55.89	184.59	644.11	57.83	184.09	645.53	59.81	183.55	646.96
3	41.58	198.90	445.21	43.08	198.84	446.69	44.56	198.80	448.16
4	26.12	214.36	230.85	27.08	214.84	231.85	28.06	215.30	232.86
5	9.45	230.85	0	9.85	231.85	0	10.21	232.86	0

	9% interest—$20.76 monthly payment			10% interest—$21.25 monthly payment			12⅜% interest—$22.43 monthly payment		
Year	Interest	Amort.	Balance	Interest	Amort.	Balance	Interest	Amort.	Balance
1	83.27	165.85	834.15	92.71	162.29	837.71	115.17	153.99	846.01
2	67.72	181.40	652.75	75.70	179.30	658.41	94.97	174.19	671.82
3	50.40	198.42	454.33	56.91	198.09	460.32	72.16	197.00	474.82
4	32.10	217.02	237.31	36.20	218.80	241.52	46.34	222.82	252.00
5	11.71	237.31	0	13.25	241.52	0	17.16	252.00	0

	12¾% interest—$22.63 monthly payment			13⅛% interest—$22.82 monthly payment			13½% interest—$23.01 monthly payment		
Year	Interest	Amort.	Balance	Interest	Amort.	Balance	Interest	Amort.	Balance
1	118.84	152.72	847.28	122.39	151.45	848.55	125.92	150.20	849.80
2	98.18	173.38	673.90	101.27	172.57	675.98	104.38	171.74	678.06
3	74.74	196.82	477.08	77.20	196.64	479.34	78.67	196.45	481.61
4	48.13	223.43	253.65	49.80	224.04	255.30	51.45	224.67	256.94
5	17.91	253.65	0	18.54	255.30	0	19.18	256.94	0

	13⅞% interest—$23.20 monthly payment			14¼% interest—$23.40 monthly payment			14⅝% interest—$23.59 monthly payment		
Year	Interest	Amort.	Balance	Interest	Amort.	Balance	Interest	Amort.	Balance
1	129.47	148.93	851.07	133.12	147.68	852.32	136.65	146.43	853.57
2	107.44	170.96	680.11	110.66	170.14	682.18	113.74	169.34	684.23
3	82.17	196.23	483.88	84.75	196.05	486.13	87.23	195.85	488.38
4	53.11	225.29	258.59	54.92	225.88	260.25	56.61	226.47	261.91
5	19.81	258.59	0	20.55	260.25	0	21.17	261.91	0

	15% interest—$23.79 monthly payment			15⅜% interest—$23.99 monthly payment			15¾% interest—$24.19 monthly payment		
Year	Interest	Amort.	Balance	Interest	Amort.	Balance	Interest	Amort.	Balance
1	140.29	145.19	854.81	143.92	143.96	856.04	147.54	142.74	857.26
2	116.94	168.54	686.27	120.16	167.72	688.32	123.37	166.91	690.35
3	89.85	195.63	490.64	92.48	195.40	492.92	95.09	195.19	495.16
4	58.41	227.07	263.57	60.19	227.69	265.23	62.03	228.25	266.91
5	21.91	263.57	0	22.65	265.23	0	23.37	266.91	0

Self-Liquidating Mortgages—Annual Interest, Amortization And Remaining Balance *(continued)*

Note: The tables for 10, 15, and 20 year terms are carried through 15¾% and the tables for 25 and 30 years terms are carried through 12%.

10-Year Term

	4% interest—$10.13 monthly payment			4½% interest—$10.37 monthly payment			5% interest—$10.61 monthly payment		
Year	Interest	Amort.	Balance	Interest	Amort.	Balance	Interest	Amort.	Balance
1	38.49	83.07	916.93	43.34	81.10	918.90	48.21	79.11	920.89
2	35.11	86.45	830.48	39.62	84.82	834.08	44.16	83.16	837.73
3	31.56	90.00	740.48	35.72	88.72	745.36	39.90	87.42	750.31
4	27.92	93.64	646.84	31.65	92.79	652.57	35.44	91.88	658.43
5	24.11	97.45	549.39	27.41	97.03	555.54	30.72	96.60	561.83
6	20.13	101.43	447.96	22.92	101.52	454.02	25.79	101.53	460.30
7	16.00	105.56	342.40	18.26	106.18	347.84	20.60	106.72	353.58
8	11.70	109.86	232.54	13.37	111.07	236.77	15.13	112.19	241.39
9	7.23	114.33	118.21	8.29	116.15	120.62	9.41	117.91	123.48
10	2.54	118.21	0	2.94	120.62	0	3.35	123.48	0

	5¼% interest—$10.73 monthly payment			5½% interest—$10.86 monthly payment			5¾% interest—$10.98 monthly payment		
Year	Interest	Amort.	Balance	Interest	Amort.	Balance	Interest	Amort.	Balance
1	50.64	78.12	921.88	53.08	77.24	922.76	55.50	76.26	923.74
2	46.43	82.33	839.55	48.72	81.60	841.16	51.01	80.75	842.99
3	42.01	86.75	752.80	44.11	86.21	754.95	46.24	85.52	757.47
4	37.35	91.41	661.39	39.24	91.08	663.87	41.19	90.57	666.90
5	32.43	96.33	565.06	34.11	96.21	567.66	35.85	95.91	570.99
6	27.25	101.51	463.55	28.68	101.64	466.02	30.20	101.56	469.43
7	21.79	106.97	356.58	22.96	107.36	358.66	24.20	107.56	361.87
8	16.04	112.72	243.86	16.89	113.43	245.23	17.83	113.93	247.94
9	9.98	118.78	125.08	10.49	119.83	125.40	11.12	120.64	127.30
10	3.58	125.08	0	3.71	125.40	0	3.98	127.30	0

	6% interest—$11.11 monthly payment			6¼% interest—$11.23 monthly payment			6½% interest—$11.36 monthly payment		
Year	Interest	Amort.	Balance	Interest	Amort.	Balance	Interest	Amort.	Balance
1	57.96	75.36	924.64	60.40	74.36	925.64	59.83	73.49	926.51
2	53.31	80.01	844.63	55.60	79.16	846.48	57.90	78.42	848.09
3	48.35	84.97	759.66	50.53	84.23	762.25	52.66	83.66	764.43
4	43.13	90.19	669.47	45.09	89.67	672.58	47.04	89.28	675.15
5	37.56	95.76	573.71	39.33	95.43	577.15	41.09	95.23	579.92
6	31.67	101.65	472.06	33.20	101.56	475.59	34.71	101.61	478.31
7	25.38	107.94	364.11	26.67	108.09	367.50	27.88	108.44	369.87
8	18.73	114.59	249.52	19.73	115.03	252.47	20.63	115.69	254.18
9	11.68	121.64	127.88	12.31	122.45	130.02	12.91	123.41	130.77
10	4.14	127.88	0	4.44	130.02	0	4.63	130.77	0

¶112

Self-Liquidating Mortgages—Annual Interest, Amortization And Remaining Balance (continued)

10-Year Term (continued)

	6¾% interest—$11.49 monthly payment			7% interest—$11.62 monthly payment			7¼% interest—$11.75 monthly payment		
Year	Interest	Amort.	Balance	Interest	Amort.	Balance	Interest	Amort.	Balance
1	65.29	72.59	927.41	67.73	71.71	928.29	70.18	70.82	929.18
2	60.23	77.65	849.76	62.55	76.89	851.40	64.86	76.14	853.04
3	54.81	83.07	766.69	57.01	82.44	768.97	59.14	81.86	771.18
4	49.04	88.84	677.85	51.02	88.41	680.55	53.02	87.98	683.20
5	42.85	95.03	582.82	44.64	94.80	585.75	45.43	94.57	588.63
6	36.23	101.65	481.17	37.79	101.65	484.10	39.36	101.64	486.99
7	29.17	108.71	372.46	30.42	109.02	375.08	31.73	109.27	377.72
8	21.59	116.29	256.17	22.55	116.89	258.19	23.50	117.50	260.22
9	13.50	124.38	131.79	14.12	125.32	132.87	14.71	126.29	133.93
10	4.82	131.79	0	5.06	132.87	0	5.27	133.93	0

	7½% interest—$11.88 monthly payment			7¾% interest—$12.01 monthly payment			8% interest—$12.14 monthly payment		
Year	Interest	Amort.	Balance	Interest	Amort.	Balance	Interest	Amort.	Balance
1	72.60	69.93	930.07	75.08	69.04	930.96	77.54	68.14	931.86
2	67.19	75.37	854.70	69.54	74.58	856.38	71.87	73.81	858.05
3	61.35	81.21	773.49	63.55	80.57	775.81	65.74	79.94	778.11
4	55.04	87.52	685.97	57.07	87.05	688.76	59.12	86.56	691.55
5	48.26	94.30	591.67	50.09	94.03	594.73	51.94	93.74	597.81
6	40.94	101.62	490.05	42.53	101.59	493.14	44.17	101.51	496.30
7	33.04	109.52	380.53	34.38	109.74	383.40	35.73	109.95	386.35
8	24.54	118.02	262.51	25.56	118.56	264.84	26.61	119.07	267.28
9	15.38	127.18	135.33	16.05	128.07	136.77	16.72	128.96	138.32
10	5.51	135.33	0	5.75	136.77	0	6.02	138.32	0

	8½% interest—$12.40 monthly payment			8¾% interest—$12.53 monthly payment			9% interest—$12.67 monthly payment		
Year	Interest	Amort.	Balance	Interest	Amort.	Balance	Interest	Amort.	Balance
1	82.46	66.34	933.66	84.91	65.45	934.55	87.36	64.68	935.32
2	76.59	72.21	861.45	78.95	71.41	863.14	81.31	70.73	864.59
3	70.21	78.59	782.86	72.45	77.91	785.23	74.65	77.39	787.20
4	63.27	85.53	697.32	65.35	85.01	700.22	67.40	84.64	702.56
5	55.71	93.09	604.23	57.61	92.75	607.47	59.46	92.58	609.98
6	47.48	101.32	502.91	49.16	101.20	506.27	50.78	101.26	508.72
7	38.52	110.28	392.63	39.94	110.42	395.85	41.30	110.74	397.98
8	28.78	120.02	272.61	29.89	120.47	275.38	30.90	121.14	276.84
9	18.17	130.63	141.98	18.92	131.44	143.94	19.56	132.48	144.36
10	6.63	142.17	0.18	6.94	143.42	0.52	7.10	144.36	0

¶112

Self-Liquidating Mortgages—Annual Interest, Amortization And Remaining Balance (continued)

10-Year Term (continued)

	9¼% interest—$12.80 monthly payment			9½% interest—$12.94 monthly payment			9¾% interest—$13.08 monthly payment		
Year	Interest	Amort.	Balance	Interest	Amort.	Balance	Interest	Amort.	Balance
1	89.84	63.76	936.23	92.31	62.97	937.03	94.76	62.20	937.80
2	83.68	69.92	866.32	86.06	69.22	867.81	88.42	68.54	869.26
3	76.93	76.67	789.65	79.19	76.09	791.71	81.43	75.53	793.74
4	69.53	84.07	705.51	71.64	83.64	708.07	73.73	83.23	710.51
5	61.42	92.18	613.40	63.34	91.94	616.13	65.25	91.71	618.79
6	52.52	101.08	512.33	54.21	101.07	515.06	55.89	101.07	517.73
7	42.77	110.83	401.50	44.18	111.10	403.97	45.59	111.37	406.36
8	32.07	121.53	279.97	33.16	122.12	281.85	34.23	122.73	283.63
9	20.34	133.26	146.72	21.04	134.24	147.61	21.72	135.24	148.38
10	7.48	146.12	0.60	7.72	147.56	0.05	7.93	149.03	0.64

	10% interest—$13.22 monthly payment			10¼% interest—$13.35 monthly payment			10½% interest—$13.49 monthly payment		
Year	Interest	Amort.	Balance	Interest	Amort.	Balance	Interest	Amort.	Balance
1	97.23	61.41	938.59	99.71	60.49	939.51	102.18	59.70	940.30
2	90.81	67.83	870.76	93.21	66.99	872.51	95.61	66.27	874.03
3	83.72	74.92	795.84	86.01	74.19	798.32	88.30	73.58	800.45
4	75.86	82.78	713.06	78.04	82.16	716.16	80.19	81.69	718.77
5	67.19	91.45	621.61	69.21	90.99	625.17	71.19	90.69	628.08
6	57.62	101.02	520.59	59.44	100.76	524.41	61.20	100.68	527.40
7	47.03	111.61	408.98	48.61	111.59	412.82	50.10	111.78	415.62
8	35.35	123.29	285.69	36.62	123.58	289.24	37.79	124.09	291.53
9	22.44	136.20	149.49	23.34	136.86	152.38	24.11	137.77	153.76
10	8.17	149.49	0	8.64	151.56	0.82	8.93	152.95	0.81

	10¾% interest—$13.63 monthly payment			11% interest—$13.77 monthly payment			11¼% interest—$13.92 monthly payment		
Year	Interest	Amort.	Balance	Interest	Amort.	Balance	Interest	Amort.	Balance
1	104.65	58.91	941.09	107.12	58.12	941.88	109.59	57.45	942.55
2	98.00	65.56	875.53	100.39	64.85	877.03	102.79	64.25	878.30
3	90.59	72.97	802.56	92.89	72.35	804.68	95.18	71.86	806.44
4	82.35	81.21	721.35	84.52	80.72	723.96	86.66	80.38	726.06
5	73.18	90.38	630.96	75.18	90.06	633.90	77.14	89.90	636.16
6	62.79	100.59	530.37	64.76	100.48	533.42	66.49	100.55	535.60
7	51.61	111.95	418.42	53.13	112.11	421.32	54.57	112.47	423.14
8	38.96	124.60	293.82	40.16	125.08	296.24	41.25	125.79	297.35
9	24.89	138.67	155.15	25.69	139.55	156.69	26.34	140.70	156.65
10	9.23	154.33	0.82	9.54	155.70	1.00	9.68	157.36	0.70

¶112

Self-Liquidating Mortgages—Annual Interest, Amortization And Remaining Balance *(continued)*

10-Year Term *(continued)*

	11½% interest—$14.06 monthly payment			11¾% interest—$14.20 monthly payment			12% interest—$14.35 monthly payment		
Year	Interest	Amort.	Balance	Interest	Amort.	Balance	Interest	Amort.	Balance
1	112.07	56.65	943.35	114.55	55.85	944.15	117.03	55.17	944.83
2	105.20	63.52	879.83	107.63	62.77	881.38	110.03	62.17	882.65
3	97.50	71.22	808.61	99.84	70.56	810.82	102.15	70.05	812.60
4	88.86	79.86	728.75	91.09	79.31	731.51	93.26	78.94	733.66
5	79.18	89.54	639.21	81.25	89.15	642.37	83.25	88.95	644.71
6	68.33	100.39	538.82	70.20	100.20	542.16	71.97	100.23	544.49
7	56.15	112.57	426.25	57.77	112.63	429.53	59.26	112.94	431.55
8	42.50	126.22	300.03	43.80	126.60	302.93	44.94	127.26	304.29
9	27.20	141.52	158.51	28.10	142.30	160.63	28.80	143.40	160.90
10	10.04	158.68	0	10.45	159.95	0.67	10.62	161.58	0

	12⅜% interest—$14.56 monthly payment			12¾% interest—$14.78 monthly payment			13⅛% interest—$15.00 monthly payment		
Year	Interest	Amort.	Balance	Interest	Amort.	Balance	Interest	Amort.	Balance
1	120.71	54.01	945.99	124.43	52.93	947.07	128.14	51.86	948.14
2	113.61	61.11	884.88	117.28	60.08	886.99	120.92	59.08	889.06
3	105.61	69.11	815.77	109.14	68.22	818.77	112.67	67.33	821.73
4	96.56	78.16	737.61	99.92	77.44	741.33	103.30	76.70	745.03
5	86.32	88.40	649.21	89.45	87.91	653.42	92.58	87.42	657.61
6	74.75	99.97	549.24	77.57	99.79	553.63	80.41	99.59	558.02
7	61.63	113.09	436.15	64.08	113.28	440.35	66.51	113.49	444.53
8	46.83	127.89	308.26	48.75	128.61	311.74	50.70	129.30	315.23
9	30.07	144.65	163.61	31.36	146.00	165.74	32.65	147.35	167.88
10	11.11	163.61	0	11.62	165.74	0	12.12	167.88	0

	13½% interest—$15.23 monthly payment			13⅞% interest—$15.45 monthly payment			14¼% interest—$15.68 monthly payment		
Year	Interest	Amort.	Balance	Interest	Amort.	Balance	Interest	Amort.	Balance
1	131.95	50.81	949.19	135.64	49.76	950.24	139.43	48.73	951.27
2	124.68	58.08	891.11	128.29	57.11	893.13	132.02	56.14	895.13
3	116.32	66.44	824.67	119.84	65.56	827.57	123.46	64.70	830.43
4	106.77	75.99	748.68	110.14	75.26	752.31	113.63	74.53	755.90
5	95.87	86.89	661.79	99.00	86.40	665.91	102.29	85.87	670.03
6	83.36	99.40	562.39	86.22	99.18	566.73	89.21	98.95	571.08
7	69.08	113.68	448.71	71.57	113.83	452.90	74.16	114.00	457.08
8	52.78	129.98	318.73	54.72	130.68	322.22	56.80	131.36	325.72
9	34.07	148.69	170.04	35.39	150.01	172.21	36.81	151.35	174.37
10	12.72	170.04	0	13.19	172.21	0	13.79	174.37	0

Self-Liquidating Mortgages—Annual Interest, Amortization And Remaining Balance *(continued)*

10-Year Term *(continued)*

Year	14⅞% interest—$15.90 monthly payment			15% interest—$16.13 monthly payment		
	Interest	Amort.	Balance	Interest	Amort.	Balance
1	143.08	47.72	952.28	146.83	46.73	953.27
2	135.62	55.18	897.10	139.32	54.24	899.03
3	126.97	63.83	833.27	130.59	62.97	836.06
4	116.99	73.81	759.46	120.49	73.07	762.99
5	105.45	85.35	674.11	108.73	84.83	678.16
6	92.10	98.70	575.41	95.10	98.46	579.70
7	76.64	114.16	461.25	79.26	114.30	465.40
8	58.78	132.02	329.23	60.89	132.67	332.73
9	38.22	152.58	176.65	39.58	153.98	178.75
10	14.15	176.65	0	14.81	178.75	0

Year	15⅜% interest—$16.36 monthly payment			15¾% interest—$16.60 monthly payment		
	Interest	Amort.	Balance	Interest	Amort.	Balance
1	150.57	45.75	954.25	154.41	44.79	955.21
2	143.02	53.30	900.95	146.82	52.38	902.83
3	134.22	62.10	838.85	137.95	61.25	841.58
4	123.95	72.37	766.48	127.57	71.63	769.95
5	112.04	84.28	682.20	115.44	83.76	686.19
6	98.10	98.22	583.98	101.26	97.94	588.25
7	81.90	114.42	469.56	84.66	114.54	473.71
8	63.01	133.31	336.25	65.66	133.54	339.77
9	41.01	155.31	180.94	42.58	156.62	183.15
10	15.38	180.94	0	16.05	183.15	0

Self-Liquidating Mortgages—Annual Interest, Amortization And Remaining Balance (continued)

15-Year Term

Year	4% interest—$7.40 monthly payment Interest	Amort.	Balance	4½% interest—$7.65 monthly payment Interest	Amort.	Balance	5% interest—$7.91 monthly payment Interest	Amort.	Balance
1	39.10	49.70	950.30	44.04	47.76	952.24	48.96	45.96	954.04
2	37.06	51.74	898.56	41.83	49.97	902.27	46.62	48.30	905.74
3	34.99	53.81	844.75	39.53	52.27	850.00	44.13	50.79	854.95
4	32.77	56.03	788.72	37.12	54.68	795.32	41.54	53.38	801.57
5	30.49	58.31	730.41	34.62	57.18	738.14	38.80	56.12	745.45
6	28.10	60.70	669.71	31.99	59.81	678.33	35.94	58.98	686.47
7	25.65	63.15	606.56	29.26	62.54	615.79	32.92	62.00	624.47
8	23.07	65.73	540.83	26.39	65.41	550.38	29.74	65.18	559.29
9	20.38	68.42	472.41	23.36	68.44	481.94	26.41	68.50	490.79
10	17.63	71.17	401.24	20.22	71.58	410.36	22.91	72.01	418.78
11	14.71	74.09	327.15	16.94	74.86	335.50	19.22	75.70	343.08
12	11.69	77.11	250.04	13.49	78.31	257.19	15.35	79.57	263.51
13	8.54	80.26	169.78	9.89	81.91	175.28	11.28	83.64	179.87
14	5.27	83.53	86.25	6.14	85.66	89.62	7.02	87.90	91.97
15	1.86	86.25	0	2.20	89.62	0	2.50	91.97	0

Year	5¼% interest—$8.04 monthly payment Interest	Amort.	Balance	5½% interest—$8.18 monthly payment Interest	Amort.	Balance	5¾% interest—$8.31 monthly payment Interest	Amort.	Balance
1	51.42	45.06	954.94	53.90	44.26	955.74	56.36	43.36	956.64
2	49.00	47.48	907.46	51.39	46.77	908.97	53.81	45.91	910.73
3	46.46	50.02	857.44	48.77	49.39	859.58	51.10	48.62	862.11
4	43.76	52.72	804.72	45.96	52.20	807.38	48.24	51.48	810.63
5	40.92	55.56	749.16	43.03	55.13	752.25	45.19	54.53	756.10
6	37.94	58.54	690.62	39.92	58.24	694.01	41.97	57.75	698.35
7	34.79	61.69	628.93	36.63	61.53	632.48	38.56	61.16	637.19
8	31.47	65.01	563.92	33.17	64.99	567.49	34.95	64.77	572.42
9	27.96	68.52	495.40	29.20	68.66	498.83	31.13	68.59	503.83
10	24.30	72.18	423.22	25.62	72.54	426.29	26.97	72.65	431.18
11	20.41	76.07	347.15	21.53	76.63	349.66	22.80	76.92	354.26
12	16.32	80.16	266.99	17.20	80.96	268.70	18.25	81.47	272.79
13	12.01	84.47	182.52	12.65	85.51	183.19	13.43	86.29	186.50
14	7.48	89.00	93.52	7.84	90.32	92.87	8.34	91.38	95.12
15	2.66	93.52	0	2.73	92.87	0	2.94	95.12	0

Self-Liquidating Mortgages—Annual Interest, Amortization And Remaining Balance (continued)

15-Year Term (continued)

	6% interest—$8.44 monthly payment			6¼% interest—$8.58 monthly payment			6½% interest—$8.72 monthly payment		
Year	Interest	Amort.	Balance	Interest	Amort.	Balance	Interest	Amort.	Balance
1	58.85	42.43	957.57	61.33	41.63	958.37	63.79	40.85	959.15
2	56.23	45.05	912.52	58.65	44.31	914.06	61.08	43.56	915.59
3	53.40	47.88	864.64	55.80	47.16	866.90	58.15	46.49	869.10
4	50.49	50.79	813.85	52.76	50.20	816.70	55.03	49.61	819.49
5	47.37	53.91	759.94	49.53	53.43	763.27	51.72	52.92	766.57
6	44.04	57.24	702.70	46.08	56.88	706.39	48.16	56.48	710.09
7	40.50	60.78	641.92	42.43	60.53	645.86	44.39	60.25	649.84
8	36.75	64.53	577.39	38.54	64.42	581.44	40.34	64.30	585.54
9	32.78	68.50	508.89	34.38	68.58	512.86	36.04	68.60	516.94
10	28.58	72.72	436.17	29.99	72.97	439.89	31.46	73.18	443.76
11	24.07	77.21	358.96	25.30	77.66	362.23	26.55	78.09	365.67
12	19.29	81.99	276.97	20.30	82.66	279.57	21.31	83.33	282.34
13	14.24	87.04	189.93	14.99	87.97	191.60	15.72	88.92	193.42
14	8.89	92.39	97.54	9.35	93.61	97.99	9.79	94.85	98.57
15	3.19	97.54	0	3.29	97.99	0	3.41	98.57	0

	6¾% interest—$8.85 monthly payment			7% interest—$8.99 monthly payment			7¼% interest—$9.13 monthly payment		
Year	Interest	Amort.	Balance	Interest	Amort.	Balance	Interest	Amort.	Balance
1	66.29	39.91	960.09	68.74	39.14	960.86	71.25	38.31	961.69
2	63.48	42.72	917.37	65.94	41.94	918.92	68.38	41.18	920.51
3	60.54	45.66	871.71	62.90	44.98	873.94	65.29	44.27	876.24
4	57.35	48.85	822.86	59.65	48.23	825.71	61.98	47.58	828.66
5	53.93	52.27	770.59	56.16	51.72	773.99	58.39	51.17	777.49
6	50.30	55.90	714.69	52.42	55.46	718.53	54.55	55.01	722.48
7	46.41	59.79	654.90	48.39	59.49	659.04	50.45	59.11	663.37
8	42.28	63.92	590.98	44.10	63.78	595.26	46.02	63.54	599.83
9	37.80	68.40	522.58	39.56	68.32	526.89	41.25	68.31	531.52
10	33.04	73.16	449.42	34.55	73.33	453.56	36.13	73.43	458.09
11	27.94	78.26	371.16	29.26	78.62	374.94	30.60	78.96	379.13
12	22.51	83.69	287.47	23.59	84.29	290.65	24.71	84.85	294.28
13	16.68	89.52	197.95	17.48	90.40	200.25	18.33	91.23	203.05
14	10.42	95.78	102.17	10.96	96.92	103.33	11.52	98.04	105.01
15	3.76	102.17	0	3.94	103.33	0	4.16	105.01	0

¶112

Self-Liquidating Mortgages—Annual Interest, Amortization And Remaining Balance (continued)

15-Year Term (continued)

Year	7½% interest—$9.28 monthly payment			7¾% interest—$9.42 monthly payment			8% interest—$9.56 monthly payment		
	Interest	Amort.	Balance	Interest	Amort.	Balance	Interest	Amort.	Balance
1	73.73	37.63	962.37	76.20	36.84	963.16	78.72	36.00	964.00
2	70.77	40.59	921.78	73.26	39.78	923.38	75.71	39.01	924.99
3	67.64	43.72	878.06	70.05	42.99	880.39	72.48	42.24	882.75
4	64.25	47.11	830.95	66.59	46.45	833.94	68.98	45.74	837.01
5	60.59	50.77	780.18	62.89	50.15	783.79	65.17	49.55	787.46
6	56.66	54.70	725.48	58.84	54.20	729.59	61.02	53.70	733.76
7	52.38	58.98	666.50	54.49	58.55	671.04	56.60	58.12	675.64
8	47.83	63.53	602.97	49.78	63.26	607.78	51.77	62.95	612.69
9	42.89	68.47	534.50	44.71	68.33	539.45	46.55	68.17	544.52
10	37.59	73.77	460.73	39.23	73.81	465.64	40.89	73.83	470.69
11	31.87	79.49	381.24	33.29	79.75	385.89	34.78	79.94	390.75
12	25.68	85.68	295.56	26.89	86.15	299.74	28.15	86.57	304.18
13	19.04	92.32	203.24	19.98	93.06	206.68	20.96	93.76	210.42
14	11.88	99.48	103.76	12.49	100.55	106.13	13.16	101.56	108.86
15	4.15	103.76	0	4.41	106.13	0	4.76	108.86	0

Year	8¼% interest—$9.70 monthly payment			8½% interest—$9.85 monthly payment			8¾% interest—$9.99 monthly payment		
	Interest	Amort.	Balance	Interest	Amort.	Balance	Interest	Amort.	Balance
1	81.19	35.22	964.79	83.68	34.53	965.48	86.17	33.72	966.29
2	78.18	38.23	926.56	80.63	37.58	927.90	83.10	36.79	929.51
3	74.90	41.51	885.06	77.31	40.90	887.01	79.75	40.14	889.38
4	71.34	45.07	840.00	73.69	44.52	842.49	76.10	43.79	845.59
5	67.48	48.93	791.07	69.76	48.45	794.05	72.11	47.78	797.82
6	63.29	53.12	737.96	65.48	52.73	741.32	67.76	52.13	745.69
7	58.74	57.67	680.29	60.82	57.39	683.93	63.01	56.88	688.81
8	53.80	62.61	617.68	55.74	62.47	621.47	57.83	62.06	626.75
9	48.43	67.98	549.71	50.22	67.99	553.48	52.17	67.72	559.04
10	42.61	73.80	475.91	44.21	74.00	479.49	46.00	73.89	485.15
11	36.28	80.13	395.79	37.67	80.54	398.96	39.27	80.62	404.54
12	29.42	86.99	308.80	30.55	87.66	311.30	31.93	87.96	316.59
13	21.96	94.45	214.36	22.81	95.40	215.90	23.92	95.97	220.62
14	13.87	102.54	111.82	14.37	103.84	112.07	15.18	104.71	115.91
15	5.08	111.33	0.50	5.20	113.01	0.94	5.65	114.25	1.66

¶112

Self-Liquidating Mortgages—Annual Interest, Amortization And Remaining Balance *(continued)*

15-Year Term *(continued)*

Year	9% interest—$10.15 monthly payment			9¼% interest—$10.29 monthly payment			9½% interest—$10.44 monthly payment		
	Interest	Amort.	Balance	Interest	Amort.	Balance	Interest	Amort.	Balance
1	88.66	33.14	966.86	91.16	32.33	967.68	93.65	31.64	968.37
2	85.55	36.25	930.61	88.04	35.45	932.23	90.51	34.78	933.60
3	82.14	39.66	890.95	84.62	38.87	893.36	87.06	38.23	895.37
4	78.43	43.37	847.58	40.86	42.63	850.74	83.27	42.02	853.35
5	74.34	47.46	800.12	76.75	46.74	804.00	79.10	46.19	807.17
6	69.92	51.88	748.24	72.24	51.25	752.76	74.51	50.78	756.39
7	65.03	56.77	691.97	67.29	56.20	696.56	69.47	55.82	700.58
8	59.72	62.08	629.39	61.87	61.62	634.94	63.93	61.36	639.23
9	53.90	67.90	561.49	55.92	67.57	567.38	57.85	67.44	571.79
10	47.53	74.27	487.22	49.40	74.09	493.29	51.15	74.14	497.66
11	40.57	81.23	405.99	42.25	81.24	412.06	43.79	81.50	416.17
12	32.92	88.88	317.11	34.41	89.08	322.98	35.71	89.58	326.59
13	24.59	97.21	219.90	25.81	97.68	225.30	26.82	98.47	228.12
14	15.50	106.30	113.60	16.38	107.11	118.19	17.04	108.25	119.88
15	5.51	113.60	0	6.04	117.45	0.75	6.30	118.99	0.89

Year	9¾% interest—$10.59 monthly payment			10% interest—$10.75 monthly payment			10¼% interest—$10.90 monthly payment		
	Interest	Amort.	Balance	Interest	Amort.	Balance	Interest	Amort.	Balance
1	96.15	30.94	969.07	98.62	30.38	969.62	101.14	29.67	970.34
2	92.99	34.10	934.97	95.46	33.54	936.08	97.95	32.86	937.48
3	89.51	37.58	897.40	91.95	37.05	899.03	94.42	36.39	901.09
4	85.68	41.41	856.00	88.06	40.94	858.09	90.51	40.30	860.80
5	81.46	45.63	810.38	83.78	45.22	812.87	86.18	44.63	816.17
6	76.81	50.28	760.10	79.04	49.96	762.91	81.38	49.43	766.75
7	71.68	55.41	704.70	73.81	55.19	707.72	76.07	54.74	712.02
8	66.03	61.06	643.65	68.03	60.97	646.75	70.19	60.62	651.40
9	59.81	67.28	576.37	61.64	67.36	579.39	63.68	67.13	584.28
10	52.95	74.14	502.23	54.30	74.40	504.99	56.46	74.35	509.93
11	45.38	81.71	420.53	46.79	82.21	422.78	48.48	82.33	427.61
12	37.05	90.04	330.50	38.18	90.82	331.96	39.63	91.18	336.43
13	27.87	99.22	231.28	28.70	100.30	231.66	29.83	100.98	235.46
14	17.75	109.34	121.95	18.20	110.80	120.86	18.98	111.83	123.64
15	6.61	120.48	1.47	6.57	120.86	0	6.97	123.84	0.20

Self-Liquidating Mortgages—Annual Interest, Amortization And Remaining Balance (continued)

15-Year Term (continued)

	10½% interest—$11.05 monthly payment			10¾% interest—$11.21 monthly payment			11% interest—$11.37 monthly payment		
Year	Interest	Amort.	Balance	Interest	Amort.	Balance	Interest	Amort.	Balance
1	103.64	28.97	971.04	106.13	28.40	971.61	108.63	27.82	972.19
2	100.44	32.17	938.88	102.93	31.60	940.01	105.41	31.04	941.16
3	96.90	35.71	903.17	99.36	35.17	904.85	101.82	34.63	906.53
4	92.97	39.64	863.53	95.38	39.15	865.71	97.81	38.64	867.90
5	88.60	44.01	819.53	90.96	43.57	822.14	93.34	43.11	824.80
6	83.75	48.86	770.67	86.04	48.49	773.66	88.36	48.09	776.71
7	78.36	54.25	716.43	80.57	53.96	719.70	82.79	53.66	723.06
8	72.39	60.22	656.21	74.47	60.06	659.65	76.58	59.87	663.19
9	65.75	66.86	589.35	67.69	66.84	592.81	69.66	66.79	596.40
10	58.38	74.23	515.13	60.14	74.39	518.42	61.93	74.52	521.89
11	50.20	82.41	432.73	51.73	82.80	435.63	53.30	83.15	438.74
12	41.12	91.49	341.24	42.38	92.15	343.49	43.68	92.77	345.98
13	31.04	101.57	239.68	31.97	102.56	240.94	32.95	103.50	242.48
14	19.85	112.76	126.92	20.39	114.14	126.80	20.97	115.48	127.01
15	7.42	125.19	1.74	7.50	127.03	0.23	7.61	128.84	1.83

	11¼% interest—$11.52 monthly payment			11½% interest—$11.68 monthly payment			11¾% interest—$11.84 monthly payment		
Year	Interest	Amort.	Balance	Interest	Amort.	Balance	Interest	Amort.	Balance
1	111.14	27.11	972.90	113.64	26.53	973.48	116.14	25.95	974.06
2	107.92	30.33	942.57	110.42	29.75	943.73	112.92	29.17	944.89
3	104.33	33.92	908.66	106.81	33.36	910.38	109.30	32.79	912.11
4	100.31	37.94	870.73	102.77	37.40	872.98	105.23	36.86	875.26
5	95.82	42.43	828.30	98.23	41.94	831.05	100.66	41.43	833.83
6	90.79	47.46	780.85	93.15	47.02	784.03	95.53	46.56	787.27
7	85.17	53.08	727.77	87.45	52.72	731.31	89.75	52.34	734.94
8	78.88	59.37	668.40	81.05	59.12	672.20	83.26	58.83	676.11
9	71.85	66.40	602.00	73.89	66.28	605.92	75.96	66.13	609.99
10	63.98	74.27	527.74	65.85	74.32	531.60	67.76	74.33	535.67
11	55.18	83.07	444.67	56.84	83.33	448.28	58.54	83.55	452.12
12	45.34	92.91	351.76	46.73	93.44	354.84	48.18	93.91	358.21
13	34.33	103.92	247.84	35.40	104.77	250.08	36.53	105.56	252.66
14	22.01	116.24	131.61	22.70	117.47	132.62	23.44	118.65	134.01
15	8.24	130.01	1.61	8.46	131.71	0.91	8.72	133.37	0.65

Self-Liquidating Mortgages—Annual Interest, Amortization And Remaining Balance *(continued)*

15-Year Term *(continued)*

Year	12% interest—$12.00 monthly payment			12⅜% interest—$12.24 monthly payment			12¾% interest—$12.49 monthly payment		
	Interest	*Amort.*	*Balance*	*Interest*	*Amort.*	*Balance*	*Interest*	*Amort.*	*Balance*
1	118.64	25.37	974.64	122.35	24.53	975.47	126.17	23.71	976.29
2	115.42	28.59	946.06	119.12	27.76	947.71	122.96	26.92	949.37
3	111.80	32.21	913.85	115.49	31.39	916.32	119.31	30.57	918.80
4	107.71	36.30	877.56	111.67	35.21	881.11	115.19	34.69	884.11
5	103.11	40.90	836.67	106.43	40.45	840.66	110.48	39.40	844.71
6	97.92	46.09	790.59	101.47	45.41	795.25	105.18	44.70	800.01
7	92.08	51.93	738.66	95.51	51.37	743.88	99.13	50.75	749.26
8	85.49	58.52	680.15	88.79	58.09	685.79	92.25	57.63	691.63
9	78.07	65.94	614.22	81.18	65.70	620.09	84.48	65.40	626.23
10	69.71	74.30	539.93	72.56	74.32	545.77	75.08	74.80	551.43
11	60.29	83.72	456.21	62.83	84.05	461.72	66.11	83.77	467.66
12	49.67	94.34	361.88	51.82	95.06	366.66	54.19	95.69	371.97
13	37.71	106.30	255.59	39.36	107.52	259.14	41.24	108.64	263.33
14	24.23	119.78	135.81	25.27	121.61	137.53	27.06	122.82	140.51
15	9.04	134.97	0.84	9.35	137.53	0	9.37	140.51	0

Year	13⅛% interest—$12.73 monthly payment			13½% interest—$12.98 monthly payment			13⅞% interest—$13.23 monthly payment		
	Interest	*Amort.*	*Balance*	*Interest*	*Amort.*	*Balance*	*Interest*	*Amort.*	*Balance*
1	129.85	22.91	977.09	133.62	22.14	977.86	136.93	21.38	978.62
2	126.65	26.11	950.98	130.45	25.31	952.55	134.22	24.54	954.08
3	123.01	29.75	921.23	126.81	28.95	923.60	130.59	28.17	925.91
4	118.87	33.89	887.34	122.65	33.11	890.49	126.42	32.34	893.57
5	114.13	38.63	848.71	117.90	37.86	852.63	121.64	37.12	856.45
6	108.75	44.01	804.70	112.44	43.32	809.31	116.14	42.62	813.83
7	102.61	50.15	754.55	106.22	49.54	759.77	109.86	48.90	764.93
8	95.63	57.13	697.42	99.13	56.63	703.14	102.60	56.16	708.77
9	87.65	65.11	632.31	90.96	64.80	638.34	93.41	64.45	644.32
10	78.58	74.18	558.13	81.66	74.10	564.24	84.77	73.99	570.33
11	68.23	84.53	473.60	71.02	84.74	479.50	73.82	84.94	485.39
12	56.44	96.32	377.28	58.84	96.92	382.58	61.25	97.51	387.88
13	43.01	109.75	267.53	44.93	110.83	271.75	46.84	111.92	275.96
14	27.72	125.04	142.49	28.99	126.77	144.98	30.29	128.47	147.49
15	10.27	142.49	0	10.78	144.98	0	11.27	147.49	0

¶112

Self-Liquidating Mortgages—Annual Interest, Amortization And Remaining Balance (continued)

15-Year Term (continued)

Year	14¼% interest—$13.49 monthly payment			14⅝% interest—$13.74 monthly payment			15% interest—$14.00 monthly payment		
	Interest	Amort.	Balance	Interest	Amort.	Balance	Interest	Amort.	Balance
1	141.24	20.64	979.36	144.95	19.93	980.07	148.76	19.24	980.76
2	138.09	23.79	955.57	141.84	23.04	957.03	145.67	22.33	958.43
3	134.47	27.41	928.16	138.21	26.67	930.36	142.08	25.92	932.51
4	130.31	31.57	896.59	134.06	30.82	899.54	137.91	30.09	902.42
5	125.50	36.38	860.21	129.23	35.65	863.89	133.08	34.92	867.50
6	119.97	41.91	818.30	123.67	41.21	822.68	127.46	40.54	826.96
7	113.58	48.30	770.00	117.20	47.68	775.00	120.95	47.05	779.91
8	106.23	55.65	714.35	109.74	55.14	719.86	113.38	54.62	725.29
9	97.76	64.12	650.23	101.21	63.67	656.19	104.61	63.39	661.90
10	88.02	73.86	576.37	91.05	73.83	582.36	94.41	73.59	588.31
11	76.76	85.12	491.25	79.60	85.28	497.08	82.59	85.41	502.90
12	63.82	98.06	393.19	66.27	98.61	398.47	68.55	99.45	403.45
13	48.88	113.00	280.19	50.82	114.06	284.41	53.21	114.79	288.66
14	31.69	130.19	150.00	33.00	131.88	152.53	34.40	133.60	155.06
15	11.88	150.00	0	12.35	152.53	0	12.94	155.06	0

Year	15⅜% interest—$14.25 monthly payment			15¾% interest—$14.51 monthly payment		
	Interest	Amort.	Balance	Interest	Amort.	Balance
1	152.43	18.57	981.43	156.20	17.92	982.08
2	149.38	21.62	959.81	153.18	20.94	961.14
3	145.80	25.20	934.61	149.62	24.50	936.64
4	141.63	29.37	905.24	145.48	28.64	908.00
5	136.80	34.20	871.04	140.62	33.50	874.50
6	131.14	39.86	831.18	134.95	39.17	835.33
7	124.58	46.42	784.76	128.31	45.81	789.52
8	116.90	54.10	730.66	120.56	53.56	735.96
9	107.98	63.02	667.64	111.48	62.64	673.32
10	97.57	73.43	594.21	100.88	73.24	600.08
11	85.46	85.54	508.67	88.46	85.66	514.42
12	71.33	99.67	409.00	73.96	100.16	414.26
13	54.89	116.11	292.89	57.00	117.12	297.14
14	35.73	135.27	157.62	37.15	136.97	160.17
15	13.38	157.62	0	13.95	160.17	0

Self-Liquidating Mortgages—Annual Interest, Amortization And Remaining Balance *(continued)*

20-Year Term

Year	4% interest—$6.06 monthly payment			4½% interest—$6.33 monthly payment			5% interest—$6.60 monthly payment		
	Interest	Amort.	Balance	Interest	Amort.	Balance	Interest	Amort.	Balance
1	39.49	33.32	966.68	44.34	31.62	968.38	49.33	29.87	970.13
2	38.03	34.69	931.99	42.90	33.06	935.32	47.79	31.41	938.72
3	36.66	36.06	895.93	41.38	34.58	900.74	46.19	33.01	905.71
4	35.15	37.57	858.36	39.80	36.16	864.58	44.50	34.70	871.01
5	33.62	39.10	819.26	38.12	37.84	826.74	42.71	36.49	834.52
6	32.04	40.68	778.58	36.40	39.56	787.18	40.83	38.37	796.15
7	30.38	42.34	736.24	34.58	41.38	745.80	38.90	40.30	755.85
8	28.64	44.08	692.16	32.68	43.28	702.52	36.84	42.36	713.49
9	26.85	45.87	646.29	30.68	45.28	657.24	34.66	44.54	668.95
10	24.99	47.73	598.56	28.61	47.35	609.89	32.39	46.81	622.14
11	23.05	49.67	548.89	26.44	49.52	560.37	29.99	49.21	572.93
12	21.01	51.71	497.18	24.16	51.80	508.57	27.47	51.73	521.20
13	18.90	53.82	443.36	21.78	54.18	454.39	24.80	54.40	466.80
14	16.72	56.00	387.36	19.29	56.67	397.72	22.09	57.11	409.69
15	14.43	58.29	329.07	16.70	59.26	338.46	19.13	60.07	349.62
16	12.06	60.66	268.41	13.95	62.01	276.45	16.05	63.15	286.47
17	9.59	63.13	205.28	11.14	64.82	211.63	12.81	66.39	220.08
18	7.01	65.71	139.57	8.13	67.83	143.80	9.42	69.78	150.30
19	4.36	68.36	71.21	5.02	70.94	72.86	5.85	73.35	76.95
20	1.56	71.21	0	1.76	72.86	0	2.10	76.95	0

Year	5¼% interest—$6.74 monthly payment			5½% interest—$6.88 monthly payment			5¾% interest—$7.03 monthly payment		
	Interest	Amort.	Balance	Interest	Amort.	Balance	Interest	Amort.	Balance
1	51.79	29.09	970.91	54.29	28.27	971.73	56.78	27.58	972.42
2	50.25	30.63	940.28	52.69	29.87	941.86	55.16	29.20	943.22
3	48.60	32.28	908.00	51.02	31.54	910.32	53.43	30.93	912.29
4	46.86	34.02	873.98	49.23	33.33	876.99	51.60	32.76	879.53
5	45.02	35.86	838.12	47.36	35.20	841.79	49.66	34.70	844.83
6	43.11	37.77	800.35	45.37	37.19	804.60	47.93	36.43	808.10
7	41.07	39.81	760.54	43.26	39.30	765.30	45.45	38.91	769.19
8	38.94	41.94	718.60	41.06	41.50	723.80	43.16	41.20	727.99
9	36.67	44.21	674.39	38.70	43.86	679.94	40.72	43.64	684.35
10	34.29	46.59	627.80	36.25	46.31	633.63	38.14	46.22	638.13
11	31.78	49.10	578.70	33.63	48.93	584.70	35.41	48.95	589.18
12	29.14	51.74	526.96	30.84	51.72	532.98	32.51	51.85	537.33
13	26.40	54.48	472.48	27.96	54.60	478.38	29.46	54.90	482.43
14	23.45	57.43	415.05	24.86	57.70	420.68	26.21	58.15	424.28
15	20.34	60.54	354.51	21.62	60.94	359.74	22.80	61.56	362.72
16	17.10	63.78	290.73	18.16	64.40	295.34	19.16	65.20	297.52
17	13.66	67.22	223.51	14.56	68.00	227.34	15.30	69.06	228.46
18	10.04	70.84	152.67	10.71	71.85	155.49	11.20	73.16	155.30
19	6.22	74.66	78.01	6.65	75.91	79.58	6.90	77.46	77.84
20	2.22	78.01	0	2.34	79.58	0	2.33	77.84	0

¶112

Self-Liquidating Mortgages—Annual Interest, Amortization And Remaining Balance (continued)

20-Year Term (continued)

	6% interest—$7.17 monthly payment			6¼% interest—$7.31 monthly payment			6½% interest—$7.46 monthly payment		
Year	Interest	Amort.	Balance	Interest	Amort.	Balance	Interest	Amort.	Balance
1	59.28	26.76	973.24	61.77	25.95	974.05	64.25	25.27	974.73
2	57.62	28.42	944.82	60.09	27.63	946.42	62.58	26.94	947.78
3	55.86	30.18	914.64	58.31	29.41	917.01	60.75	28.77	919.01
4	54.01	32.03	882.61	56.44	31.28	885.73	58.83	30.69	888.32
5	52.03	34.01	848.60	54.42	33.30	852.43	56.76	32.76	855.56
6	49.92	36.12	812.48	52.25	35.47	816.96	54.57	34.95	820.61
7	47.72	38.32	774.16	50.00	37.72	779.24	52.25	37.27	783.34
8	45.32	40.72	733.44	47.58	40.14	739.10	49.76	39.76	743.58
9	42.83	43.21	690.23	44.97	42.75	696.35	47.09	42.43	701.15
10	40.16	45.88	644.35	42.25	45.47	650.88	44.26	45.26	655.89
11	37.33	48.71	595.64	39.30	48.42	602.46	41.21	48.31	607.58
12	34.34	51.70	543.94	36.19	51.53	550.93	37.99	51.53	556.05
13	31.14	54.90	489.04	32.90	54.82	496.11	34.52	55.00	501.05
14	27.77	58.27	430.77	29.35	58.37	437.74	30.82	58.70	442.35
15	24.16	61.88	368.89	25.58	62.14	375.60	26.91	62.61	379.74
16	20.35	65.69	303.20	21.62	66.10	309.50	22.74	66.78	312.96
17	16.30	69.74	233.46	17.33	70.39	239.11	18.25	71.27	241.69
18	12.01	74.03	159.43	12.82	74.90	164.21	13.47	76.05	165.64
19	7.44	78.60	80.83	8.03	79.69	84.52	8.39	81.13	84.51
20	2.54	80.83	0	2.88	84.52	0	2.95	84.51	0

	6¾% interest—$7.61 monthly payment			7% interest—$7.76 monthly payment			7¼% interest—$7.91 monthly payment		
Year	Interest	Amort.	Balance	Interest	Amort.	Balance	Interest	Amort.	Balance
1	66.75	24.57	975.43	69.24	23.88	976.12	71.74	23.18	976.82
2	65.03	26.29	949.14	67.52	25.60	950.52	70.01	24.91	951.91
3	63.20	28.12	921.02	65.67	27.45	923.07	68.13	26.79	925.12
4	61.25	30.07	890.95	63.68	29.44	893.63	66.13	28.79	896.33
5	59.16	32.16	858.79	61.54	31.58	862.05	63.96	30.96	865.37
6	56.92	34.40	824.39	59.28	33.84	828.21	61.65	33.27	832.10
7	54.52	36.80	787.59	56.83	36.29	791.92	59.16	35.76	796.34
8	51.97	39.35	748.24	54.19	38.93	752.99	56.47	38.45	757.89
9	49.20	42.12	706.12	51.37	41.75	711.24	53.61	41.31	716.58
10	46.28	45.04	661.08	48.38	44.74	666.50	50.50	44.42	672.16
11	43.15	48.17	612.91	45.12	48.00	618.50	47.17	47.75	624.41
12	39.81	51.51	561.40	41.67	51.45	567.05	43.59	51.33	573.08
13	36.20	55.12	506.28	37.95	55.17	511.88	39.74	55.18	517.90
14	32.36	58.96	447.32	33.97	59.15	452.73	35.58	59.34	458.56
15	28.26	63.06	384.26	29.68	63.44	389.29	31.16	63.76	394.80
16	23.87	67.45	316.81	25.10	68.02	321.27	26.37	68.55	326.25
17	19.18	72.14	244.67	20.19	72.93	248.34	21.25	73.67	252.58
18	14.15	77.17	167.50	14.90	78.22	170.12	15.72	79.20	173.38
19	8.80	82.52	84.98	9.25	83.87	86.25	9.79	85.13	88.25
20	3.05	84.98	0	3.19	86.25	0	3.40	88.25	0

¶112

Self-Liquidating Mortgages—Annual Interest, Amortization And Remaining Balance (continued)

20-Year Term (continued)

	7½% interest—$8.06 monthly payment			7¾% interest—$8.21 monthly payment			8% interest—$8.37 monthly payment		
Year	Interest	Amort.	Balance	Interest	Amort.	Balance	Interest	Amort.	Balance
1	74.24	22.48	977.52	76.74	21.78	978.22	79.24	21.20	978.80
2	72.49	24.23	953.29	75.00	23.52	954.70	77.47	22.97	955.83
3	70.63	26.09	927.20	73.10	25.42	929.28	75.58	24.86	930.97
4	68.58	28.14	899.06	71.05	27.47	901.81	73.49	26.95	904.02
5	66.40	30.32	868.74	68.84	29.68	872.13	71.29	29.15	874.87
6	64.04	32.68	836.06	66.47	32.05	840.08	68.85	31.59	843.28
7	61.51	35.21	800.85	63.89	34.63	805.45	66.21	34.23	809.05
8	58.81	37.91	762.94	61.09	37.43	768.02	63.37	37.07	771.98
9	55.83	40.89	722.05	58.11	40.41	727.61	60.32	40.12	731.86
10	52.66	44.06	677.99	54.86	43.66	683.95	56.99	43.45	688.41
11	49.25	47.47	630.52	51.35	47.17	636.78	53.36	47.08	641.33
12	45.56	51.16	579.36	47.56	50.96	585.82	49.46	50.98	590.35
13	41.58	55.14	524.22	43.49	55.03	530.79	45.26	55.18	535.17
14	37.32	59.40	464.82	39.05	59.47	471.32	40.67	59.77	475.40
15	32.69	64.03	400.79	34.28	64.24	407.08	35.67	64.77	410.63
16	27.73	68.99	331.80	29.12	69.40	337.68	30.31	70.13	340.50
17	22.37	74.35	257.45	23.52	75.00	262.68	24.51	75.93	264.57
18	16.61	80.11	177.34	17.53	80.99	181.69	18.20	82.24	182.33
19	10.37	86.35	90.99	11.01	87.51	94.18	11.33	89.11	93.22
20	3.67	90.99	0	3.99	94.18	0	3.97	93.22	0

	8¼% interest—$8.52 monthly payment			8½% interest—$8.68 monthly payment			8¾% interest—$8.84 monthly payment		
Year	Interest	Amort.	Balance	Interest	Amort.	Balance	Interest	Amort.	Balance
1	81.74	20.51	979.50	84.24	19.93	980.08	86.74	19.35	980.66
2	79.98	22.27	957.24	82.48	21.69	958.40	84.98	21.11	959.56
3	78.08	24.17	933.07	80.56	23.61	934.79	83.06	23.03	936.53
4	76.01	26.24	906.83	78.48	25.69	909.10	80.96	25.13	911.40
5	73.76	28.49	878.34	76.21	27.96	881.15	78.67	27.42	883.99
6	71.32	30.93	847.42	73.73	30.44	850.71	76.17	29.92	854.07
7	68.67	33.58	813.84	71.04	33.13	817.59	73.45	32.64	821.44
8	65.79	36.46	777.38	68.12	36.05	781.55	70.47	35.62	785.83
9	62.66	39.59	737.80	64.93	39.24	742.31	67.23	38.86	746.97
10	59.27	42.98	694.83	61.46	42.71	699.61	63.69	42.40	704.58
11	55.59	46.66	648.17	57.69	46.48	653.13	59.83	46.26	658.32
12	51.59	50.66	597.52	53.58	50.59	602.55	55.62	50.47	607.85
13	47.25	55.00	542.52	49.11	55.06	547.49	51.02	55.07	552.79
14	42.54	59.71	482.82	44.24	59.93	487.57	46.00	60.09	492.70
15	37.42	64.83	417.99	38.95	65.22	422.35	40.53	65.56	427.15
16	31.87	70.38	347.62	33.18	70.99	351.37	34.56	71.53	355.62
17	25.84	76.41	271.21	26.91	77.26	274.11	28.04	78.05	277.57
18	19.29	82.96	188.25	20.08	84.09	190.02	20.93	85.16	192.42
19	12.18	90.07	98.19	12.65	91.52	98.51	13.17	92.92	99.51
20	4.46	97.79	0.40	4.56	99.61	1.10	4.71	101.38	1.87

¶112

Self-Liquidating Mortgages—Annual Interest, Amortization
And Remaining Balance (continued)

20-Year Term (continued)

	9% interest—$9.00 monthly payment			9¼% interest—$9.16 monthly payment			9½% interest—$9.32 monthly payment		
Year	Interest	Amort.	Balance	Interest	Amort.	Balance	Interest	Amort.	Balance
1	89.24	18.76	981.24	91.75	18.18	981.83	94.25	17.60	982.41
2	87.47	20.53	960.71	89.99	19.94	961.89	92.51	19.34	963.07
3	85.56	22.44	938.27	88.07	21.86	940.04	90.59	21.26	941.81
4	83.45	24.55	913.72	85.96	23.97	916.07	88.48	23.37	918.45
5	81.14	26.86	886.86	83.65	26.28	889.79	86.16	25.69	892.76
6	78.64	29.36	857.50	81.11	28.82	860.98	83.61	28.24	864.52
7	75.83	32.17	825.33	78.33	31.60	829.38	80.81	31.04	833.48
8	72.85	35.15	790.18	75.28	34.65	794.73	77.73	34.12	799.36
9	69.56	38.44	751.74	71.93	38.00	756.74	74.34	37.51	761.86
10	65.96	42.04	709.70	68.27	41.66	715.08	70.62	41.23	720.63
11	62.01	45.99	663.71	64.25	45.68	669.40	66.52	45.33	675.31
12	57.70	50.30	613.41	59.84	50.09	619.31	62.03	49.82	625.49
13	52.97	55.03	558.38	55.00	54.93	564.39	57.08	54.77	570.73
14	47.80	60.20	498.18	49.70	60.23	504.16	51.65	60.20	510.53
15	42.19	65.81	432.37	43.89	66.04	438.13	45.67	66.18	444.36
16	35.99	72.01	360.36	37.51	72.42	365.71	39.11	72.74	371.62
17	29.20	78.80	281.56	30.52	79.41	286.31	31.89	79.96	291.66
18	21.86	86.14	195.42	22.86	87.07	199.24	23.95	87.90	203.76
19	13.76	94.24	101.18	14.45	95.48	103.77	15.23	96.62	107.15
20	4.93	101.18	0	5.24	104.69	0.91	5.64	106.21	0.94

	9¾% interest—$9.49 monthly payment			10% interest—$9.66 monthly payment			10¼% interest—$9.82 monthly payment		
Year	Interest	Amort.	Balance	Interest	Amort.	Balance	Interest	Amort.	Balance
1	96.75	17.14	982.87	99.25	16.67	983.33	101.76	16.09	983.92
2	95.01	18.88	963.99	97.49	18.43	964.90	100.04	17.81	966.11
3	93.08	20.81	943.19	95.59	20.33	944.57	98.12	19.73	946.39
4	90.96	22.93	920.26	93.44	22.48	922.09	96.00	21.85	924.55
5	88.62	25.27	895.00	91.09	24.83	897.26	93.66	24.19	900.36
6	86.04	27.85	867.16	88.51	27.41	869.85	91.06	26.79	873.57
7	83.21	30.68	836.48	85.62	30.30	839.55	88.18	29.67	843.90
8	80.08	33.81	802.67	82.45	33.47	806.08	84.99	32.86	811.05
9	76.63	37.26	765.42	78.94	36.98	769.10	81.46	36.39	774.66
10	72.83	41.06	724.36	75.08	40.84	728.26	77.55	40.30	734.36
11	68.64	45.25	679.12	70.81	45.11	683.15	73.22	44.63	689.74
12	64.03	49.86	629.26	66.07	49.85	633.30	68.42	49.43	640.31
13	58.95	54.94	574.32	60.86	55.06	578.24	63.11	54.74	585.58
14	53.34	60.55	513.78	55.11	60.81	517.43	57.23	60.62	524.97
15	47.17	66.72	447.07	48.71	67.21	450.22	50.72	67.13	457.84
16	40.37	73.52	373.55	41.68	74.24	375.98	43.51	74.34	383.50
17	32.87	81.02	292.53	33.90	82.02	293.96	35.52	82.33	301.18
18	24.61	89.28	203.25	25.32	90.60	203.36	26.67	91.18	210.00
19	15.50	98.39	104.87	15.83	100.09	103.27	16.87	100.98	109.03
20	5.47	108.42	3.55	5.35	103.27	0	6.02	111.83	2.79

¶112

Self-Liquidating Mortgages—Annual Interest, Amortization And Remaining Balance (continued)

20-Year Term (continued)

Year	10½% interest—$9.98 monthly payment			10¾% interest—$10.15 monthly payment			11% interest—$10.32 monthly payment		
	Interest	Amort.	Balance	Interest	Amort.	Balance	Interest	Amort.	Balance
1	104.27	15.50	984.51	106.78	15.03	984.98	109.29	14.56	985.45
2	102.57	17.20	967.31	105.08	16.73	968.26	107.60	16.25	969.20
3	100.67	19.10	948.22	103.19	18.62	949.64	105.72	18.13	951.08
4	98.57	21.20	927.02	101.09	20.72	928.93	103.62	20.23	930.86
5	96.23	23.54	903.49	98.75	23.06	905.87	101.28	22.57	908.29
6	93.64	26.13	877.36	96.15	25.66	880.21	98.67	25.18	883.12
7	90.76	29.01	848.35	93.25	28.56	851.66	95.76	28.09	855.04
8	87.56	32.21	816.15	90.02	31.79	819.87	92.51	31.34	823.70
9	84.01	35.76	780.40	86.43	35.38	784.50	88.88	34.97	788.74
10	80.07	39.70	740.70	82.44	39.37	745.13	84.84	39.01	749.73
11	75.70	44.07	696.64	77.99	43.82	701.32	80.32	43.53	706.21
12	70.84	48.93	647.71	73.04	48.77	652.55	75.29	48.56	657.66
13	65.45	54.32	593.40	67.53	54.28	598.28	69.67	54.18	603.48
14	59.46	60.31	533.09	61.40	60.41	537.87	63.40	60.45	543.04
15	52.82	66.95	466.15	54.58	67.23	470.65	56.41	67.44	475.60
16	45.44	74.33	391.82	46.98	74.83	395.82	48.60	75.25	400.35
17	37.25	82.52	309.31	38.53	83.28	312.55	39.89	83.96	316.40
18	28.16	91.61	217.70	29.13	92.68	219.87	30.18	93.67	222.74
19	18.06	101.71	116.00	18.66	103.15	116.72	19.34	104.51	118.23
20	6.85	112.92	3.08	7.01	114.80	1.92	7.25	116.60	1.64

Year	11¼% interest—$10.49 monthly payment			11½% interest—$10.66 monthly payment			11¾% interest—$10.84 monthly payment		
	Interest	Amort.	Balance	Interest	Amort.	Balance	Interest	Amort.	Balance
1	111.79	14.10	985.91	114.30	13.63	986.38	116.80	13.29	986.72
2	110.12	15.77	970.15	112.65	15.28	971.11	115.16	14.93	971.80
3	108.26	17.63	952.52	110.80	17.13	953.98	113.31	16.78	955.02
4	106.17	19.72	932.80	108.72	19.21	934.77	111.22	18.87	936.16
5	103.83	22.06	910.75	106.39	21.54	913.24	108.89	21.20	914.96
6	101.22	24.67	886.08	103.78	24.15	889.10	106.26	23.83	891.13
7	98.29	27.60	858.49	100.85	27.08	862.03	103.30	26.79	864.34
8	95.03	30.86	827.63	97.57	30.36	831.67	99.98	30.11	834.24
9	91.37	34.52	793.12	93.89	34.04	797.64	96.24	33.85	800.40
10	87.28	38.61	754.51	89.76	38.17	759.47	92.05	38.04	762.36
11	82.71	43.18	711.33	85.14	42.79	716.68	87.33	42.76	719.60
12	77.59	48.30	663.04	79.95	47.98	668.71	82.02	48.07	671.54
13	71.87	54.02	609.02	74.13	53.80	614.91	76.06	54.03	617.51
14	65.47	60.42	548.60	67.61	60.32	554.59	69.36	60.73	556.79
15	58.31	67.58	481.03	60.29	67.64	486.96	61.83	68.26	488.53
16	50.30	75.59	405.44	52.09	75.84	411.12	53.36	76.73	411.81
17	41.35	84.54	320.90	42.89	85.04	326.09	43.85	86.24	325.57
18	31.33	94.56	226.35	32.58	95.35	230.75	33.15	96.94	228.64
19	20.13	105.76	120.59	21.02	106.91	123.85	21.13	108.96	119.68
20	7.59	118.30	2.30	8.06	119.87	3.98	7.61	122.48	2.80

¶112

Self-Liquidating Mortgages—Annual Interest, Amortization And Remaining Balance (continued)

20-Year Term (continued)

	12% interest—$11.01 monthly payment			12⅜% interest—$11.27 monthly payment			12¾% interest—$11.54 monthly payment		
Year	Interest	Amort.	Balance	Interest	Amort.	Balance	Interest	Amort.	Balance
1	119.32	12.81	987.20	123.04	12.20	987.80	126.86	11.62	988.38
2	117.69	14.44	972.76	121.43	13.81	973.99	125.29	13.19	975.19
3	115.86	16.27	956.50	119.62	15.62	958.37	123.50	14.98	960.21
4	113.80	18.33	938.17	117.57	17.67	940.70	121.48	17.00	943.21
5	111.47	20.66	917.52	115.27	19.97	920.73	119.17	19.31	923.90
6	108.85	23.28	894.25	112.65	22.59	898.14	116.58	21.90	902.00
7	105.90	26.23	868.03	109.68	25.56	872.58	113.61	24.87	877.13
8	102.58	29.55	838.48	106.34	28.90	843.68	110.24	28.24	848.89
9	98.83	33.30	805.18	101.56	33.68	811.00	106.43	32.05	816.84
10	94.61	37.52	767.67	98.26	36.98	774.02	102.09	36.39	780.45
11	89.85	42.28	725.39	93.43	41.81	732.21	97.16	41.32	739.13
12	84.49	47.64	677.75	87.95	47.29	684.92	91.59	46.89	692.24
13	78.45	53.68	624.07	81.75	53.49	631.43	85.25	53.23	639.01
14	71.64	60.49	563.59	74.73	60.51	570.92	78.05	60.43	578.58
15	63.97	68.16	495.43	66.83	68.41	502.51	69.87	68.61	509.97
16	55.32	76.81	418.63	57.84	77.40	425.11	60.59	77.89	432.08
17	45.58	86.55	332.08	47.73	87.51	337.60	50.07	88.41	343.67
18	34.61	97.52	234.57	36.24	99.00	238.60	38.10	100.38	243.29
19	22.24	109.89	124.68	23.27	111.97	126.63	24.54	113.94	129.35
20	8.30	123.83	0.86	8.61	126.63	0	9.13	129.35	0

	13⅛% interest—$11.80 monthly payment			13½% interest—$12.07 monthly payment			13⅝% interest—$12.34 monthly payment		
Year	Interest	Amort.	Balance	Interest	Amort.	Balance	Interest	Amort.	Balance
1	130.54	11.06	988.94	134.31	10.53	989.47	138.07	10.01	989.99
2	129.00	12.60	976.34	132.82	12.02	977.45	136.60	11.48	978.51
3	127.24	14.36	961.98	131.08	13.76	963.69	134.90	13.18	965.33
4	125.25	16.35	945.63	129.10	15.74	947.95	132.94	15.14	950.19
5	122.95	18.65	926.98	126.85	17.99	929.96	130.71	17.37	932.82
6	120.36	21.24	905.74	124.25	20.59	909.37	128.13	19.95	912.87
7	117.40	24.20	881.54	121.29	23.55	885.82	125.20	22.88	889.99
8	114.03	27.57	853.97	117.93	26.91	858.91	121.80	26.28	863.71
9	110.17	31.43	822.54	114.04	30.80	828.11	117.92	30.16	833.55
10	105.80	35.80	786.74	109.62	35.22	792.89	113.45	34.63	798.92
11	100.81	40.79	745.95	104.57	40.27	752.62	108.33	39.75	759.17
12	95.11	46.49	699.46	98.77	46.07	706.55	102.44	45.64	713.53
13	88.63	52.97	646.49	92.17	52.67	653.88	95.71	52.37	661.16
14	81.25	60.35	586.14	84.59	60.25	593.63	87.96	60.12	601.04
15	72.83	68.77	517.37	75.93	68.91	524.72	79.05	69.03	532.01
16	62.89	78.71	438.66	66.03	78.81	445.91	68.85	79.23	452.78
17	52.67	88.93	349.73	54.72	90.12	355.79	57.13	90.95	361.83
18	39.87	101.73	248.00	41.76	103.08	252.71	43.68	104.40	257.43
19	25.68	115.92	132.08	26.96	117.88	134.83	28.22	119.86	137.57
20	9.52	132.08	0	10.01	134.83	0	10.51	137.57	0

¶112

Self-Liquidating Mortgages—Annual Interest, Amortization And Remaining Balance *(continued)*

20-Year Term *(continued)*

Year	14¼% interest—$12.62 monthly payment			14⅝% interest—$12.89 monthly payment			15% interest—$13.17 monthly payment		
	Interest	*Amort.*	*Balance*	*Interest*	*Amort.*	*Balance*	*Interest*	*Amort.*	*Balance*
1	141.93	9.51	990.49	145.64	9.04	990.96	149.45	8.59	991.41
2	140.48	10.96	979.53	144.23	10.45	980.51	148.07	9.97	981.44
3	138.81	12.63	966.90	142.58	12.10	968.41	146.46	11.58	969.86
4	136.89	14.55	952.35	140.70	13.98	954.43	144.61	13.43	956.43
5	134.68	16.76	935.59	138.51	16.17	938.26	142.45	15.59	940.84
6	132.13	19.31	916.28	135.99	18.69	919.57	139.94	18.10	922.74
7	129.19	22.25	894.03	133.05	21.63	897.94	137.03	21.01	901.73
8	125.79	25.65	868.38	129.67	25.01	872.93	133.65	24.39	877.34
9	121.90	29.54	838.84	125.85	28.83	844.10	129.74	28.30	849.04
10	117.41	34.03	804.81	121.14	33.54	810.56	125.18	32.86	816.18
11	112.22	39.22	765.59	116.00	38.68	771.88	119.91	38.13	778.05
12	106.26	45.18	720.41	109.95	44.73	727.15	113.47	44.57	733.48
13	99.37	52.07	668.34	102.94	51.74	675.41	106.95	51.09	682.39
14	91.45	59.99	608.35	94.86	59.82	615.59	98.39	59.65	622.74
15	82.33	69.11	539.24	85.50	69.18	546.41	88.80	69.24	553.50
16	71.81	79.63	459.61	74.66	80.02	466.39	77.68	80.36	473.14
17	59.68	91.76	367.85	62.16	92.52	373.87	64.76	93.28	379.86
18	45.73	105.71	262.14	47.67	107.01	266.86	49.76	108.28	271.58
19	29.64	121.80	140.34	30.94	123.74	143.12	32.35	125.69	145.89
20	11.11	140.34	0	11.56	143.12	0	12.15	145.89	0

Year	15⅜% interest—$13.45 monthly payment			15¾% interest—$13.73 monthly payment		
	Interest	*Amort.*	*Balance*	*Interest*	*Amort.*	*Balance*
1	153.24	8.16	991.84	157.01	7.75	992.25
2	151.90	9.50	982.34	155.70	9.06	983.19
3	150.33	11.07	971.27	154.17	10.59	972.60
4	148.48	12.92	958.35	152.37	12.39	960.21
5	146.38	15.02	943.33	150.27	14.49	945.72
6	143.88	17.52	925.81	147.82	16.94	928.78
7	141.00	20.40	905.41	144.95	19.81	908.97
8	137.62	23.78	881.63	141.59	23.17	885.80
9	133.71	27.69	853.94	137.67	27.09	858.71
10	129.13	32.27	821.67	133.09	31.67	827.04
11	123.81	37.59	784.08	127.71	37.05	789.99
12	117.60	43.80	740.28	121.44	43.32	746.67
13	110.37	51.03	689.25	114.11	50.65	696.02
14	101.96	59.44	629.81	105.52	59.24	636.78
15	92.13	69.27	560.54	95.48	69.28	567.50
16	80.70	80.70	479.84	83.76	81.00	486.50
17	67.39	94.01	385.83	70.04	94.72	391.78
18	51.86	109.54	276.29	53.99	110.77	281.01
19	33.79	127.61	148.68	35.23	129.53	151.48
20	12.72	148.68	0	13.28	151.48	0

¶112

Self-Liquidating Mortgages—Annual Interest, Amortization and Remaining Balance (continued)

25-Year Term

	4% interest—$5.28 monthly payment			4½% interest—$5.56 monthly payment			5% interest—$5.85 monthly payment		
Year	Interest	Amort.	Balance	Interest	Amort.	Balance	Interest	Amort.	Balance
1.	39.56	23.80	976.20	44.54	22.18	977.82	49.54	20.66	979.34
2	38.60	24.76	951.44	43.54	23.18	954.64	48.48	21.72	957.62
3	37.57	25.79	925.65	42.44	24.28	930.36	47.36	22.84	934.78
4	36.55	26.81	898.84	41.35	25.37	904.99	46.19	24.01	910.77
5	35.45	27.91	870.93	40.18	26.54	878.45	44.98	25.22	885.55
6	34.31	29.05	841.88	38.97	27.75	850.70	43.67	26.53	859.02
7	33.13	30.23	811.65	37.70	29.02	821.68	42.30	27.90	831.12
8	31.89	31.47	780.18	36.34	30.38	791.30	40.87	29.33	801.79
9	30.61	32.75	747.43	34.98	31.74	759.56	39.39	30.81	770.98
10	29.28	34.08	713.35	33.53	33.19	726.37	37.81	32.39	738.59
11	27.91	35.45	677.90	31.96	34.76	691.61	36.16	34.04	704.55
12	26.45	36.91	640.99	30.38	36.34	655.27	34.41	35.79	668.76
13	24.94	38.42	602.57	28.71	38.01	617.26	32.59	37.61	631.15
14	23.38	39.98	562.59	26.96	39.76	577.50	30.66	39.54	591.61
15	21.75	41.61	520.98	25.14	41.58	535.92	28.65	41.55	550.06
16	20.05	43.31	477.67	23.22	43.50	492.42	26.52	43.68	506.38
17	18.29	45.07	432.60	21.23	45.49	446.93	24.27	45.93	460.45
18	16.46	46.90	385.70	19.14	47.58	399.35	21.93	48.27	412.18
19	14.54	48.82	336.88	16.97	49.75	349.60	19.46	50.74	361.44
20	12.55	50.81	286.07	14.66	52.06	297.54	16.86	53.34	308.10
21	10.49	52.87	233.20	12.28	54.44	243.10	14.15	56.05	252.05
22	8.32	55.04	178.16	9.77	56.95	186.15	11.57	58.93	193.12
23	6.10	57.26	120.90	7.15	59.57	126.58	8.24	61.96	131.16
24	3.73	59.63	61.27	4.44	62.28	64.30	5.07	65.13	66.03
25	1.31	61.27	0	1.56	64.30	0	1.75	66.03	0

	5¼% interest—$6.00 monthly payment			5½% interest—$6.15 monthly payment			5¾% interest—$6.30 monthly payment		
Year	Interest	Amort.	Balance	Interest	Amort.	Balance	Interest	Amort.	Balance
1	52.03	19.97	980.03	54.51	19.29	980.71	57.02	18.58	981.42
2	50.96	21.04	958.99	53.44	20.36	960.35	55.90	19.70	961.72
3	49.82	22.18	936.81	52.27	21.53	938.82	54.77	20.83	940.89
4	48.63	23.37	913.44	51.07	22.73	916.09	53.52	22.08	918.82
5	47.38	24.62	888.82	49.79	24.01	892.08	52.23	23.37	895.44
6	46.01	25.99	862.83	48.42	25.38	866.70	50.82	24.78	870.66
7	44.70	27.30	835.53	46.98	26.82	839.88	49.39	26.21	844.45
8	43.16	28.84	806.69	45.50	28.30	811.58	47.83	27.77	816.68
9	41.63	30.37	776.32	43.88	29.92	781.66	46.18	29.42	787.26
10	40.00	32.00	744.32	42.20	31.60	750.06	44.48	31.12	756.14
11	38.27	33.73	710.59	40.42	33.38	716.68	42.61	32.99	723.15
12	36.47	35.53	675.06	38.55	35.25	681.43	40.67	34.93	688.22
13	34.55	37.45	637.61	36.55	37.25	644.18	38.62	36.98	651.24
14	32.54	39.46	598.15	34.44	39.36	604.82	36.41	39.18	612.06
15	30.42	41.58	556.57	32.22	41.58	563.24	34.11	41.49	570.57

Self-Liquidating Mortgages—Annual Interest, Amortization and Remaining Balance *(continued)*

25-Year Term *(continued)*

	5¼% interest—$6.00 monthly payment			5½% interest—$6.15 monthly payment			5¾% interest—$6.30 monthly payment		
Year	Interest	Amort.	Balance	Interest	Amort.	Balance	Interest	Amort.	Balance
16	28.16	43.84	512.73	29.90	43.90	519.34	31.65	43.95	526.62
17	25.82	46.18	466.55	27.41	46.39	472.95	29.07	46.53	480.09
18	23.34	48.66	417.89	24.79	49.01	423.94	26.27	49.33	430.76
19	20.71	51.29	366.60	22.04	51.76	372.18	23.41	52.19	378.57
20	17.98	54.02	312.58	19.12	54.68	317.50	20.33	55.27	323.30
21	15.06	56.94	255.64	16.01	57.79	259.71	17.05	58.55	264.75
22	11.99	60.01	195.63	12.76	61.04	198.67	13.61	61.99	202.76
23	8.76	63.24	132.39	9.32	64.48	134.19	9.94	65.66	137.10
24	5.36	66.64	65.75	5.70	68.10	66.09	6.07	69.53	67.57
25	1.78	65.75	0	1.83	66.09	0	1.97	67.57	0

	6% interest—$6.45 monthly payment			6¼% interest—$6.60 monthly payment			6½% interest—$6.76 monthly payment		
Year	Interest	Amort.	Balance	Interest	Amort.	Balance	Interest	Amort.	Balance
1	59.52	17.88	982.12	62.01	17.19	982.81	64.50	16.62	983.38
2	58.41	18.99	963.13	60.91	18.29	964.52	63.41	17.71	965.67
3	57.24	20.16	942.97	59.73	19.47	945.05	62.21	18.91	946.76
4	55.99	21.41	921.56	58.48	20.72	924.33	60.95	20.17	926.59
5	54.66	22.74	898.82	57.17	22.03	902.30	59.58	21.54	905.05
6	53.23	24.17	874.65	55.73	23.47	878.83	58.14	22.98	882.07
7	51.78	25.62	849.03	54.22	24.98	853.85	56.62	24.50	857.57
8	50.20	27.20	821.83	52.61	26.59	827.26	54.97	26.15	831.42
9	48.53	28.87	792.96	50.89	28.31	798.95	53.23	27.89	803.53
10	46.74	30.66	762.30	49.09	30.11	768.84	51.37	29.75	773.78
11	44.84	32.56	729.74	47.14	32.06	736.78	49.34	31.78	742.00
12	42.84	34.56	695.18	45.08	34.12	702.66	47.22	33.90	708.10
13	40.72	36.68	658.50	42.88	36.32	666.34	44.98	36.14	671.96
14	38.44	38.96	619.54	40.54	38.66	627.68	42.54	38.58	633.38
15	36.05	41.35	578.19	38.06	41.14	586.54	39.95	41.17	592.21
16	33.50	43.90	534.29	35.42	43.78	542.76	37.21	43.91	548.30
17	30.80	46.60	487.69	32.64	46.56	496.20	34.26	46.86	501.44
18	27.93	49.47	438.22	29.60	49.60	446.60	31.12	50.00	451.44
19	24.87	52.53	385.69	26.40	52.80	393.80	27.17	53.34	398.10
20	21.62	55.78	329.91	23.03	56.17	337.63	24.19	56.93	341.17
21	18.18	59.22	270.69	19.41	59.79	277.84	20.40	60.72	280.45
22	14.53	62.87	207.82	15.56	63.64	214.20	16.32	64.80	215.65
23	10.64	66.76	141.06	11.46	67.74	146.46	12.00	69.12	146.53
24	6.53	70.87	70.19	7.10	72.10	74.36	7.34	73.78	72.75
25	2.17	70.19	0	2.48	74.36	0	2.40	72.75	0

¶112

Self-Liquidating Mortgages—Annual Interest, Amortization And Remaining Balance (continued)

25-Year Term (continued)

	6¾% interest—$6.91 monthly payment			7% interest—$7.07 monthly payment			7¼% interest—$7.23 monthly payment		
Year	Interest	Amort.	Balance	Interest	Amort.	Balance	Interest	Amort.	Balance
1	67.02	15.90	984.10	69.51	15.33	984.67	72.02	14.74	985.26
2	65.92	17.00	967.10	68.40	16.44	968.23	70.91	15.85	969.41
3	64.72	18.20	948.90	67.20	17.64	950.59	69.72	17.04	952.37
4	63.45	19.47	929.43	65.94	18.90	931.69	68.44	18.32	934.05
5	62.10	20.82	908.61	64.62	20.22	911.47	67.01	19.75	914.30
6	60.66	22.26	886.35	63.12	21.72	889.75	65.58	21.18	893.12
7	59.09	23.83	862.52	61.55	23.29	866.46	64.01	22.75	870.37
8	57.44	25.48	837.04	59.85	24.99	841.47	62.30	24.46	845.91
9	55.68	27.24	809.80	58.06	26.78	814.69	60.47	26.29	819.62
10	53.77	29.15	780.65	56.12	28.72	785.97	58.49	28.27	791.35
11	51.74	31.18	749.47	54.05	30.79	755.18	56.39	30.37	760.98
12	49.57	33.35	716.12	51.81	33.03	722.15	54.10	32.66	728.32
13	47.24	35.68	680.44	49.44	35.40	686.75	51.66	35.10	693.22
14	44.78	38.14	642.30	46.87	37.97	648.78	49.02	37.74	655.48
15	42.10	40.82	601.48	44.14	40.70	608.08	46.20	40.56	614.92
16	39.24	43.68	557.80	41.20	43.64	564.44	43.15	43.61	571.31
17	36.22	46.70	511.10	38.03	46.81	517.63	39.89	46.87	524.44
18	32.98	49.94	461.16	34.65	50.19	467.44	36.35	50.41	474.03
19	29.51	53.41	407.75	31.02	53.82	413.62	32.59	54.17	419.86
20	25.79	57.13	350.62	27.12	57.72	355.90	28.53	58.23	361.63
21	21.79	61.13	289.49	22.99	61.85	294.05	24.15	62.61	299.02
22	17.56	65.36	224.13	18.48	66.36	227.69	19.47	67.29	231.73
23	12.99	69.93	154.20	13.67	71.17	156.52	14.43	72.33	159.40
24	8.11	74.81	79.39	8.52	76.32	80.20	8.98	77.78	81.62
25	2.92	79.39	0	3.01	80.20	0	3.17	81.62	0

	7½% interest—$7.39 monthly payment			7¾% interest—$7.56 monthly payment			8% interest—$7.72 monthly payment		
Year	Interest	Amort.	Balance	Interest	Amort.	Balance	Interest	Amort.	Balance
1	74.52	14.16	985.84	77.02	13.70	986.30	79.53	13.11	986.89
2	73.42	15.26	970.58	75.92	14.80	971.50	78.43	14.21	972.68
3	72.23	16.45	954.13	74.72	16.00	955.50	77.26	15.38	957.30
4	70.96	17.72	936.41	73.45	17.27	938.23	75.99	16.65	940.65
5	69.54	19.14	917.27	72.06	18.66	919.57	74.58	18.06	922.59
6	68.11	20.57	896.70	70.57	20.15	899.42	73.10	19.54	903.05
7	66.48	22.20	874.50	68.95	21.77	877.65	71.49	21.15	881.90
8	64.77	23.91	850.59	67.19	23.53	854.12	69.72	22.92	858.98
9	62.92	25.76	824.83	65.30	25.42	828.70	67.83	24.81	834.17
10	60.93	27.75	797.08	63.28	27.44	801.26	65.76	26.88	807.29
11	58.78	29.90	767.18	61.05	29.67	771.59	63.52	29.12	778.17
12	56.44	32.24	734.94	58.67	32.05	739.54	61.11	31.53	746.64
13	53.94	34.74	700.20	56.10	34.62	704.92	58.50	34.14	712.50
14	51.23	37.45	662.75	53.29	37.43	667.49	55.67	36.97	675.53
15	48.33	40.35	622.40	50.31	40.41	627.08	52.59	40.05	635.48

¶112

Self-Liquidating Mortgages—Annual Interest, Amortization And Remaining Balance *(continued)*

25-Year Term *(continued)*

	7½% interest—$7.39 monthly payment			7¾% interest—$7.56 monthly payment			8% interest—$7.72 monthly payment		
Year	Interest	Amort.	Balance	Interest	Amort.	Balance	Interest	Amort.	Balance
16	45.21	43.47	578.98	47.06	43.66	583.42	49.26	43.38	592.10
17	41.83	46.85	532.08	43.57	47.15	536.27	45.68	46.96	545.14
18	38.20	50.48	481.60	39.77	50.95	485.32	41.76	50.88	494.26
19	34.28	54.40	427.20	35.69	55.03	430.29	37.57	55.07	439.19
20	30.05	58.63	368.57	31.26	59.46	370.83	32.98	59.66	379.53
21	25.51	63.17	305.40	26.51	64.21	306.62	28.03	64.61	314.92
22	20.59	68.09	237.31	21.33	69.39	237.23	22.65	69.99	244.93
23	15.30	73.38	163.93	15.72	75.00	162.23	16.85	75.79	169.14
24	9.59	79.09	84.84	9.74	80.98	81.25	10.53	82.11	87.03
25	3.47	84.84	0	3.22	81.25	0	3.74	87.03	0

	8¼% interest—$7.88 monthly payment			8½% interest—$8.05 monthly payment			8¾% interest—$8.22 monthly payment		
Year	Interest	Amort.	Balance	Interest	Amort.	Balance	Interest	Amort.	Balance
1	82.04	12.53	987.48	84.54	12.07	987.94	87.05	11.60	988.41
2	80.96	13.61	973.88	83.48	13.13	974.81	85.99	12.66	975.75
3	79.80	14.77	959.11	82.32	14.29	960.52	84.84	13.81	961.95
4	78.53	16.04	943.08	81.05	15.56	944.97	83.58	15.07	946.88
5	77.16	17.41	925.68	79.68	16.93	928.04	82.21	16.44	930.44
6	75.67	18.90	906.78	78.18	18.43	909.62	80.71	17.94	912.51
7	74.05	20.52	886.27	76.55	20.06	889.57	79.08	19.57	892.94
8	72.29	22.28	863.99	74.78	21.83	867.74	77.29	21.36	871.59
9	70.38	24.19	839.81	72.85	23.76	843.99	75.35	23.30	848.29
10	68.31	26.26	813.56	70.75	25.86	818.14	73.23	25.42	822.88
11	66.06	28.51	785.05	68.47	28.14	790.00	70.91	27.74	795.14
12	63.62	30.95	754.11	65.98	30.63	759.37	68.39	30.26	764.88
13	60.97	33.60	720.51	63.27	33.34	726.04	65.63	33.02	731.87
14	58.09	36.48	684.03	60.33	36.28	689.76	62.62	36.03	695.84
15	54.96	39.61	644.43	57.12	39.49	650.28	59.34	39.31	656.54
16	51.57	43.00	601.43	53.63	42.98	607.30	55.76	42.89	613.65
17	47.88	46.69	554.75	49.83	46.78	560.53	51.85	46.80	566.86
18	43.88	50.69	504.07	45.70	50.91	509.62	47.59	51.06	515.80
19	39.54	55.03	449.05	41.20	55.41	454.21	42.94	55.71	460.09
20	34.83	59.74	389.31	36.30	60.31	393.91	37.86	60.79	399.31
21	29.71	64.86	324.45	30.97	65.64	328.27	32.33	66.32	332.99
22	24.15	70.42	254.03	25.17	71.44	256.83	26.29	72.36	260.64
23	18.11	76.46	177.58	18.85	77.76	179.08	19.69	78.96	181.68
24	11.56	83.01	94.58	11.98	84.63	94.45	12.50	86.15	95.54
25	4.45	90.12	4.46	4.50	92.11	2.35	4.65	94.00	1.55

Self-Liquidating Mortgages—Annual Interest, Amortization And Remaining Balance (continued)

25-Year Term (continued)

Year	9% interest—$8.40 monthly payment			9¼% interest—$8.56 monthly payment			9½% interest—$8.74 monthly payment		
	Interest	Amort.	Balance	Interest	Amort.	Balance	Interest	Amort.	Balance
1	89.54	11.26	988.74	92.06	10.67	989.34	94.56	10.33	989.68
2	88.49	12.31	976.43	91.03	11.70	977.65	93.54	11.35	978.34
3	87.34	13.46	962.97	89.90	12.83	964.82	92.41	12.48	965.86
4	86.07	14.73	948.24	88.66	14.07	950.76	91.17	13.72	952.15
5	84.66	16.14	932.10	87.31	15.42	935.35	89.81	15.08	937.08
6	83.18	17.62	914.48	85.82	16.91	918.44	88.32	16.57	920.52
7	81.52	19.28	895.20	84.19	18.54	899.90	86.67	18.22	902.31
8	79.72	21.08	874.12	82.40	20.33	879.58	84.87	20.02	882.29
9	77.73	23.07	851.05	80.44	22.29	857.29	82.88	22.01	860.28
10	75.58	25.22	825.83	78.28	24.45	832.85	80.70	24.19	836.10
11	74.10	27.60	798.23	75.93	26.80	806.05	78.30	26.59	809.51
12	70.62	30.18	768.05	73.34	29.39	776.66	75.66	29.23	780.28
13	67.79	33.01	735.04	70.50	32.23	744.44	72.76	32.13	748.15
14	64.69	36.11	698.93	67.39	35.34	709.10	69.57	35.32	712.83
15	61.29	39.51	659.42	63.98	38.75	670.36	66.06	38.83	674.01
16	57.60	43.20	616.22	60.24	42.49	627.88	62.21	42.68	631.33
17	53.56	47.24	568.98	56.14	46.59	581.29	57.97	46.92	584.42
18	49.12	51.68	517.30	51.64	51.09	530.21	53.32	51.57	532.85
19	44.27	56.53	460.77	46.71	56.02	474.20	48.20	56.69	476.17
20	38.96	61.84	398.93	41.31	61.42	412.78	42.57	62.32	413.85
21	33.17	67.63	331.30	35.38	67.35	345.44	36.39	68.50	345.36
22	26.81	73.99	257.31	28.88	73.85	271.59	29.59	75.30	270.07
23	19.86	80.94	176.37	21.75	80.98	190.62	22.12	82.77	187.30
24	12.27	88.53	87.84	13.94	88.79	101.83	13.90	90.99	96.32
25	3.96	87.84	0	5.36	97.37	4.47	4.87	100.02	0

Year	9¾% interest—$8.91 monthly payment			10% interest—$9.09 monthly payment			10¼% interest—$9.26 monthly payment		
	Interest	Amort.	Balance	Interest	Amort.	Balance	Interest	Amort.	Balance
1	97.07	9.86	990.15	99.56	9.52	990.48	102.09	9.04	990.97
2	96.07	10.86	979.30	98.57	10.51	979.97	101.12	10.01	980.96
3	94.96	11.97	967.33	97.48	11.60	968.37	100.04	11.09	969.88
4	93.74	13.19	954.15	96.25	12.83	955.54	98.85	12.28	957.60
5	92.40	14.53	939.62	94.93	14.15	941.39	97.53	13.60	944.01
6	90.91	16.02	923.61	93.44	15.64	925.75	96.07	15.06	928.96
7	89.28	17.65	905.96	91.78	17.30	908.45	94.45	16.68	912.29
8	87.48	19.45	886.52	90.01	19.07	889.38	92.66	18.47	893.82
9	85.50	21.43	865.09	87.99	21.09	868.29	90.68	20.45	873.38
10	83.31	23.62	841.48	85.78	23.30	844.99	88.48	22.65	850.73
11	80.91	26.02	815.47	83.33	25.75	819.24	86.05	25.08	825.66
12	78.25	28.68	786.79	80.63	28.45	790.79	83.35	27.78	797.88
13	75.33	31.60	755.20	77.68	31.40	759.39	80.37	30.76	767.13
14	72.11	34.82	720.38	74.39	34.69	724.70	77.06	34.07	733.07
15	68.56	38.37	682.01	70.74	38.34	686.36	73.40	37.73	695.35

Self-Liquidating Mortgages—Annual Interest, Amortization And Remaining Balance (continued)

25-Year Term (continued)

Year	9¾% interest—$8.91 monthly payment			10% interest—$9.09 monthly payment			10¼% interest—$9.26 monthly payment		
	Interest	Amort.	Balance	Interest	Amort.	Balance	Interest	Amort.	Balance
16	64.64	42.29	639.73	66.74	42.34	644.02	69.35	41.78	653.57
17	60.33	46.60	593.14	62.29	46.79	597.23	64.86	46.27	607.31
18	55.58	51.35	541.80	57.39	51.69	545.54	59.89	51.24	556.08
19	50.35	56.58	485.22	51.96	57.12	488.42	54.39	56.74	499.34
20	44.58	62.35	422.87	46.01	63.07	425.35	48.29	62.84	436.50
21	38.22	68.71	354.16	39.37	69.71	355.64	41.54	69.59	366.92
22	31.21	75.72	278.45	32.09	76.99	278.65	34.06	77.07	289.85
23	23.49	83.44	195.02	24.06	85.02	193.63	25.78	85.35	204.50
24	14.98	91.95	103.07	15.13	93.95	99.68	16.61	94.52	109.99
25	5.61	101.32	1.75	5.29	99.68	0	6.45	104.68	5.31

Year	10½% interest—$9.44 monthly payment			10¾% interest—$9.62 monthly payment			11% interest—$9.80 monthly payment		
	Interest	Amort.	Balance	Interest	Amort.	Balance	Interest	Amort.	Balance
1	104.59	8.70	991.31	107.10	8.35	991.66	109.61	8.00	992.01
2	103.64	9.65	981.67	106.16	9.29	982.38	108.68	8.93	983.09
3	102.57	10.72	970.96	105.11	10.34	972.04	107.65	9.96	973.14
4	101.39	11.90	959.06	103.94	11.51	960.54	106.50	11.11	962.03
5	100.08	13.21	945.86	102.64	12.81	947.74	105.22	12.39	949.64
6	98.63	14.66	931.20	101.20	14.25	933.49	103.78	13.83	935.82
7	97.01	16.28	914.93	99.59	15.86	917.64	102.18	15.43	920.40
8	95.22	18.07	896.87	97.80	17.65	899.99	100.40	17.21	903.19
9	93.23	20.06	876.81	95.80	19.65	880.35	98.41	19.20	883.99
10	91.02	22.27	854.54	93.59	21.86	858.49	96.19	21.42	862.57
11	88.56	24.73	829.82	91.12	24.33	834.16	93.71	23.90	838.67
12	85.84	27.45	802.38	88.37	27.08	807.08	90.94	26.67	812.01
13	82.82	30.47	771.91	85.31	30.14	776.95	87.86	29.75	782.26
14	79.46	33.83	738.08	81.91	33.54	743.41	84.41	33.20	749.07
15	75.73	37.56	700.52	78.12	37.33	706.08	80.57	37.04	712.03
16	71.59	41.70	658.83	73.90	41.55	664.54	76.29	41.32	670.72
17	67.00	46.29	612.54	69.21	46.24	618.30	71.50	46.11	624.62
18	61.89	51.40	561.15	63.99	51.46	566.84	66.17	51.44	573.18
19	56.23	57.06	504.10	58.17	57.28	509.57	60.22	57.39	515.79
20	49.94	63.35	440.76	51.70	63.75	445.82	53.58	64.03	451.77
21	42.96	70.33	370.44	44.50	70.95	374.88	46.17	71.44	380.33
22	35.21	78.08	292.36	36.49	78.96	295.93	37.90	79.71	300.63
23	26.61	86.68	205.69	27.57	87.88	208.05	28.68	88.93	211.70
24	17.06	96.23	109.46	17.64	97.81	110.25	18.39	99.22	112.49
25	6.45	106.84	2.63	6.60	108.85	1.40	6.91	110.70	1.79

¶112

Self-Liquidating Mortgages—Annual Interest, Amortization And Remaining Balance (continued)

25-Year Term (continued)

Year	11¼% interest—$9.98 monthly payment			11½% interest—$10.16 monthly payment			11¾% interest—$10.35 monthly payment		
	Interest	Amort.	Balance	Interest	Amort.	Balance	Interest	Amort.	Balance
1	112.12	7.65	992.36	114.63	7.30	992.71	117.13	7.08	992.93
2	111.21	8.56	983.81	113.74	8.19	984.53	116.25	7.96	984.98
3	110.20	9.57	974.24	112.75	9.18	975.35	115.27	8.94	976.05
4	109.07	10.70	963.54	111.64	10.29	965.07	114.16	10.05	966.00
5	107.80	11.97	951.58	110.39	11.54	953.53	112.91	11.30	954.71
6	106.38	13.39	938.19	108.99	12.94	940.60	111.51	12.70	942.02
7	104.79	14.98	923.22	107.43	14.50	926.10	109.94	14.27	927.75
8	103.02	16.75	906.48	105.67	16.26	909.85	108.17	16.04	911.72
9	101.04	18.73	887.75	103.70	18.23	891.62	106.18	18.03	893.69
10	98.82	20.95	866.80	101.49	20.44	871.18	103.95	20.26	873.44
11	96.34	23.43	843.37	99.01	22.92	848.26	101.43	22.78	850.66
12	93.56	26.21	817.17	96.23	25.70	822.56	98.61	25.60	825.07
13	90.45	29.32	787.86	93.11	28.82	793.75	95.43	28.78	796.29
14	86.98	32.79	755.07	89.62	32.31	761.44	91.86	32.35	763.95
15	83.10	36.67	718.41	85.70	36.23	725.21	87.85	36.36	727.60
16	78.75	41.02	677.39	81.31	40.62	684.60	83.34	40.87	686.74
17	73.89	45.88	631.52	76.38	45.55	639.05	78.27	45.94	640.81
18	68.46	51.31	580.22	70.86	51.07	587.99	72.58	51.63	589.18
19	62.38	57.39	522.83	64.67	57.26	530.73	66.17	58.04	531.15
20	55.58	64.19	458.65	57.72	64.21	466.53	58.98	65.23	465.92
21	47.98	71.79	386.86	49.94	71.99	394.54	50.88	73.33	392.60
22	39.47	80.30	306.56	41.21	80.72	313.83	41.79	82.42	310.19
23	29.96	89.81	216.75	31.42	90.51	223.33	31.57	92.64	217.55
24	19.31	100.46	116.30	20.45	101.48	121.85	20.08	104.13	113.42
25	7.41	112.36	3.95	8.14	113.79	8.07	7.16	117.05	0

12% interest—$10.35 monthly payment

Year	Interest	Amort.	Balance	Year	Interest	Amort.	Balance
1	119.64	6.73	993.28	13	98.20	28.17	802.73
2	118.79	7.58	985.71	14	94.62	31.75	770.99
3	117.83	8.54	977.17	15	90.60	35.77	735.23
4	116.75	9.62	967.56	16	86.06	40.31	694.93
5	115.53	10.84	956.72	17	80.95	45.42	649.51
6	114.15	12.22	944.51	18	75.19	51.18	598.34
7	112.60	13.77	930.75	19	68.70	57.67	540.68
8	110.86	15.51	915.24	20	61.39	64.98	475.70
9	108.89	17.48	897.77	21	53.15	73.22	402.48
10	106.68	19.69	878.08	22	43.86	82.51	319.98
11	104.18	22.19	855.90	23	33.40	92.97	227.01
12	101.37	25.00	830.90	24	21.61	104.76	122.26
				25	8.32	118.05	4.22

¶112

Self-Liquidating Mortgages—Annual Interest, Amortization And Remaining Balance (continued)

30-Year Term

Year	4% interest—$4.77 monthly payment			4¼% interest—$4.92 monthly payment			4½% interest—$5.07 monthly payment		
	Interest	Amort.	Balance	Interest	Amort.	Balance	Interest	Amort.	Balance
1	39.69	17.56	982.45	42.18	16.87	983.14	44.67	16.18	983.83
2	38.97	18.28	964.17	41.45	17.60	965.54	43.93	16.92	966.92
3	38.23	19.02	945.15	40.69	18.36	947.18	43.15	17.70	949.23
4	37.45	19.80	925.36	39.89	19.16	928.03	42.34	18.51	930.73
5	36.64	20.61	904.76	39.06	19.99	908.04	41.49	19.36	911.37
6	35.80	21.45	883.32	38.19	20.86	887.19	40.60	20.25	891.13
7	34.93	22.32	861.00	37.29	21.76	865.44	39.67	21.18	869.96
8	34.02	23.23	837.78	36.35	22.70	842.74	38.70	22.15	847.81
9	33.08	24.17	813.61	35.36	23.69	819.06	37.68	23.17	824.65
10	32.09	25.16	788.46	34.34	24.71	794.35	36.62	24.23	800.42
11	31.07	26.18	762.28	33.27	25.78	768.57	35.51	25.34	775.08
12	30.00	27.25	735.03	32.15	26.90	741.67	34.34	26.51	748.58
13	28.89	28.36	706.68	30.98	28.07	713.61	33.12	27.73	720.86
14	27.73	29.52	677.17	29.77	29.28	684.33	31.85	29.00	691.86
15	26.53	30.72	646.45	28.50	30.55	653.79	30.52	30.33	661.54
16	25.28	31.97	614.49	27.17	31.88	621.92	29.13	31.72	629.82
17	23.98	33.27	581.23	25.79	33.26	588.67	27.67	33.18	596.64
18	22.62	34.63	546.60	24.35	34.70	553.97	26.14	34.71	561.94
19	21.21	36.04	510.57	22.85	36.20	517.78	24.55	36.30	525.64
20	19.75	37.50	473.07	21.28	37.77	480.01	22.88	37.97	487.68
21	18.22	39.03	434.04	19.64	39.41	440.61	21.14	39.71	447.97
22	16.63	40.62	393.43	17.94	41.11	399.50	19.31	41.54	406.44
23	14.97	42.28	351.15	16.15	42.90	356.61	17.41	43.44	363.00
24	13.25	44.00	307.16	14.30	44.75	311.86	15.41	45.44	317.57
25	11.46	45.79	261.37	12.36	46.69	265.17	13.32	47.53	270.05
26	9.59	47.66	213.72	10.33	48.72	216.46	11.14	49.71	220.34
27	7.65	49.60	164.13	8.22	50.83	165.64	8.86	51.99	168.35
28	5.63	51.62	112.51	6.02	53.03	112.61	6.47	54.38	113.98
29	3.53	53.72	58.79	3.72	55.33	57.29	3.97	56.88	57.10
30	1.34	55.91	2.89	1.32	57.73	0	1.36	59.49	0

Self-Liquidating Mortgages—Annual Interest, Amortization And Remaining Balance (continued)

30-Year Term (continued)

Year	4¾% interest—$5.22 monthly payment			5% interest—$5.37 monthly payment			5¼% interest—$5.52 monthly payment		
	Interest	Amort.	Balance	Interest	Amort.	Balance	Interest	Amort.	Balance
1	47.17	15.48	984.53	49.67	14.78	985.23	52.17	14.08	985.93
2	46.42	16.23	968.31	48.91	15.54	969.70	51.41	14.84	971.10
3	45.63	17.02	951.29	48.12	16.33	953.37	50.61	15.64	955.47
4	44.81	17.84	933.45	47.28	17.17	936.21	49.77	16.48	939.00
5	43.94	18.71	914.75	46.41	18.04	918.17	48.89	17.36	921.64
6	43.03	19.62	895.14	45.48	18.97	899.21	47.95	18.30	903.35
7	42.08	20.57	874.57	44.51	19.94	879.28	46.97	19.28	884.08
8	41.08	21.57	853.01	43.49	20.96	858.32	45.93	20.32	863.76
9	40.03	22.62	830.40	42.42	22.03	836.30	44.84	21.41	842.36
10	38.94	23.71	806.69	41.29	23.16	813.15	43.69	22.56	819.81
11	37.79	24.86	781.83	40.11	24.34	788.81	42.48	23.77	796.04
12	36.58	26.07	755.76	38.86	25.59	763.23	41.20	25.05	771.00
13	35.31	27.34	728.43	37.56	26.89	736.34	39.85	26.40	744.60
14	33.99	28.66	699.77	36.18	28.27	708.08	38.43	27.82	716.79
15	32.60	30.05	669.72	34.73	29.72	678.37	36.94	29.31	687.49
16	31.14	31.51	638.22	33.21	31.24	647.14	35.36	30.89	656.60
17	29.61	33.04	605.18	31.62	32.83	614.31	33.70	32.55	624.06
18	28.00	34.65	570.54	29.94	34.51	579.80	31.95	34.30	589.76
19	26.32	36.33	534.21	28.17	36.28	543.52	30.11	36.14	553.62
20	24.56	38.09	496.12	26.32	38.13	505.39	28.16	38.09	515.54
21	22.71	39.94	456.19	24.36	40.09	465.31	26.11	40.14	475.41
22	20.77	41.88	414.31	22.31	42.14	423.18	23.96	42.29	433.12
23	18.74	43.91	370.41	20.16	44.29	378.90	21.68	44.57	388.56
24	16.61	46.04	324.37	17.89	46.56	332.34	19.28	46.97	341.60
25	14.37	48.28	276.09	15.51	48.94	283.41	16.76	49.49	292.11
26	12.03	50.62	225.47	13.01	51.44	231.97	14.10	52.15	239.97
27	9.57	53.08	172.40	10.38	54.07	177.90	11.29	54.96	185.01
28	6.99	55.66	116.74	7.61	56.84	121.06	8.34	57.91	127.11
29	4.29	58.36	58.39	4.70	59.75	61.32	5.22	61.03	66.08
30	1.46	61.19	0	1.64	62.81	0	1.94	64.31	1.78

Self-Liquidating Mortgages—Annual Interest, Amortization And Remaining Balance (continued)

30-Year Term (continued)

Year	5½% interest—$5.68 monthly payment			5¾% interest—$5.84 monthly payment			6% interest—$6.00 monthly payment		
	Interest	Amort.	Balance	Interest	Amort.	Balance	Interest	Amort.	Balance
1	54.67	13.50	986.51	57.17	12.92	987.09	59.67	12.34	987.67
2	53.91	14.26	972.25	56.41	13.68	973.41	58.91	13.10	974.57
3	53.10	15.07	957.19	55.60	14.49	958.92	58.10	13.91	960.67
4	52.25	15.92	941.28	54.74	15.35	943.58	57.24	14.77	945.91
5	51.36	16.81	924.47	53.84	16.25	927.33	56.33	15.68	930.23
6	50.41	17.76	906.71	52.88	17.21	910.12	55.37	16.64	913.60
7	49.41	18.76	887.95	51.86	18.23	891.90	54.34	17.67	895.93
8	48.35	19.82	868.13	50.79	19.30	872.60	53.25	18.76	877.18
9	47.23	20.94	847.19	49.65	20.44	852.16	52.09	19.92	857.27
10	46.05	22.12	825.08	48.44	21.65	830.52	50.87	21.14	836.13
11	44.80	23.37	801.71	47.16	22.93	807.59	49.56	22.45	813.68
12	43.48	24.69	777.03	45.81	24.28	783.32	48.18	23.83	789.85
13	42.09	26.08	750.96	44.37	25.72	757.61	46.71	25.30	764.56
14	40.62	27.55	723.41	42.86	27.23	730.38	45.15	26.86	737.70
15	39.07	29.10	694.31	41.25	28.84	701.54	43.49	28.52	709.19
16	37.42	30.75	663.57	39.55	30.54	671.00	41.73	30.28	678.91
17	35.69	32.48	631.10	37.74	32.35	638.66	39.87	32.14	646.77
18	33.86	34.31	596.79	35.83	34.26	604.41	37.88	34.13	612.65
19	31.92	36.25	560.55	33.81	36.28	568.14	35.78	36.23	576.43
20	29.88	38.29	522.27	31.67	38.42	529.72	33.54	38.47	537.96
21	27.72	40.45	481.82	29.40	40.69	489.04	31.17	40.84	497.13
22	25.44	42.73	439.10	27.00	43.09	445.96	28.65	43.36	453.78
23	23.03	45.14	393.96	24.46	45.63	400.33	25.98	46.03	407.75
24	20.48	47.69	346.28	21.76	48.33	352.01	23.14	48.87	358.89
25	17.79	50.38	295.90	18.91	51.18	300.84	20.13	51.88	307.01
26	14.95	53.22	242.69	15.89	54.20	246.64	16.93	55.08	251.93
27	11.95	56.22	186.47	12.69	57.40	189.24	13.53	58.48	193.46
28	8.78	59.39	127.09	9.30	60.79	128.46	9.92	62.09	131.38
29	5.43	62.74	64.35	5.71	64.38	64.09	6.09	65.92	65.47
30	1.89	66.28	0	1.91	68.18	0	2.03	69.98	0

¶112

Self-Liquidating Mortgages—Annual Interest, Amortization And Remaining Balance (continued)

30-Year Term (continued)

Year	6¼% interest—$6.16 monthly payment			6½% interest—$6.32 monthly payment			6¾% interest—$6.49 monthly payment		
	Interest	Amort.	Balance	Interest	Amort.	Balance	Interest	Amort.	Balance
1	62.17	11.76	988.25	64.68	11.17	988.84	67.18	10.71	989.30
2	61.42	12.51	975.74	63.93	11.92	976.92	66.43	11.46	977.85
3	60.61	13.32	962.43	63.13	12.72	964.20	65.63	12.26	965.59
4	59.76	14.17	948.26	62.28	13.57	950.64	64.78	13.11	952.49
5	58.84	15.09	933.18	61.37	14.48	936.16	63.87	14.02	938.48
6	57.87	16.06	917.13	60.40	15.45	920.72	62.89	15.00	923.48
7	56.84	17.09	900.04	59.37	16.48	904.24	61.85	16.04	907.45
8	55.74	18.19	881.86	58.26	17.59	886.66	60.73	17.16	890.30
9	54.57	19.36	862.51	57.09	18.76	867.90	59.54	18.35	871.95
10	53.33	20.60	841.91	55.83	20.02	847.88	58.26	19.63	852.33
11	52.00	21.93	819.99	54.49	21.36	826.52	56.90	20.99	831.34
12	50.59	23.34	796.66	53.06	22.79	803.74	55.43	22.46	808.89
13	49.09	24.84	771.83	51.53	24.32	779.42	53.87	24.02	784.87
14	47.50	26.43	745.40	49.90	25.95	753.48	52.20	25.69	759.19
15	45.80	28.13	717.27	48.17	27.68	725.80	50.41	27.48	731.71
16	43.99	29.94	687.33	46.31	29.54	696.27	48.50	29.39	702.32
17	42.06	31.87	655.46	44.33	31.52	664.76	46.45	31.44	670.89
18	40.01	33.92	621.55	42.22	33.63	631.14	44.26	33.63	637.27
19	37.83	36.10	585.45	39.97	35.88	595.27	41.92	35.97	601.30
20	35.51	38.42	547.04	37.57	38.28	556.99	39.42	38.47	562.84
21	33.04	40.89	506.15	35.01	40.84	516.15	36.74	41.15	521.69
22	30.41	43.52	462.63	32.27	43.58	472.58	33.87	44.02	477.68
23	27.61	46.32	416.31	29.35	46.50	426.09	30.81	47.08	430.60
24	24.63	49.30	367.01	26.24	49.61	376.48	27.53	50.36	380.25
25	21.46	52.47	314.54	22.92	52.93	323.56	24.03	53.86	326.39
26	18.08	55.85	258.70	19.37	56.48	267.08	20.28	57.61	268.78
27	14.49	59.44	199.27	15.59	60.26	206.83	16.26	61.63	207.16
28	10.67	63.26	136.01	11.56	64.29	142.54	11.97	65.92	141.25
29	6.60	67.33	68.68	7.25	68.60	73.95	7.38	70.51	70.75
30	2.27	71.66	0	2.66	73.19	0.76	2.48	75.41	0

¶112

Self-Liquidating Mortgages—Annual Interest, Amortization And Remaining Balance (continued)

30-Year Term (continued)

Year	7% interest—$6.65 monthly payment			7¼% interest—$6.82 monthly payment			7½% interest—$6.99 monthly payment		
	Interest	Amort.	Balance	Interest	Amort.	Balance	Interest	Amort.	Balance
1	69.68	10.13	989.88	72.19	9.66	990.35	74.69	9.20	990.81
2	68.95	10.86	979.03	71.46	10.39	979.97	73.98	9.91	980.91
3	68.17	11.64	967.40	70.69	11.16	968.81	73.21	10.68	970.23
4	67.33	12.48	954.92	69.85	12.00	956.81	72.38	11.51	958.73
5	66.43	13.38	941.54	68.95	12.90	943.92	71.49	12.40	946.33
6	65.46	14.35	927.19	67.98	13.87	930.06	70.53	13.36	932.98
7	64.42	15.39	911.81	66.95	14.90	915.16	69.49	14.40	918.58
8	63.31	16.50	895.31	65.83	16.02	899.14	68.37	15.52	903.07
9	62.12	17.69	877.62	64.63	17.22	881.93	67.17	16.72	886.35
10	60.84	18.97	858.65	63.34	18.51	863.42	65.87	18.02	868.34
11	59.47	20.34	838.31	61.95	19.90	843.52	64.47	19.42	848.92
12	58.00	21.81	816.50	60.46	21.39	822.14	62.96	20.93	828.00
13	56.42	23.39	793.12	58.86	22.99	799.15	61.34	22.55	805.46
14	54.73	25.08	768.04	57.13	24.72	774.43	59.59	24.30	781.16
15	52.92	26.89	741.15	55.28	26.57	747.87	57.70	26.19	754.98
16	50.97	28.84	712.32	53.29	28.56	719.31	55.67	28.22	726.77
17	48.89	30.92	681.40	51.15	30.70	688.62	53.48	30.41	696.36
18	46.65	33.16	648.25	48.85	33.00	655.62	51.12	32.77	663.60
19	44.26	35.55	612.70	46.37	35.48	620.15	48.58	35.31	628.29
20	41.69	38.12	574.58	43.71	38.14	582.02	45.84	38.05	590.24
21	38.93	40.88	533.71	40.86	40.99	541.03	42.88	41.01	549.24
22	35.98	43.83	489.88	37.78	44.07	496.97	39.70	44.19	505.06
23	32.81	47.00	442.88	34.48	47.37	449.60	36.27	47.62	457.44
24	29.41	50.40	392.48	30.93	50.92	398.69	32.57	51.32	406.13
25	25.77	54.04	338.45	27.11	54.74	343.96	28.59	55.30	350.83
26	21.86	57.95	280.50	23.01	58.84	285.13	24.30	59.59	291.24
27	17.67	62.14	218.37	18.60	63.25	221.88	19.67	64.22	227.03
28	13.18	66.63	151.74	13.86	67.99	153.90	14.69	69.20	157.83
29	8.36	71.45	80.30	8.77	73.08	80.82	9.31	74.58	83.26
30	3.20	76.61	3.70	3.29	78.56	2.26	3.52	80.37	2.90

¶112

Self-Liquidating Mortgages—Annual Interest, Amortization And Remaining Balance (continued)

30-Year Term (continued)

Year	7¾% interest—$7.16 monthly payment			8% interest—$7.34 monthly payment			8¼% interest—$7.51 monthly payment		
	Interest	Amort.	Balance	Interest	Amort.	Balance	Interest	Amort.	Balance
1	77.20	8.73	991.28	79.70	8.39	991.62	82.21	7.92	992.09
2	76.50	9.43	981.85	79.01	9.08	982.54	81.53	8.60	983.50
3	75.74	10.19	971.67	78.25	9.84	972.71	80.80	9.33	974.17
4	74.92	11.01	960.67	77.44	10.65	962.06	80.00	10.13	964.04
5	74.04	11.89	948.78	76.55	11.54	950.53	79.13	11.00	953.04
6	73.09	12.84	935.94	75.60	12.49	938.04	78.19	11.94	941.10
7	72.05	13.88	922.07	74.56	13.53	924.52	77.16	12.97	928.14
8	70.94	14.99	907.09	73.44	14.65	909.87	76.05	14.08	914.07
9	69.74	16.19	890.90	72.22	15.87	894.00	74.85	15.28	898.79
10	68.44	17.49	873.41	70.90	17.19	876.82	73.54	16.59	882.20
11	67.03	18.90	854.52	69.48	18.61	858.21	72.12	18.01	864.19
12	65.52	20.41	834.11	67.93	20.16	838.06	70.57	19.56	844.64
13	63.88	22.05	812.06	66.26	21.83	816.24	68.90	21.23	823.41
14	62.11	23.82	788.24	64.45	23.64	792.60	67.08	23.05	800.36
15	60.19	25.74	762.51	62.49	25.60	767.01	65.10	25.03	775.34
16	58.13	27.80	734.71	60.36	27.73	739.28	62.96	27.17	748.17
17	55.89	30.04	704.68	58.06	30.03	709.26	60.63	29.50	718.68
18	53.48	32.45	672.23	55.57	32.52	676.75	58.10	32.03	686.65
19	50.87	35.06	637.18	52.87	35.22	641.53	55.36	34.77	651.89
20	48.06	37.87	599.32	49.95	38.14	603.40	52.38	37.75	614.14
21	45.02	40.91	558.41	46.78	41.31	562.10	49.14	40.99	573.16
22	41.73	44.20	514.22	43.36	44.73	517.37	45.63	44.50	528.67
23	38.18	47.75	466.48	39.64	48.45	468.93	41.82	48.31	480.36
24	34.35	51.58	414.90	35.62	52.47	416.46	37.68	52.45	427.92
25	30.21	55.72	359.19	31.27	56.82	359.65	33.19	56.94	370.98
26	25.73	60.20	299.00	26.55	61.54	298.11	28.31	61.82	309.16
27	20.90	65.03	233.97	21.45	66.64	231.48	23.01	67.12	242.05
28	15.68	70.25	163.72	15.91	72.18	159.30	17.26	72.87	169.18
29	10.03	75.90	87.83	9.92	78.17	81.14	11.02	79.11	90.07
30	3.94	81.99	5.85	3.44	84.65	0	4.24	85.89	4.19

Self-Liquidating Mortgages—Annual Interest, Amortization And Remaining Balance (continued)

30-Year Term (continued)

Year	8½% interest—$7.69 monthly payment			8¾% interest—$7.87 monthly payment			9% interest—$8.05 monthly payment		
	Interest	Amort.	Balance	Interest	Amort.	Balance	Interest	Amort.	Balance
1	84.71	7.58	992.43	87.22	7.23	992.78	89.73	6.88	993.13
2	84.05	8.24	984.20	86.56	7.89	984.90	89.08	7.53	985.60
3	83.32	8.97	975.23	85.84	8.61	976.29	88.37	8.24	977.37
4	82.52	9.77	965.47	85.06	9.39	966.91	87.60	9.01	968.37
5	81.66	10.63	954.84	84.20	10.25	956.67	86.76	9.85	958.52
6	80.72	11.57	943.28	83.27	11.18	945.50	85.83	10.78	947.75
7	79.70	12.59	930.70	82.25	12.20	933.31	84.82	11.79	935.97
8	78.59	13.70	917.00	81.14	13.31	920.01	83.72	12.89	923.08
9	77.38	14.91	902.09	79.93	14.52	905.49	82.51	14.10	908.99
10	76.06	16.23	885.87	78.61	15.84	889.66	81.19	15.42	893.57
11	74.63	17.66	868.21	77.17	17.28	872.38	79.74	16.87	876.71
12	73.06	19.23	848.99	75.59	18.86	853.53	78.16	18.45	858.26
13	71.37	20.92	828.07	73.88	20.57	832.96	76.43	20.18	838.09
14	69.52	22.77	805.30	72.00	22.45	810.52	74.54	22.07	816.02
15	67.50	24.79	780.52	69.96	24.49	786.03	72.47	24.14	791.88
16	65.31	26.98	753.55	67.73	26.72	759.31	70.20	26.41	765.48
17	62.93	29.36	724.20	65.29	29.16	730.16	67.73	28.88	736.60
18	60.34	31.95	692.25	62.64	31.81	698.36	65.02	31.59	705.01
19	57.51	34.78	657.47	59.74	34.71	663.65	62.05	34.56	670.46
20	54.44	37.85	619.63	56.58	37.87	625.79	58.81	37.80	632.67
21	51.09	41.20	578.43	53.13	41.32	584.47	55.27	41.34	591.33
22	47.45	44.84	533.60	49.37	45.08	539.39	51.39	45.22	546.11
23	43.49	48.80	484.80	45.26	49.19	490.21	47.15	49.46	496.66
24	39.18	53.11	431.69	40.78	53.67	436.54	42.51	54.10	442.56
25	34.48	57.81	373.89	35.89	58.56	377.99	37.43	59.18	383.39
26	29.37	62.92	310.98	30.56	63.89	314.10	31.88	64.73	318.67
27	23.81	68.48	242.50	24.74	69.71	244.39	25.81	70.80	247.87
28	17.76	74.53	167.98	18.39	76.06	168.33	19.17	77.44	170.44
29	11.17	81.12	86.86	11.46	82.99	85.34	11.91	84.70	85.74
30	4.00	88.29	0	3.90	90.55	0	3.96	92.65	0

Self-Liquidating Mortgages—Annual Interest, Amortization And Remaining Balance (continued)

30-Year Term (continued)

Year	9¼% interest—$8.23 monthly payment Interest	Amort.	Balance	9½% interest—$8.41 monthly payment Interest	Amort.	Balance	9¾% interest—$8.59 monthly payment Interest	Amort.	Balance
1	92.23	6.54	993.47	94.74	6.19	993.82	97.25	5.84	994.17
2	91.60	7.17	986.31	94.13	6.80	987.02	96.65	6.44	987.74
3	90.91	7.86	978.46	93.45	7.48	979.55	96.00	7.09	980.65
4	90.15	8.62	969.84	92.71	8.22	971.33	95.28	7.81	972.84
5	89.32	9.45	960.40	91.89	9.04	962.30	94.48	8.61	964.23
6	88.41	10.36	950.04	91.00	9.93	952.38	93.60	9.49	954.75
7	87.41	11.36	938.69	90.01	10.92	941.47	92.63	10.46	944.30
8	86.31	12.46	926.24	88.93	12.00	929.47	91.57	11.52	932.78
9	85.11	13.66	912.59	87.74	13.19	916.29	90.39	12.70	920.09
10	83.79	14.98	897.62	86.43	14.50	901.79	89.10	13.99	906.10
11	82.35	16.42	881.20	84.99	15.94	885.86	87.67	15.42	890.69
12	80.77	18.00	863.20	83.41	17.52	868.35	86.10	16.99	873.71
13	79.03	19.74	843.46	81.67	19.26	849.09	84.37	18.72	854.99
14	77.12	21.65	821.82	79.76	21.17	827.93	82.46	20.63	834.37
15	75.03	23.74	798.09	77.66	23.27	804.67	80.36	22.73	811.64
16	72.74	26.03	772.07	75.35	25.58	779.10	78.04	25.05	786.60
17	70.23	28.54	743.53	72.82	28.11	750.99	75.49	27.60	759.00
18	67.48	31.29	712.25	70.03	30.90	720.09	72.67	30.42	728.58
19	64.46	34.31	677.94	66.96	33.97	686.13	69.57	33.52	695.07
20	61.15	37.62	640.32	63.59	37.34	648.79	66.15	36.94	658.14
21	57.51	41.26	599.07	59.88	41.05	607.75	62.39	40.70	617.44
22	53.53	45.24	553.83	55.81	45.12	562.63	58.24	44.85	572.59
23	49.17	49.60	504.24	51.33	49.60	513.04	53.66	49.43	523.16
24	44.38	54.39	449.85	46.41	54.52	458.52	48.62	54.47	468.70
25	39.13	59.64	390.21	41.00	59.93	398.60	43.07	60.02	408.68
26	33.37	65.40	324.82	35.05	65.88	332.73	36.95	66.14	342.55
27	27.06	71.71	253.12	28.51	72.42	260.32	30.20	72.89	269.66
28	20.14	78.63	174.49	21.33	79.60	180.72	22.77	80.32	189.35
29	12.55	86.22	88.28	13.43	87.50	93.22	14.58	88.51	100.84
30	4.23	94.54	0	4.74	96.19	0	5.55	97.54	3.31

¶112

Self-Liquidating Mortgages—Annual Interest, Amortization And Remaining Balance (continued)

30-Year Term (continued)

Year	10% interest—$8.78 monthly payment			10¼% interest—$8.96 monthly payment			10½% interest—$9.15 monthly payment		
	Interest	Amort.	Balance	Interest	Amort.	Balance	Interest	Amort.	Balance
1	99.75	5.62	994.39	102.26	5.27	994.74	104.77	5.04	994.97
2	99.16	6.21	988.19	101.70	5.83	988.91	104.21	5.60	989.37
3	98.52	6.85	981.34	101.07	6.46	982.46	103.60	6.21	983.16
4	97.80	7.57	973.78	100.38	7.15	975.31	102.91	6.90	976.27
5	97.01	8.36	965.42	99.61	7.92	967.40	102.15	7.66	968.62
6	96.13	9.24	956.18	98.76	8.77	958.63	101.31	8.50	960.12
7	95.16	10.21	945.98	97.82	9.71	948.92	100.37	9.44	950.69
8	94.10	11.27	934.71	96.77	10.76	938.17	99.33	10.48	940.21
9	92.92	12.45	922.26	95.62	11.91	926.26	98.18	11.63	928.59
10	91.61	13.76	908.51	94.34	13.19	913.07	96.90	12.91	915.68
11	90.17	15.20	893.31	92.92	14.61	898.47	95.47	14.34	901.35
12	88.58	16.79	876.53	91.35	16.18	882.30	93.89	15.92	885.44
13	86.82	18.55	857.99	89.61	17.92	864.39	92.14	17.67	867.77
14	84.88	20.49	837.50	87.69	19.84	844.55	90.19	19.62	848.16
15	82.74	22.63	814.87	85.56	21.97	822.58	88.03	21.78	826.39
16	80.37	25.00	789.88	83.20	24.33	798.26	85.63	24.18	802.22
17	77.75	27.62	762.26	80.58	26.95	771.31	82.97	26.84	775.39
18	74.86	30.51	731.75	77.69	29.84	741.48	80.01	29.80	745.60
19	71.66	33.71	698.05	74.48	33.05	708.43	76.73	33.08	712.52
20	68.13	37.24	660.82	70.93	36.60	671.84	73.09	36.72	675.80
21	64.24	41.13	619.69	67.00	40.53	631.31	69.04	40.77	635.04
22	59.93	45.44	574.25	62.64	44.89	586.43	64.55	45.26	589.78
23	55.17	50.20	524.06	57.82	49.71	536.73	59.56	50.25	539.53
24	49.91	55.46	468.61	52.48	55.05	481.69	54.02	55.79	483.75
25	44.11	61.26	407.35	46.57	60.96	420.73	47.87	61.94	421.82
26	37.69	67.68	339.68	40.02	67.51	353.22	41.05	68.76	353.06
27	30.61	74.76	264.92	32.76	74.77	278.46	33.47	76.34	276.73
28	22.78	82.59	182.34	24.73	82.80	195.66	25.06	84.75	191.98
29	14.13	91.24	91.11	15.83	91.70	103.97	15.72	94.09	97.90
30	4.58	100.79	0	5.98	101.55	2.42	5.35	104.46	0

Self-Liquidating Mortgages—Annual Interest, Amortization And Remaining Balance (continued)

30-Year Term (continued)

Year	10¾% interest—$9.33 monthly payment			11% interest—$9.52 monthly payment			11¼% interest—$9.71 monthly payment		
	Interest	Amort.	Balance	Interest	Amort.	Balance	Interest	Amort.	Balance
1	107.28	4.69	995.32	109.78	4.47	995.54	112.29	4.24	995.77
2	106.75	5.22	990.10	109.27	4.98	990.57	111.79	4.74	991.04
3	106.16	5.81	984.30	108.69	5.56	985.02	111.23	5.30	985.74
4	105.50	6.47	977.84	108.05	6.20	978.82	110.60	5.93	979.81
5	104.77	7.20	970.65	107.33	6.92	971.91	109.90	6.63	973.19
6	103.96	8.01	962.64	106.53	7.72	964.20	109.11	7.42	965.78
7	103.06	8.91	953.74	105.64	8.61	955.59	108.24	8.29	957.49
8	102.05	9.92	943.82	104.65	9.60	945.99	107.25	9.28	948.22
9	100.93	11.04	932.79	103.53	10.72	935.28	106.15	10.38	937.85
10	99.69	12.28	920.51	102.29	11.96	923.33	104.93	11.60	926.25
11	98.30	13.67	906.85	100.91	13.34	910.00	103.55	12.98	913.27
12	96.76	15.21	891.64	99.37	14.88	895.12	102.01	14.52	898.76
13	95.04	16.93	874.71	97.65	16.60	878.53	100.30	16.23	882.53
14	93.13	18.84	855.87	95.73	18.52	860.01	98.37	18.16	864.38
15	91.00	20.97	834.90	93.58	20.67	839.35	96.22	20.31	844.08
16	88.63	23.34	811.57	91.19	23.06	816.30	93.82	22.71	821.37
17	85.99	25.98	785.60	88.53	25.72	790.58	91.13	25.40	795.97
18	83.06	28.91	756.69	85.55	28.70	761.88	88.12	28.41	767.56
19	79.79	32.18	724.52	82.23	32.02	729.87	84.75	31.78	735.79
20	76.16	35.81	688.71	78.52	35.73	694.15	80.98	35.55	700.24
21	72.12	39.85	648.87	74.39	39.86	654.29	76.77	39.76	660.49
22	67.61	44.36	604.52	69.78	44.47	609.83	72.06	44.47	616.03
23	62.60	49.37	555.16	64.63	49.62	560.21	66.80	49.73	566.30
24	57.03	54.94	500.22	58.89	55.36	504.86	60.90	55.63	510.68
25	50.82	61.15	439.08	52.49	61.76	443.10	54.31	62.22	448.47
26	43.92	68.05	371.03	45.34	68.91	374.20	46.94	69.59	378.89
27	36.23	75.74	295.30	37.37	76.88	297.32	38.70	77.83	301.06
28	27.68	84.29	211.01	28.47	85.78	211.55	29.48	87.05	214.01
29	18.15	93.82	117.20	18.55	95.70	115.85	19.16	97.37	116.65
30	7.56	104.41	12.79	7.47	106.78	9.07	7.63	108.90	7.75

Self-Liquidating Mortgages—Annual Interest, Amortization And Remaining Balance (continued)

30-Year Term (continued)

Year	11½% interest—$9.90 monthly payment			11¾% interest—$10.09 monthly payment			12% interest—$10.29 monthly payment		
	Interest	Amort.	Balance	Interest	Amort.	Balance	Interest	Amort.	Balance
1	114.80	4.01	996.00	117.31	3.78	996.23	119.81	3.68	996.33
2	114.31	4.50	991.51	116.84	4.25	991.98	119.34	4.15	992.18
3	113.77	5.04	986.47	116.31	4.78	987.20	118.82	4.67	987.51
4	113.16	5.65	980.82	115.72	5.37	981.84	118.22	5.27	982.25
5	112.47	6.34	974.49	115.05	6.04	975.80	117.56	5.93	976.32
6	111.70	7.11	967.39	114.30	6.79	969.02	116.80	6.69	969.64
7	110.84	7.97	959.42	113.46	7.63	961.40	115.96	7.53	962.11
8	109.88	8.93	950.50	112.52	8.57	952.83	115.00	8.49	953.63
9	108.79	10.02	940.48	111.45	9.64	943.20	113.93	9.56	944.07
10	107.58	11.23	929.26	110.26	10.83	932.38	112.71	10.78	933.29
11	106.22	12.59	916.68	108.92	12.17	920.21	111.35	12.14	921.16
12	104.69	14.12	902.56	107.41	13.68	906.53	109.81	13.68	907.48
13	102.98	15.83	886.74	105.71	15.38	891.16	108.07	15.42	892.06
14	101.06	17.75	869.00	103.80	17.29	873.88	106.12	17.37	874.70
15	98.91	19.90	849.11	101.66	19.43	854.45	103.91	19.58	855.13
16	96.50	22.31	826.80	99.25	21.84	832.62	101.43	22.06	833.07
17	93.80	25.01	801.79	96.54	24.55	808.08	98.64	24.85	808.23
18	90.76	28.05	773.75	93.50	27.59	780.49	95.48	28.01	780.23
19	87.36	31.45	742.31	90.08	31.01	749.48	91.93	31.56	748.67
20	83.55	35.26	707.06	86.23	34.86	714.63	87.93	35.56	713.12
21	79.28	39.53	667.53	81.91	39.18	675.45	83.42	40.07	673.06
22	74.48	44.33	623.20	77.05	44.04	631.42	78.34	45.15	627.92
23	69.11	49.70	573.51	71.59	49.50	581.92	72.62	50.87	577.05
24	63.08	55.73	517.78	65.45	55.64	526.28	66.17	57.32	519.73
25	56.32	62.49	455.30	58.55	62.54	463.74	58.90	64.59	455.14
26	48.75	70.06	385.24	50.79	70.30	393.44	50.70	72.79	382.36
27	40.25	78.56	306.69	42.07	79.02	314.42	41.47	82.02	300.35
28	30.73	88.08	218.61	32.27	88.82	225.61	31.07	92.42	207.94
29	20.05	98.76	119.86	21.25	99.84	125.77	19.35	104.14	103.80
30	8.07	110.74	9.12	8.87	112.22	13.55	6.15	117.34	0

[¶113]

Mortgage Yield Table

When you purchase a mortgage which has already begun to run, you are buying, in fact, a right to receive a certain number of monthly payments. The table below is designed to help you determine how much you should pay for a mortgage in order to get the various yields. To find the price, multiply the number you get from the table times the amount of the monthly payment.

Example: Find the price to pay for a mortgage with 6 years to run, monthly payments of $249.75, if the desired yield is 12%:

From the table at 6 years and 12% — 51.1504
Monthly payment ($249.75) times (51.1504) — $12,774.81

Years to Run	6%	7%	8%	9%
1	11.6189	11.5571	11.4958	11.4349
2	22.5629	22.3351	22.1105	21.8891
3	32.8710	32.3865	31.9118	31.4468
4	42.5803	41.7602	40.9619	40.1848
5	51.7256	50.5020	49.3184	48.1734
6	60.3395	58.6544	57.0345	55.4768
7	68.4530	66.2573	64.2593	62.1540
8	76.0952	73.3476	70.7380	68.2548
9	83.2934	75.9598	76.8125	73.8394
10	90.0735	86.1264	82.4215	78.9417
11	96.4596	91.8771	87.6006	83.6064
12	102.4747	97.2402	92.3828	87.8711
13	108.1404	102.2417	96.7985	91.7700
14	113.4770	106.9061	100.8758	95.3346
15	118.5035	111.2560	104.6406	98.5934
16	123.2380	115.3126	108.1169	101.5728
17	127.6975	119.0957	111.3267	104.2966
18	131.8979	122.6238	114.2906	106.7869
19	135.8542	125.9141	117.0273	109.0635
20	139.5808	128.9825	119.5543	111.1450

Years to Run	10%	11%	12%	15%	18%
1	11.3745	11.3146	11.2551	11.0793	10.9075
2	21.6709	21.4556	21.2434	20.6242	20.0304
3	30.9912	30.5449	30.1075	28.8473	27.6607
4	39.4282	38.6914	37.9740	35.9315	34.0426
5	47.0654	45.9930	44.9550	42.0346	39.3803
6	53.9787	52.5373	51.1504	47.2925	43.8447
7	60.2367	58.4029	56.6485	51.8222	47.5786
8	65.9015	63.6601	61.5277	55.7246	50.7017
9	71.0294	68.3720	65.8578	59.0865	53.3137
10	75.6712	72.5953	69.7005	61.9828	55.4985
11	79.8730	76.3805	73.1108	64.4781	57.3257
12	83.6765	79.7731	76.1372	66.6277	58.8540
13	87.1195	82.8139	78.8229	68.4797	60.1323
14	90.2362	85.5392	81.2064	70.0751	61.2014
15	93.0574	87.9819	83.3217	71.4496	62.0956
16	95.6113	90.1713	85.1988	72.6338	62.8435
17	97.9230	92.1336	86.8647	73.6540	63.4690
18	100.0156	93.8923	88.3431	74.5328	63.9922
19	101.9099	95.4687	89.6551	75.2900	64.4297
20	103.6246	96.8815	90.8194	75.9423	64.7957

[¶ 114]
Real Estate Investment Rules of Thumb

The net yield from your real estate investment in rental property probably should exceed the prime mortgage interest rate, and the amount it exceeds the prime should depend on the risk of declining value.

Another way of looking at yield is to break it into its components. The percentage rates of these components will, of course, vary with economic conditions, quality of the property, and your own judgment. But they can be broken down as follows:

1. Safe rate (that is, current rate of return on investments having the greatest liquidity and safety, such as long-term United States Government bonds) 8.0%
2. Risk rate (reasonable allowance for continued ability of property to earn current income) 3.0%
3. Penalty for nonliquidity (relative salability, rentability, and collateral value of the property) 2.5%
4. Burden of management of funds rate (cost involved in managing the investment) 2.5%
            Total rate of capitalization 16%

These factors might be offset (to make a lower yield acceptable) by these additional considerations:
5. Tax protection—depreciation making a large portion of the income tax free for a period of years 1%
6. Amortization return—the scaling down of mortgage debt, which, though taxable, is offset by depreciation and promises an early refinancing to produce tax-free cash or an increased yield through lower financing charge 1%

Still another way to look at yield is to consider the difference in the rent you can get on the property under an ordinary lease and under a net lease where your tenant assumes all the costs and expenses of taxes, upkeep, and services for the property. Then, if you discount this annual difference using the present worth table for periodic future payments, you will get another measure of what a renter thinks is the risk factor on the property over the period of a lease.

Finally, yield also depends on the amount of cash involved. If you have a 75% mortgage, and the property appreciates, your gain over your cash basis will be much greater than the gain you would have if you had no mortgage leverage. However, if the property goes down in value, your decline will also be leveraged. But you must also consider the additional depreciation thrown off by mortgaged property as well as the fact that the interest payments are deductible.

Inflation is another factor to consider, since the yield from long-term rentals will be steadily eroded by inflation. However, this is offset by the increase in the value of the property as a result of inflation. As a rule of thumb, you should only look for properties which will increase in value at a rate greater than the prevailing rate of inflation.

Once you have decided on a real estate investment, the following tables may prove useful. The first table is a state-by-state guide to the increase in value of the average price of farmland. Since farmland generates few expenses, it is a useful table for comparison with developed land in any state because the potential income plus appreciation of developed property, discounted for inflation, should at least equal the gain to be made on farmland. Otherwise, the risks of developed property are too great.

The second and third tables are conversion tables. They convert prices per square foot into prices for lots and acres, and yearly rentals into monthly and daily rentals for purposes of prorating.

The fourth table is a depreciation table. It shows current and cumulative amounts for depreciation which can be taken on property with various useful lives, using different methods. Since depreciation is deductible, this table can be used to determine the tax-shelter potential of real estate. But you should remember that this depreciation is subject to recapture under Code §1250.

[¶115]

State-by-State Table of Farmland Value

The following table is based on prices as of November 1973. It is based on data developed by the U.S. Department of Agriculture. No data were developed for Alaska, Hawaii, and Florida.

State	Average Value Per Acre	Increase Over 1972	Increase Over Past 5 Years
Ala.	$ 325	30%	67%
Ariz.	82	11	48
Ark.	371	18	47
Calif.	566	12	17
Colo.	147	33	70
Conn.	1,380	15	84
Del.	732	19	64
Ga.	404	25	103
Ida.	232	13	42
Ill.	657	21	42
Ind.	587	28	41
Ia.	563	30	52
Kans.	212	22	41
Ky.	374	19	57
La.	451	10	49
Me.	231	15	87
Md.	943	27	80
Mass.	818	15	90
Mich.	497	24	53
Minn.	310	21	45
Miss.	317	16	41
Mo.	325	15	58
Mont.	83	24	54
Nebr.	226	23	49
Nev.	79	20	95
N. H.	357	15	75
N. J.	1,743	24	95
N. M.	60	26	53
N. Y.	410	19	59
N. C.	516	21	66
N. D.	132	25	49
Ohio	559	18	48
Okla.	251	23	47
Oreg.	195	12	53
Pa.	594	31	103
R. I.	1,056	15	84
S. C.	446	29	83
S. D.	113	24	41
Tenn.	427	24	71
Tex.	202	16	44
Utah	123	17	61
Vt.	330	15	69
Va.	443	21	80
Wash.	284	19	28
W. Va.	200	16	82
Wis.	370	23	78
Wyo.	58	22	50

[¶116]

Land Value Per Square Foot

The following table converts value per square foot into both value for a lot 25' x 100' and value for an acre of land.

Value per Sq. Ft. of Land	Value of a Lot 25' x 100'	Value of an Acre	Value per Sq. Ft. of Land	Value of a Lot 25' x 100'	Value of an Acre
$.01	$ 25.00	$ 435.60	$ .51	$1,275.00	$22,215.60
.02	50.00	871.20	.52	1,300.00	22,651.20
.03	75.00	1,306.80	.53	1,325.00	23,086.80
.04	100.00	1,742.40	.54	1,350.00	23,522.40
.05	125.00	2,178.00	.55	1,375.00	23,958.00
.06	150.00	2,613.60	.56	1,400.00	24.393.60
.07	175.00	3,049.20	.57	1,425.00	24,829.20
.08	200.00	3,484.80	.58	1,450.00	25,264.80
.09	225.00	3,920.40	.59	1,475.00	25,700.40
.10	250.00	4,356.00	.60	1,500.00	26,136.00
.11	275.00	4,791.60	.61	1,525.00	26,571.60
.12	300.00	5,227.20	.62	1,550.00	27,007.20
.13	325.00	5,662.80	.63	1,575.00	27,442.80
.14	350.00	6,098.40	.64	1,600.00	27,878.40
.15	375.00	6,534.00	.65	1,625.00	28,314.00
.16	400.00	6,969.60	.66	1,650.00	28,749.60
.17	425.00	7,405.20	.67	1,675.00	29,185.20
.18	450.00	7,840.80	.68	1,700.00	29,620.80
.19	475.00	8,276.40	.69	1,725.00	30,056.40
.20	500.00	8,712.00	.70	1,750.00	30,492.00
.21	525.00	9,147.60	.71	1,775.00	30,927.60
.22	550.00	9,583.20	.72	1,800.00	31,363.20
.23	575.00	10,018.80	.73	1,825.00	31,798.80
.24	600.00	10,454.40	.74	1,850.00	32,234.40
.25	625.00	10,890.00	.75	1,875.00	32,670.00
.26	650.00	11,325.60	.76	1,900.00	33,105.60
.27	675.00	11,761.20	.77	1,925.00	33,541.20
.28	700.60	12,196.80	.78	1,950.00	33,976.80
.29	725.00	12,632.40	.79	1,975.00	34,412.40
.30	750.00	13,068.00	.80	2,000.00	34,848.00
.31	775.00	13,503.60	.81	2,025.00	35,283.60
.32	800.00	13,939.20	.82	2,050.00	35,719.20
.33	825.00	14.374.80	.83	2,075.00	36,154.80
.34	850.00	14,810.40	.84	2,100.00	36,590.40
.35	875.00	15,246.00	.85	2,125.00	37,026.00
.36	900.00	15,681.60	.86	2,150.00	37,461.60
.37	925.00	16,117.20	.87	2,175.00	37,897.20
.38	950.00	16,552.80	.88	2,200.00	38,332.80
.39	975.00	16,988.40	.89	2,225.00	38,768.40
.40	1,000.00	17,424.00	.90	2,250.00	39.204.00
.41	1,025.00	17,859.60	.91	2,275.00	39,639.60
.42	1,050.00	18,295.20	.92	2,300.00	40,075.20
.43	1,075.00	18,730.80	.93	2,325.00	40,510.80
.44	1,100.00	19,166.40	.94	2,350.00	40,946.40
.45	1,125.00	19,602.00	.95	2,375.00	41,382.00
.46	1,150.00	20,037.60	.96	2,400.00	41,817.60
.47	1,175.00	20,473.20	.97	2,425.00	42,253.20
.48	1,200.00	20.908.80	.98	2,450.00	42,688.80
.49	1,225.00	21,344.40	.99	2,475.00	43,124.40
.50	1,250.00	21,780.00	1.00	2,500.00	43,560.00

[¶117]

Rent Prorating Table

Not infrequently, in commencing or terminating tenancies, it becomes necessary to calculate daily rent charges for short periods. This table may be used where only the annual rental is known. Note that the table indicates daily rental for both 30-day and 31-day months.

Rent per Year	Rent per Month	Rent per Day (30 days)	Rent per Day (31 days)	Rent per Year	Rent per Month	Rent per Day (30 days)	Rent per Day (31 days)
$ 1	$ .09	$ .003	$ .0029	$ 725	$ 60.42	$ 2.01	$ 1.950
2	.17	.0056	.0054	750	62.50	2.084	2.017
3	.25	.0083	.008	775	64.59	2.153	2.084
4	.34	.011	.01	800	66.67	2.223	2.151
5	.42	.014	.013	825	68.75	2.292	2.218
6	.50	.016	.016	850	70.84	2.361	2.286
7	.59	.019	.019	875	72.92	2.431	2.352
8	.67	.022	.021	900	75.00	2.50	2.420
9	.75	.025	.024	925	77.09	2.57	2.487
10	.84	.028	.027	950	79.17	2.639	2.554
20	1.67	.056	.054	975	81.25	2.708	2.621
25	2.09	.070	.068	1,000	83.34	2.778	2.689
30	2.50	.084	.081	1,025	85.42	2.847	2.756
40	3.34	.111	.108	1,050	87.50	2.917	2.823
50	4.17	.139	.134	1,075	89.59	2.986	2.880
60	5.00	.166	.161	1,100	91.67	3.056	2.958
70	5.84	.195	.189	1,125	93.75	3.125	3.025
75	6.25	.208	.202	1,150	95.84	3.195	3.092
80	6.67	.222	.215	1,175	97.92	3.264	3.159
90	7.50	.25	.242	1,200	100.00	3.334	3.226
100	8.34	.278	.269	1,300	108.34	3.611	3.495
125	10.42	.347	.336	1,400	116.67	3.889	3.764
150	12.50	.417	.403	1,500	125.00	4.167	4.033
175	14.59	.486	.470	1,600	133.34	4.445	4.301
200	16.67	.556	.538	1,700	141.67	4.723	4.570
225	18.75	.625	.605	1,800	150.00	5.00	4.84
250	20.84	.695	.672	1,900	158.34	5.278	5.108
275	22.92	.764	.740	2,000	166.67	5.56	5.377
300	25.00	.834	.807	3,000	250.00	8.334	8.065
325	27.09	.903	.874	4,000	333.34	11.111	10.753
350	29.17	.972	.941	5,000	416.67	13.889	13.441
375	31.25	1.042	1.009	6,000	500.00	16.667	16.13
400	33.34	1.112	1.076	7,000	583.34	19.445	18.818
425	35.42	1.181	1.143	8,000	666.67	22.223	21.506
450	37.50	1.25	1.21	9,000	750.00	25.00	24.194
475	39.59	1.32	1.277	10,000	833.34	27.778	26.882
500	41.67	1.389	1.344	11,000	916.67	30.556	29.57
525	43.75	1.458	1.412	12,000	1,000.00	33.34	29.889
550	45.84	1.528	1.479	13,000	1,083.34	36.111	34.947
575	47.92	1.598	1.546	14,000	1,166.67	38.889	37.635
600	50.00	1.667	1.613	15,000	1,250.00	41.667	40.323
625	52.09	1.737	1.68	16,000	1,333.34	44.445	43.811
650	54.17	1.806	1.748	17,000	1,416.67	47.223	44.690
675	56.25	1.875	1.815	18,000	1,500,00	50.00	48.387
700	58.34	1.945	1.882	19,000	1,583.34	52.778	50.070

Depreciation

Here is a rundown on the different methods of depreciation available with respect to various types of real estate where the depreciation rules set up by the Tax Reform Act of 1969 apply:

(1) Straight-Line: This method of depreciation is available for all types of new and used properties.

(2) 125%-Declining-Balance: This depreciation method is available for (a) used real property acquired before July 25, 1969, and (b) used residential rental property acquired after July 24, 1969, with a useful life of 20 years or more.

(3) 150%-Declining-Balance: This method is the fastest available for (a) used real property acquired before July 25, 1969, and (b) new nonresidential rental properties constructed after July 24, 1969.

(4) Sum-of-the-Years-Digits: This method of depreciation is available only for (a) new real estate constructed before July 25, 1969, and (b) new residential rental properties.

(5) 200%-Declining-Balance: This depreciation method is available only for (a) new real estate constructed before July 25, 1969, and (b) new residential rental properties.

(6) Straight-Line Using Short Useful Life: This method of depreciation is available only for rehabilitation expenses for low- or moderate-income residential rental properties. Expenditures are depreciated on a straight-line basis using a 60-month useful life.

The following tables give the straight-line; the 200%-, 150%- and the 125%-declining-balance; and sum-of-the-years-digits annual depreciation amounts and cumulative totals for assets having useful lives of from 3 to 50 years.

Computing Depreciation

Straight-Line: The cost or other basis of the asset is reduced by the anticipated salvage value. The remainder is then divided by the remaining useful life to find the annual straight-line deduction.

200%-Declining-Balance: The straight-line rate is first determined by dividing the useful life into 100. For example, if the useful life is 10 years, the straight-line rate is 10% (100 divided by 10). This rate is then doubled. The doubled rate is applied to the cost for the first year. The second year, the doubled rate is applied to the remaining (or declining) cost. This process is repeated each year.

At any time when the *remaining balance* divided by the *remaining useful life* (straight-line) gives a larger deduction than would result from continuing to apply the 200%-declining-balance rate against the declining balance, you can shift to straight-line. For example, with an asset having a useful life of 20 years, the 200%-declining-balance rate is 10%. If the asset cost $100, $68.62 will have been recovered via depreciation after 11 years, leaving a balance of $31.38. If we continue to apply the 10% rate, the 12th year's depreciation will be $3.14 (10% of $31.38). But if we used straight-line, dividing the remaining balance of $31.38 by the remaining useful life of 9 years, the result would be an annual depreciation for the last 9 years of $3.49.

In computing 200%-declining-balance, you need not consider salvage in determining the annual deductions. But, you cannot depreciate below the salvage value. When you switch to straight-line, you must take salvage into account in computing the straight-line deductions.

150%-Declining-Balance: You determine the straight-line rate as above. Then, multiply it by 1.5.

125%-Declining-Balance: First, determine the straight-line rate. Then multiply it by 1.25.

Sum-of-the-Years-Digits: Total the years that make up the useful life. For example, if the useful life is 5 years, add together 1, 2, 3, 4 and 5, for a total of 15. Then, the first year's depreciation is 5/15 of your cost, say, $1,500, or $500. The second year, deduct 4/15 of $1,500, then 3/15 of $1,500, etc. Alternatively, starting with the second year, you can reduce the denominator by the previous year's numerator and apply the fraction to the reduced balance. Thus, in the second year, you'd take 4/10 of $1,000 and in the third year, 3/6 of $600. You cannot switch to straight-line without permission.

[¶120]   **First-Year Depreciation**

Personal tangible property (i.e., non-real estate items) get a special first-year writeoff under the tax law. Twenty percent of the cost up to $10,000 ($20,000 on a joint return) can be written off immediately. Then, the balance of the cost is subject to depreciation under the usual rules. Thus, if an asset costs $10,000, $2,000 is written off in the year of acquisition. The cost then becomes $8,000, and that is subject to depreciation under any approved method from the time of acquisition as if the original cost were $8,000.

[¶121]   **Comparative Depreciation Table**

25-Year Life

Year	Straight-Line		200%-Declining Balance		150%-Declining Balance		125%-Declining Balance		Sum-of-Digits	
	Annual %	Cum. %	Annual %	Cum. %	Annual %	Cum. %	Annual %	Cum. %	Annual %	Cum. %
1	4.00	4.00	8.00	8.00	6.00	6.00	5.00	5.00	7.69	7.69
2	4.00	8.00	7.36	15.36	5.64	11.64	4.75	9.75	7.39	15.08
3	4.00	12.00	6.77	22.13	5.30	16.94	4.51	14.26	7.07	22.15
4	4.00	16.00	6.23	28.36	4.98	21.92	4.29	18.55	6.77	28.92
5	4.00	20.00	5.73	34.09	4.68	26.60	4.07	22.62	6.47	35.39
6	4.00	24.00	5.27	39.36	4.40	31.00	3.87	26.49	6.15	41.54
7	4.00	28.00	4.86	44.22	4.14	35.14	3.68	30.17	5.85	47.39
8	4.00	32.00	4.46	48.68	3.89	39.03	3.49	33.66	5.53	52.92
9	4.00	36.00	4.10	52.78	3.66	42.69	3.32	36.98	5.23	58.15
10	4.00	40.00	3.78	56.56	3.43	46.12	3.15	40.13	4.93	63.08
11	4.00	44.00	3.48	60.04	3.23	49.35	2.99	43.12	4.61	67.69
12	4.00	48.00	3.19	63.23	3.03	52.38	2.84	45.96	4.31	72.00
13	4.00	52.00	2.94	66.17	2.86	55.24	2.70	48.97	4.00	76.00
14	4.00	56.00	2.71	68.88	2.68	57.92	2.57	51.23	3.69	79.69
15	4.00	60.00	2.49	71.37	2.52	60.44	2.44	53.67	3.39	83.08
16	4.00	64.00	2.29	73.66	2.37	62.81	2.32	55.99	3.07	86.15
17	4.00	68.00	2.11	75.77	2.23	65.04	2.20	58.19	2.77	88.92
18	4.00	72.00	1.94	77.71	2.10	67.14	2.09	60.28	2.47	91.39
19	4.00	76.00	1.78	79.49	1.97	69.11	1.99	62.26	2.15	93.54
20	4.00	80.00	1.64	81.13	1.85	70.96	1.89	64.15	1.85	95.39
21	4.00	84.00	1.51	82.64	1.74	72.70	1.79	65.94	1.53	96.92
22	4.00	88.00	1.39	84.03	1.64	74.34	1.70	67.65	1.23	98.15
23	4.00	92.00	1.28	85.31	1.54	75.88	1.62	69.26	.93	99.08
24	4.00	96.00	1.17	86.48	1.45	77.33	1.54	70.80	.61	99.69
25	4.00	100.00	1.08	87.56	1.36	78.69	1.46	72.26	.31	100.00

¶119

Comparative Depreciation Table *(continued)*

30-Year Life

Year	Straight-Line		200%-Declining Balance		150%-Declining Balance		125%-Declining Balance		Sum-of-Digits	
	Annual %	Cum. %	Annual %	Cum. %	Annual %	Cum. %	Annual %	Cum. %	Annual %	Cum. %
1	3.33	3.33	6.67	6.67	5.00	5.00	4.16	4.16	6.45	6.45
2	3.34	6.67	6.22	12.89	4.75	9.75	3.99	8.16	6.24	12.69
3	3.33	10.00	5.81	18.70	4.51	14.26	3.83	11.99	6.02	18.71
4	3.33	13.33	5.42	24.12	4.29	18.55	3.67	15.65	5.81	24.52
5	3.34	16.67	5.06	29.18	4.07	22.62	3.51	19.17	5.59	30.11
6	3.33	20.00	4.72	33.90	3.87	26.49	3.37	22.54	5.37	35.48
7	3.33	23.33	4.40	38.30	3.68	30.17	3.23	25.76	5.17	40.65
8	3.34	26.67	4.12	42.42	3.49	33.66	3.09	28.86	4.94	45.59
9	3.33	30.00	3.84	46.26	3.32	36.98	2.96	31.82	4.73	50.32
10	3.33	33.33	3.58	49.84	3.15	40.13	2.84	34.66	4.52	54.84
11	3.34	36.67	3.34	53.18	2.99	43.12	2.72	37.38	4.30	59.14
12	3.33	40.00	3.12	56.30	2.84	45.96	2.61	39.99	4.09	63.23
13	3.33	43.33	2.92	59.22	2.70	48.66	2.50	42.49	3.87	67.10
14	3.34	46.67	2.72	61.94	2.57	51.23	2.40	44.89	3.65	70.75
15	3.33	50.00	2.53	64.47	2.44	53.67	2.30	47.19	3.44	74.19
16	3.33	53.33	2.37	66.84	2.32	55.99	2.20	49.39	3.23	77.42
17	3.34	56.67	2.21	69.05	2.20	58.19	2.11	51.50	3.01	80.43
18	3.33	60.00	2.07	71.12	2.09	60.28	2.02	53.52	2.80	83.23
19	3.33	63.33	1.92	73.04	1.99	62.27	1.94	55.45	2.58	85.81
20	3.34	66.67	1.80	74.84	1.89	64.16	1.86	57.31	2.36	88.17
21	3.33	70.00	1.68	76.52	1.80	65.96	1.78	59.09	2.15	90.32
22	3.33	73.33	1.56	78.08	1.70	67.66	1.70	60.79	1.94	92.26
23	3.34	76.67	1.46	79.54	1.62	69.28	1.63	62.43	1.72	93.98
24	3.33	80.00	1.37	80.91	1.54	70.82	1.57	63.99	1.61	95.49
25	3.33	83.33	1.27	82.18	1.46	72.28	1.50	65.49	1.29	96.78
26	3.34	86.67	1.19	83.37	1.39	73.67	1.44	66.93	1.07	97.85
27	3.33	90.00	1.11	84.48	1.32	74.99	1.38	68.31	.86	98.71
28	3.33	93.33	1.03	85.51	1.25	76.24	1.32	69.63	.65	99.36
29	3.34	96.67	.97	86.48	1.19	77.43	1.27	70.89	.43	99.79
30	3.33	100.00	.90	87.38	1.13	78.56	1.21	72.11	.21	100.00

Comparative Depreciation Table *(continued)*

35-Year Life

Year	Straight-Line		200%-Declining Balance		150%-Declining Balance		125%-Declining Balance		Sum-of-Digits	
	Annual %	Cum. %	Annual %	Cum. %	Annual %	Cum. %	Annual %	Cum. %	Annual %	Cum. %
1	2.86	2.86	5.71	5.71	4.29	4.29	3.57	3.57	5.56	5.56
2	2.86	5.72	5.38	11.09	4.10	8.39	3.44	7.02	5.40	10.96
3	2.85	8.57	5.07	16.16	3.93	12.32	3.32	10.34	5.24	16.20
4	2.86	11.43	4.78	20.94	3.76	16.08	3.20	13.54	5.08	21.28
5	2.86	14.29	4.51	25.45	3.60	19.68	3.09	16.63	4.92	26.20
6	2.85	17.14	4.25	29.70	3.44	23.12	2.98	19.60	4.76	30.96
7	2.86	20.00	4.01	33.71	3.29	26.41	2.87	22.48	4.60	35.56
8	2.86	22.86	3.78	37.49	3.15	29.56	2.77	25.24	4.44	40.00
9	2.85	25.71	3.56	41.05	3.02	32.58	2.67	27.91	4.29	44.29
10	2.86	28.57	3.36	44.41	2.89	35.47	2.57	30.49	4.13	48.42
11	2.86	31.43	3.17	47.58	2.77	38.24	2.48	32.97	3.97	52.39
12	2.85	34.28	2.99	50.57	2.65	40.89	2.39	35.36	3.81	56.20
13	2.86	37.14	2.82	53.39	2.53	43.42	2.31	37.67	3.65	59.85
14	2.86	40.00	2.66	56.05	2.42	45.84	2.23	39.90	3.49	63.34
15	2.85	42.85	2.51	58.56	2.32	48.16	2.15	42.05	3.33	66.67
16	2.86	45.71	2.37	60.93	2.22	50.38	2.07	44.12	3.18	69.85
17	2.86	48.57	2.23	63.16	2.13	52.51	2.00	46.11	3.02	72.87
18	2.85	51.42	2.10	65.26	2.03	54.54	1.92	48.04	2.86	75.73
19	2.86	54.28	1.98	67.24	1.95	56.49	1.86	49.89	2.70	78.43
20	2.86	57.14	1.87	69.11	1.86	58.35	1.79	51.68	2.54	80.97
21	2.86	60.00	1.76	70.87	1.79	60.14	1.73	53.41	2.38	83.35
22	2.86	62.86	1.66	72.53	1.71	61.85	1.66	55.07	2.22	85.57
23	2.86	65.72	1.57	74.10	1.64	63.49	1.60	56.68	2.06	87.63
24	2.85	68.57	1.48	75.58	1.56	65.05	1.55	58.22	1.90	89.53
25	2.86	71.43	1.40	76.98	1.50	66.55	1.49	59.72	1.75	91.28
26	2.86	74.29	1.32	78.30	1.43	67.98	1.44	61.15	1.59	92.87
27	2.85	77.14	1.24	79.54	1.37	69.35	1.39	62.54	1.43	94.30
28	2.86	80.00	1.17	80.71	1.31	70.66	1.34	63.88	1.27	95.57
29	2.86	82.86	1.10	81.81	1.26	71.92	1.29	65.17	1.11	96.68
30	2.85	85.71	1.04	82.85	1.20	73.12	1.24	66.41	.95	97.63
31	2.86	88.57	.98	83.83	1.15	74.27	1.20	67.61	.79	98.42
32	2.86	91.43	.92	84.75	1.10	75.37	1.16	68.77	.63	99.05
33	2.85	94.28	.87	85.62	1.06	76.43	1.12	69.88	.47	99.52
34	2.86	97.14	.82	86.44	1.01	77.44	1.08	70.96	.32	99.84
35	2.86	100.00	.77	87.21	.97	78.41	1.04	72.00	.16	100.00

Note: For additional Comparative Depreciation Tables see ¶228 in Section B—Tax Planning Aids.

¶121

[¶122]

Percentage Leases for Various Retail Establishments

The key to the profit in an investment in retail property lies in getting a minimum rental which will carry your land and building cost and support mortgage financing and, then, a percentage rental which will generate your profit. The real estate investment, in the long run, is no better than the retail business. In negotiating the percentage part of your lease you should have a rough idea of what percentages are customary in the particular business, what ratio of rent to sales the business can stand and how much sales volume a successful operation should get out of a square foot of space. These factors will vary. We have collected the figures in the following table as a guide only to the percentages paid, the minimum rental commitments made, and the sales volume per square foot achieved in different lines of business.

	Percentage Rental Rate	Minimum Rental per Square Foot	Yearly Gross Sales Per Square Foot
Appliances, Radio & TV Store	4-6	$ 1.60	$ 41.00
Art, Cards & Gifts	7-10	2.56	30.00
Automotive Parts, Retail	3½-6	1.85	50.00
Bakery	4-6	4.50	45.00
Barber Shop	10	3.50	45.00
Beauty Shop	8-11	2.20	40.00
Book & Stationery Store	7-10	2.50	---
Camera Store	6-8	3.75	55.00
Candy	6-10	10.00	35.00
Cocktail Bars & Taverns	5-7	2.60	65.00
Delicatessen	4-6	2.47	50.00
Department Stores	2-3½	1.02	60.00
Drug Stores—Independent	5-7	2.00	55.00
Drug Stores—National Chain	2½-4	2.08	70.00
Dry Cleaners & Laundry Pick-Up Stations	6-10	2.25	40.00
Florist	6-10	2.50	50.00
Furniture	4-7	1.60	40.00
Gift Shops	5-9	2.70	38.00
Hardware	4-7	2.40	40.00
Ice Cream and kindred lines	5-6	4.00	60.00
Jewelry Store	6-10	8.00	50.00
Junior Department Store	2½-5	1.00	48.00
Ladies Apparel—Independent	6-8	3.60	34.00
Liquor Store, Package—Regional Center	4-6	3.00	60.00
Luggage—Independent—Downtown	6-8	3.60	53.00
Meats & Poultry	3-5	4.00	60.00
Men's Apparel—Independent	5-8	2.85	30.00
Millinery	8-12	3.00	30.00
Music & Musical Instruments (including records)	3-7	1.10	55.00
Paint & Wallpaper, Regional Center	3-7	3.00	40.00
Reducing Salon	6-10	2.50	42.00
Restaurants	5-9	3.00	40.00
Service Stations	1-2 cts. per gal.	0.36	19.00
Shoe Repair	8-12	4.00	---
Shoe Stores, Family	5-6	2.30	50.00
Sporting Goods	5-8	3.00	50.00
Supermarkets	1-1½	1.50	100.00
Variety Stores	3-6	1.45	40.00

Here are rules of thumb as to what real estate may be worth in relation to the net income and gross rents it throws off. These rules of thumb should be used only as rough indicators. They should be modified and refined in the light of the specific circumstances involved in each piece of property, each negotiation and each deal.

How Many Times Annual Rent

Residences............................	9 to 10
Apartment Houses	7 to 9
Stores (Services)	7 to 9
Stores (Self-Operating)	8 to 10
Net Leases............................	9 to 15 (depending chiefly on credit rating)
Warehouses............................	10 to 12
Loft Buildings.........................	10 to 12
Office Buildings	8 to 12

[¶123]

How to Save Tax Dollars by Converting "Real Estate" Into Personal Property

As we've seen, the 1969 tax law bars the use of the 200%-declining-balance or sum-of-the-years-digits depreciation method on commercial real estate and it limits you to 150%-declining-balance depreciation on new and straight-line on used commercial property.

But, 200%-declining-balance depreciation continues to be available for new personal property, and you can still use 150%-declining-balance depreciation for used personal property (if it has a useful life of three or more years).

Obviously, whether you can use the depreciation rules applicable to real estate or those applicable to personal property is going to make a big difference in your tax liability and your cash flow. What's more, since items of personal property generally have shorter useful lives than real estate assets, depreciation deductions are greatly increased in the early years.

For personal, but not for real property, you get a special 20% first-year depreciation allowance, limited to $4,000 in one year.

The 1969 tax law requires you to pay an additional 10% tax on so-called tax-preference (including accelerated depreciation on real estate) income in excess of $30,000 plus the amount of your regular income taxes. But fast depreciation on personal property is not within reach of the special 10% tax with one exception—personal property subject to a net lease.

Example: You put up a small new building with air conditioning. The air-conditioning equipment costs $20,000. If it becomes part of the building, the building has a useful life of 40 years, and you use composite depreciation, this $20,000 item will, in the first year, give you all of $750 in depreciation (using 150%-declining-balance depreciation) and will save you half that amount if you're in the 50% tax bracket. On top of that, you'll add $250 to your tax-preference account. If the equipment retains its character as personal property, you get that special $4,000 first-year deduction. This reduces your basis to $16,000; but using 200%-declining-balance and a useful life of 10 years, you come out with $7,200 in first-year depreciation ($4,000 plus $3,200). So, you save $3,600 in taxes if you're in that same 50% tax bracket.

[¶124]

State Transfer Taxes

Only two states have stock transfer taxes. They are New York and Florida. The rate of the taxes is listed in the table below.

New York

2. 5¢ per share for transfers other than sale.
If by sale, the rate is:
1. 25¢ per share for shares less than $5
2. 5¢ per share for shares $5-$10
3. 75¢ per share for shares $10-$20
5¢ per share for shares over $20
Nonresidents—50% of above taxes
Maximum tax—$350

Florida

Stock: .015¢ per $100 par value; if no par, .015¢ per share

[¶125]

Price-Earnings Ratio Projector

How to Use the Projector: Let's say you are thinking about buying 100 shares of stock for $4,500. The price-earnings ratio of the stock is now 16 to 1, and you will estimate that the annual growth of the company over the next five years will be 15% a year. From the table at 16 and 15%, you get the figure 8. This represents the ratio of the price you paid to the earnings of the company in five years. If the actual price-earnings ratio at that time was 20, you could sell your stock for $11,250 ($4,500 x 20 ÷ 8), and, if it was 12, you could sell your stock for $6,750 ($4,500 x 12 ÷ 8).

Current Price-Earnings Ratio	Projected Annual Growth for 5 Years						
	10%	*15%*	*20%*	*25%*	*30%*	*40%*	*50%*
2	1.2	1.0	0.8	0.7	0.5	0.4	0.3
4	2.5	2.0	1.6	1.3	1.1	0.7	0.5
6	3.7	3.0	2.4	2.0	1.6	1.1	0.8
8	5.0	4.0	3.2	2.6	2.2	1.5	1.1
10	6.2	5.0	4.0	3.3	2.7	1.9	1.3
12	7.5	6.0	4.8	3.9	3.2	2.2	1.6
14	8.7	7.0	5.6	4.6	3.8	2.6	1.8
16	9.9	8.0	6.5	5.2	4.3	3.0	2.1
18	11.2	9.0	7.3	5.9	4.9	3.3	2.4
20	12.4	10.0	8.1	6.6	5.4	3.7	2.6
22	13.7	10.9	8.9	7.2	5.9	4.1	2.9
24	14.9	11.9	9.7	7.9	6.5	4.5	3.2
26	16.1	12.9	10.5	8.5	7.0	4.8	3.4
28	17.4	13.9	11.3	9.2	7.5	5.2	3.7
30	18.6	14.9	12.1	9.8	8.1	5.6	3.9
32	19.9	15.9	12.9	10.5	8.6	5.9	4.2
34	21.1	16.9	13.7	11.1	9.2	6.3	4.5
36	22.4	17.9	14.5	11.8	9.7	6.7	4.7
38	23.6	18.9	15.3	12.5	10.2	7.1	5.0
40	24.8	19.9	16.1	13.1	10.8	7.4	5.3
42	26.1	20.9	16.9	13.8	11.3	7.8	5.5
44	27.3	21.9	17.7	14.4	11.9	8.2	5.8
46	28.6	22.9	18.5	15.1	12.4	8.6	6.1
48	29.8	23.9	19.4	15.7	12.9	8.9	6.3
50	31.1	24.9	20.2	16.4	13.5	9.3	6.6

[¶ 126]

Gross Profit Percentage Calculator

The following table is useful in determining what percent of cost must be added to cost to get a desired gross profit percentage. For example, to get a gross profit on sales of 20% you have to add 25% of the cost to the cost to determine the selling price.

Add Following Percentage of Cost to Cost	To Get Following Gross Profit Percentage on Sales
5	4.75
7.5	7
10.00	9
11.11	10
12.36	11
12.5	11.125
13.63	12
14.94	13
16.28	14
16.43	14.25
17.65	15
19.05	16
20.00	16.67
20.49	17
21.96	18
23.46	19
25.00	20
26.58	21
28.21	22
29.88	23
31.58	24
33.33	25
35.00	26
37.5	27.25
40.00	28.5
42.86	30
45.00	31
47.00	32
50.00	33.33
53.85	35
55.00	35.5
60.00	37.5
65.00	39.5
66.67	40
70.00	41
75.00	42.75
80.00	44.5
85.00	46
90.00	47.5
100.00	50

Key Business Ratios for 125 Industries

The following tables list 14 key business ratios in 125 different industries. These ratios are based on the actual experience of major companies in each industry and are a useful aid in evaluating the performance of any company in any of the industries covered.

In the tables, there are three figures for each ratio. The middle figure is the median—the figure for the company which is in the middle of the sample. The upper figure is the median of the upper half of the sample, or the stronger companies, and the lower figure is the median for the lower half, or the weaker companies. Although none of the figures are actual averages, they are fair approximations of the ratios for strong, weak, and average companies. By comparing the figures for the company you are interested in with the three figures for the industry, you can get a fair idea of its strength, weakness, or any special problems.

The 14 ratios are listed below with an explanation for each:

Current Assets to Current Debt:

Current assets are divided by total current debt. Current assets are the sum of cash, notes, and accounts receivable (less reserves for bad debt), advances on merchandise, merchandise inventories, and listed federal, state, and municipal securities not in excess of market value. Current debt is the total of all liabilities falling due within one year. This is one test of solvency.

Net Profits on Net Sales:

Obtained by dividing the net earnings of the business, after taxes, by net sales (the dollar volume less returns, allowances, and cash discounts). This important yardstick in measuring profitability should be related to the ratio which follows.

Net Profits on Tangible Net Worth:

Tangible net worth is the equity of stockholders in the business, as obtained by subtracting total liabilities from total assets and then deducting intangibles. The ratio is obtained by dividing net profits after taxes by tangible net worth. Tendency is to look increasingly to this ratio as a final criterion of profitability. Generally, a relationship of at least 10% is regarded as a desirable objective for providing dividends plus funds for future growth.

Net Profits on Net Working Capital:

Net working capital represents the excess of current assets over current debt. This margin represents the cushion available to the business for carrying inventories and receivables and for financing day-to-day operations. The ratio is obtained by dividing net profits after taxes by net working capital.

Net Sales to Tangible Net Worth:

Net sales are divided by tangible net worth. This gives a measure of relative turnover of invested capital.

Net Sales to Net Working Capital:

Net sales are divided by net working capital. This provides a guide as to the extent the company is turning its working capital and the margin of operating funds.

Key Business Ratios *(continued)*

Collection Period:

Annual net sales are divided by 365 days to obtain average daily credit sales, and then the average daily credit sales are divided into notes and accounts receivable, including any discounted. This ratio is helpful in analyzing the collectibility of receivables. Many feel the collection period should not exceed the net maturity indicated by selling terms by more than 10 to 15 days. When comparing the collection period of one comcern with that of another, allowances should be made for possible variations in selling terms.

Net Sales to Inventory:

Divide annual net sales by merchandise inventory as carried on the balance sheet. This quotient does not yield an actual physical turnover. It provides a yardstick for comparing stock-to-sales ratios of one concern with another or with those for the industry.

Fixed Assets to Tangible Net Worth:

Fixed assets are divided by tangible net worth. Fixed assets represent depreciated book values of building, leasehold improvements, machinery, furniture, fixtures, tools, and other physical equipment, plus land, if any, and valued at cost or appraised market value. Ordinarily, this relationship should not exceed 100% for a manufacturer and 75% for a wholesaler or retailer.

Current Debt to Tangible Net Worth:

Derived by dividing current debt by tangible net worth. Ordinarily, a business begins to pile up trouble when this relationship exceeds 80%.

Total Debt to Tangible Net Worth:

Obtained by dividing total current plus long-term debts by tangible net worth. When this relationship exceeds 100%, the equity of creditors in the assets of the business exceeds that of owners.

Inventory to Net Working Capital:

Merchandise inventory is divided by net working capital. This is an additional measure of inventory balance. Ordinarily, the relationship should not exceed 80%.

Current Debt to Inventory:

Dividing the current debt by inventory yields yet another indication of the extent to which the business relies on funds from disposal of unsold inventories to meet its debts.

Funded Debts to Working Capital:

Funded debts are all long-term obligations, as represented by mortgages, bonds, debentures, term loans, serial notes, and other types of liabilities maturing more than one year from statement date. This ratio is obtained by dividing funded debt by net working capital. Analysts tend to compare funded debts with net working capital in determining whether or not long-term debts are in proper proportion. Ordinarily, this relationship should not exceed 100%.

Key Business Ratios (continued)
Retailing

Line of Business (and number of concerns reporting)	Current assets to current debt (Times)	Net profits on net sales (Per cent)	Net profits on tangible net worth (Per cent)	Net profits on net working capital (Per cent)	Net sales to tangible net worth (Times)	Net sales to net working capital (Times)	Collection period (Days)	Net sales to inventory (Times)	Fixed assets to tangible net worth (Per cent)	Current debt to tangible net worth (Per cent)	Total debt to tangible net worth (Per cent)	Inventory to net working capital (Per cent)	Current debt to inventory (Per cent)	Funded debts to net working capital (Per cent)
5641 Children's & Infants' Wear Stores (43)	3.21	3.85	14.15	15.85	6.23	9.46	**	6.8	5.9	36.7	88.8	83.7	39.4	23.3
	2.58	2.49	9.18	10.17	4.44	5.50	**	4.9	11.6	57.9	106.0	124.8	68.3	27.6
	1.63	0.77	3.04	4.31	3.70	4.19	**	3.8	22.2	94.8	188.3	195.1	96.2	72.1
5611 Clothing & Furnishings, Men's & Boys' (205)	4.81	4.02	13.15	15.18	4.56	5.08	**	5.9	5.0	24.5	60.9	60.2	36.9	14.9
	2.84	2.36	6.62	7.29	3.11	3.54	**	4.0	11.3	44.8	107.1	94.2	63.6	27.7
	1.99	0.60	2.24	2.41	2.15	2.63	**	2.8	21.3	92.8	169.3	123.0	89.7	42.7
5311 Department Stores (243)	4.37	3.11	10.39	13.29	4.65	5.98	**	7.1	11.7	22.5	49.1	58.3	46.0	14.6
	2.82	1.91	6.04	7.69	3.30	4.15	**	5.6	25.4	40.7	78.1	80.2	69.3	37.3
	1.92	0.85	3.02	3.72	2.40	3.14	**	4.3	48.6	65.4	135.4	112.6	100.0	77.7
Discount Stores (224)	2.56	2.86	17.01	23.94	8.80	11.96	**	7.2	12.4	56.8	83.7	101.4	56.5	11.6
	1.87	1.63	11.54	15.67	6.10	7.64	**	5.3	24.0	88.1	128.3	157.9	75.0	29.0
	1.49	0.76	5.87	7.04	4.49	5.28	**	3.9	46.6	130.9	209.3	218.2	94.0	70.3
Discount Stores, Leased Departments (57)	2.14	2.70	15.76	22.53	9.19	10.73	**	7.0	13.2	64.8	82.3	108.7	58.0	7.7
	1.83	1.22	9.86	10.95	6.50	8.22	**	4.9	22.7	108.2	120.7	173.0	75.0	25.8
	1.45	0.16	2.97	3.47	5.16	5.53	**	3.0	39.0	169.7	247.3	237.8	97.4	42.3
5651 Family Clothing Stores (94)	4.56	3.92	14.53	15.08	5.68	6.83	**	6.4	6.7	23.4	54.4	59.2	38.7	17.1
	2.94	2.48	7.87	8.26	3.48	4.05	**	4.5	12.9	43.7	122.0	94.4	61.5	38.0
	1.91	0.80	3.53	4.79	2.38	2.79	**	3.4	33.4	89.2	172.2	151.4	85.2	59.6
5252 Farm Equipment Dealers (92)	2.50	3.07	15.67	19.29	7.06	8.50	17	4.6	7.8	65.0	119.5	96.2	62.6	17.5
	1.67	1.70	8.01	8.69	4.78	5.53	30	2.9	18.1	137.3	227.4	174.5	81.3	28.6
	1.35	0.81	2.50	3.46	2.64	3.15	55	2.1	28.7	242.8	374.8	294.2	100.7	60.7
5969 Farm & Garden Supply Stores (76)	3.95	4.49	16.13	28.27	4.91	12.27	**	17.1	20.0	21.0	55.0	45.2	54.5	19.9
	2.23	2.51	9.12	16.18	3.36	6.45	**	9.1	43.9	46.9	89.8	75.8	120.7	53.3
	1.50	1.42	4.29	7.56	2.30	4.06	**	5.7	64.1	73.4	116.4	114.9	213.3	124.2
5712 Furniture Stores (183)	6.33	4.38	10.10	12.34	4.37	5.58	46	6.2	3.6	19.7	52.6	34.3	50.2	8.5
	2.81	2.22	5.85	6.46	2.45	2.75	92	4.6	10.6	52.4	87.2	63.0	90.3	18.5
	1.76	0.64	1.77	2.03	1.67	1.69	202	3.6	21.8	91.8	163.9	121.8	129.5	39.3

**Not computed. Necessary information as to the division between cash sales and credit sales was available in two few cases to obtain an average collection period usable as a broad guide.

¶127

Key Business Ratios (continued)

Retailing (continued)

Line of Business (and number of concerns reporting)	Current assets to current debt	Net profits on net sales	Net profits on tangible net worth	Net profits on net working capital	Net sales to tangible net worth	Net sales to net working capital	Collection period	Net sales to inventory	Fixed assets to tangible net worth	Current debt to tangible net worth	Total debt to tangible net worth	Inventory to net working capital	Current debt to inventory	Funded debts to net working capital
	Times	Per cent	Per cent	Per cent	Times	Times	Days	Times	Per cent	Per cent	Per cent	Per cent	Per cent	Per cent
5541 Gasoline Service Stations (70)	3.84	4.14	11.23	33.04	5.22	13.13	**	23.9	23.9	17.0	34.2	41.9	69.9	22.6
	2.33	2.51	7.82	19.93	3.22	6.93	**	9.9	46.6	32.4	61.8	82.6	128.2	56.2
	1.47	1.13	4.77	10.88	1.83	4.54	**	6.1	66.4	66.5	113.3	120.8	191.9	139.7
5411 Grocery Stores (137)	2.35	1.71	15.47	48.19	13.51	39.52	**	22.9	38.7	34.2	64.2	85.5	67.9	27.4
	1.66	0.99	10.30	23.12	9.70	21.65	**	16.8	67.4	58.3	106.4	136.1	94.0	72.6
	1.27	0.63	7.28	15.51	6.89	13.00	**	12.8	90.9	84.1	169.5	233.5	124.1	184.0
5251 Hardware Stores (97)	7.82	4.35	12.45	14.54	4.30	4.59	**	5.4	5.8	11.1	44.1	62.1	25.2	15.3
	3.81	2.52	6.94	8.80	2.59	3.15	**	3.9	14.7	25.3	63.0	82.7	46.0	31.4
	2.19	0.98	2.83	3.76	1.71	2.24	**	3.0	35.3	74.3	113.3	113.2	76.1	66.3
5722 Household Appliance Stores (92)	3.13	2.80	12.50	15.47	7.04	9.15	22	8.4	7.9	36.2	72.4	61.6	64.5	10.7
	2.12	1.61	8.09	9.80	4.21	5.44	35	5.6	15.7	77.8	144.1	101.7	103.6	23.7
	1.52	0.52	3.54	4.37	2.97	3.66	66	4.2	32.3	138.7	254.8	155.5	126.8	65.9
5971 Jewelry Stores (71)	5.93	5.08	11.82	12.41	3.06	3.20	**	4.0	2.8	19.4	40.6	59.0	30.4	10.3
	3.09	2.98	6.53	6.88	2.09	2.44	***	2.9	10.7	36.0	57.1	82.4	57.2	27.9
	2.34	1.34	3.87	4.18	1.53	1.73	**	2.2	28.7	73.4	124.0	118.8	81.5	45.4
5211 Lumber & Other Bldg. Mtls. Dealers (197)	5.61	3.28	8.98	12.56	4.15	5.86	41	8.5	13.7	16.9	55.9	49.5	38.9	15.0
	3.03	1.43	4.67	6.19	2.89	3.81	56	5.4	25.0	37.1	88.1	69.8	71.9	32.7
	2.10	0.41	1.21	1.71	1.91	2.58	79	4.0	47.5	66.4	126.8	98.4	119.7	61.7
5399 Miscellaneous General Mdse. Stores (84)	5.34	5.40	16.37	21.69	5.71	7.15	**	6.3	8.1	22.6	55.1	58.8	32.6	16.2
	2.91	2.92	10.16	12.63	3.46	3.91	**	4.3	16.0	41.3	101.1	89.3	60.5	34.5
	1.92	1.06	4.02	4.69	1.79	2.44	**	3.4	39.0	77.2	160.9	148.7	94.0	51.1
5511 Motor Vehicle Dealers (102)	2.53	1.60	12.44	20.33	12.19	17.99	**	12.6	9.3	50.3	79.3	93.1	66.5	10.4
	1.82	0.86	6.77	11.66	8.23	11.53	**	7.6	27.2	76.1	126.9	138.5	83.6	42.4
	1.40	0.21	2.15	3.27	5.34	8.37	**	5.6	53.3	157.3	278.8	228.5	104.7	87.0
5231 Paint, Glass & Wallpaper Stores (36)	4.64	3.38	12.83	17.54	5.15	6.22	**	10.7	11.4	18.9	36.2	34.3	44.4	12.8
	2.99	2.41	8.08	10.00	3.14	4.23	**	6.3	17.8	38.8	58.9	63.8	75.0	30.9
	2.28	0.80	3.23	5.61	2.34	3.13	**	5.2	33.0	62.1	86.6	98.9	126.6	49.6

**Not computed. Necessary information as to the division between cash sales and credit sales was available in too few cases to obtain an average collection period usable as a broad guide.

¶127

Key Business Ratios (continued)

Retailing (continued).

Line of Business (and number of concerns reporting)	Current assets to current debt	Net profits on net sales	Net profits on tangible net worth	Net profits on net working capital	Net sales to tangible net worth	Net sales to net working capital	Collection period	Net sales to inventory	Fixed assets to tangible net worth	Current debt to tangible net worth	Total debt to tangible net worth	Inventory to net working capital	Current debt to inventory	Funded debts to net working capital
	Times	Per cent	Per cent	Per cent	Times	Times	Days	Times	Per cent	Per cent	Per cent	Per cent	Per cent	Per cent
5661 Shoe Stores (102)	6.04	4.44	12.16	13.84	4.45	5.66	**	5.0	4.3	15.7	50.2	71.9	26.4	6.6
	3.52	2.10	6.26	7.55	3.18	3.73	**	3.9	10.3	33.2	75.3	102.3	45.3	16.8
	2.38	0.78	2.85	3.28	2.14	2.65	**	2.9	23.1	61.5	141.4	134.1	69.1	38.3
5531 Tire, Battery & Accessory Stores (67)	3.28	4.50	15.63	19.13	5.38	9.48	**	8.9	13.6	27.5	68.1	64.0	51.7	19.4
	2.21	1.89	8.42	10.83	4.10	5.69	**	6.1	29.0	65.1	120.0	88.8	102.7	32.6
	1.31	0.86	4.01	4.59	2.48	3.24	**	4.4	57.1	141.6	199.8	160.1	169.0	57.8
5331 Variety Stores (68)	4.78	3.63	13.42	17.60	5.38	7.16	**	5.7	10.7	21.2	35.8	95.8	31.8	6.7
	2.97	2.12	8.55	9.83	3.75	4.67	**	4.5	22.2	39.1	61.1	113.2	45.7	17.6
	2.32	1.14	3.77	4.56	2.61	3.42	**	3.5	34.9	60.8	94.7	136.6	60.0	45.0
5621 Women's Ready-to-Wear Stores (181)	4.26	3.71	14.45	18.71	6.17	7.74	**	9.4	7.4	24.1	67.1	45.2	58.1	13.0
	2.38	1.86	7.14	9.98	3.76	4.90	**	6.7	17.5	54.5	124.1	71.1	93.9	34.0
	1.74	0.44	2.53	3.36	2.52	3.23	**	5.2	40.2	98.0	172.4	121.3	129.5	66.7

**Not computed. Necessary information as to the division between cash sales and credit sales was available in too few cases to obtain an average collection period usable as a broad guide.

Key Business Ratios (continued)

Wholesaling

Line of Business (and number of concerns reporting)	Current assets to current debt (Times)	Net profits on net sales (Per cent)	Net profits on tangible net worth (Per cent)	Net profits on net working capital (Per cent)	Net sales to tangible net worth (Times)	Net sales to net working capital (Times)	Collection period (Days)	Net sales to inventory (Times)	Fixed assets to tangible net worth (Per cent)	Current debt to tangible net worth (Per cent)	Total debt to tangible net worth (Per cent)	Inventory to net working capital (Per cent)	Current debt to inventory (Per cent)	Funded debts to net working capital (Per cent)
5077 Air Condtg. & Refrigtn. Equipt. & Supplies (53)	3.34	4.35	18.56	19.22	8.20	8.87	37	9.0	4.2	39.2	58.6	63.8	57.6	10.5
	2.13	2.78	13.91	14.23	4.64	5.60	50	6.3	11.3	66.4	99.9	90.9	88.6	23.8
	1.57	1.58	7.89	9.84	3.25	3.84	64	3.9	25.6	184.4	270.8	134.7	165.6	57.0
5013 Automotive Equipment (184)	3.90	3.46	15.18	18.52	5.34	6.49	28	6.7	6.3	30.9	54.8	72.3	46.6	12.3
	2.62	2.20	9.21	11.28	3.92	4.59	34	5.1	13.1	50.4	90.4	92.6	65.2	28.1
	1.95	1.15	4.76	5.78	2.90	3.52	45	3.9	26.9	81.0	145.8	129.0	97.4	53.2
5095 Beer, Wine & Alcoholic Beverages (88)	3.07	2.00	14.24	27.66	10.97	18.91	7	15.3	10.2	30.9	59.7	69.2	62.2	21.7
	1.87	1.11	8.21	13.19	7.38	10.94	24	9.6	26.7	77.9	138.7	114.1	99.7	41.2
	1.41	0.43	3.29	5.67	5.11	7.58	36	6.3	50.4	162.5	241.3	197.3	132.3	61.8
5029 Chemicals & Allied Products (49)	3.08	2.88	15.77	28.89	8.27	17.21	37	15.9	12.7	33.9	95.0	46.1	94.7	32.6
	1.74	1.39	8.42	15.63	5.90	8.52	51	11.7	28.2	60.9	162.8	82.1	138.6	57.9
	1.43	0.42	4.78	5.76	3.02	5.95	65	7.7	48.4	140.1	234.0	123.9	236.1	96.3
5037 Clothing & Accessories, Women's & Child's (64)	4.32	2.43	12.97	15.87	8.60	11.11	30	20.5	1.9	26.8	84.1	40.3	60.8	14.7
	2.17	1.23	8.05	11.01	5.78	5.97	45	9.7	6.6	76.8	124.0	69.3	131.7	34.9
	1.73	0.23	2.07	2.34	3.65	3.88	59	5.6	19.7	128.8	196.8	104.6	212.0	47.3
5036 Clothing & Furnishings, Men's & Boys' (51)	3.51	1.67	8.37	8.99	7.37	8.32	34	8.0	2.1	36.6	49.0	73.4	56.5	9.3
	2.34	0.53	2.59	2.80	5.35	6.00	48	5.2	6.4	70.8	119.5	87.7	80.8	21.0
	1.79	0.04	0.16	0.39	2.85	3.34	71	3.8	15.7	124.6	148.4	125.1	120.0	24.2
5081 Commercial Machines & Equipment (54)	3.66	2.60	14.36	15.74	7.49	7.68	45	10.3	4.8	30.9	78.0	57.6	66.1	6.7
	1.98	1.36	6.47	8.32	4.98	5.82	59	7.7	12.6	77.2	101.7	73.2	109.0	18.6
	1.58	0.37	1.10	2.12	3.04	3.79	75	5.4	26.6	128.5	212.0	107.1	180.5	48.3
5045 Confectionery (35)	3.87	1.67	10.76	14.72	16.13	19.32	11	19.2	6.9	38.1	111.1	57.6	57.0	14.5
	2.26	0.43	7.53	9.52	10.68	11.99	18	12.8	13.4	70.9	158.4	98.1	84.1	33.8
	1.73	0.25	3.89	4.01	6.39	7.35	27	10.3	30.0	125.4	227.7	129.4	124.8	74.3
5043 Dairy Products (54)	3.10	2.00	14.87	26.81	11.88	26.22	21	56.4	11.2	27.5	52.7	23.1	101.5	18.2
	2.06	0.97	9.99	15.60	8.09	11.84	28	25.7	36.0	58.5	139.9	36.7	161.8	29.8
	1.41	0.56	3.89	6.44	5.05	9.23	33	15.6	60.6	104.5	187.8	92.0	459.5	94.2

Key Business Ratios (continued)

Wholesaling (continued)

Line of Business (and number of concerns reporting)	Current assets to current debt (Times)	Net profits on net sales (Per cent)	Net profits on tangible net worth (Per cent)	Net profits on net working capital (Per cent)	Net sales to tangible net worth (Times)	Net sales to net working capital (Times)	Collection period (Days)	Net sales to inventory (Times)	Fixed assets to tangible net worth (Per cent)	Current debt to tangible net worth (Per cent)	Total debt to tangible net worth (Per cent)	Inventory to net working capital (Per cent)	Current debt to inventory (Per cent)	Funded debts to net working capital (Per cent)
5022 Drugs & Druggists' Sundries (103)	3.17	1.70	10.33	12.29	8.53	8.75	27	8.0	7.0	40.3	76.5	75.8	58.6	15.1
	2.22	1.17	6.78	7.52	6.19	6.60	35	6.7	14.7	74.9	123.1	102.3	77.5	25.0
	1.72	0.65	3.95	4.77	4.54	5.14	44	5.8	31.5	122.8	194.8	129.0	104.1	51.4
5064 Electrical Appliances, TV & Radio Sets (94)	2.70	1.97	9.88	11.50	7.70	9.83	33	7.2	4.2	48.3	95.4	76.9	74.2	12.2
	1.96	1.16	6.35	7.16	5.85	6.45	42	5.8	9.0	90.3	152.0	115.0	101.5	27.7
	1.54	0.56	3.25	4.00	4.23	4.88	54	5.0	20.4	156.6	224.2	159.2	120.5	48.3
5063 Electrical Apparatus & Equipment (150)	3.01	2.27	13.29	15.65	8.33	9.49	35	9.9	5.3	42.9	72.7	72.5	71.2	6.6
	2.14	1.27	7.38	8.81	5.98	6.56	43	7.6	12.7	79.2	112.3	89.5	104.2	17.5
	1.65	0.68	4.01	4.61	3.72	4.58	54	5.5	25.8	133.3	163.6	115.6	138.3	33.4
5065 Electronic Parts & Equipment (47)	2.96	3.55	12.48	13.00	6.30	7.06	30	5.5	6.1	47.8	85.2	81.5	52.1	8.8
	2.32	1.86	7.38	9.03	4.50	4.79	42	4.6	12.9	77.2	124.1	105.7	74.3	24.6
	1.90	0.95	3.83	4.23	3.18	3.47	52	3.3	24.5	111.0	174.5	138.0	92.0	54.5
5083 Farm Machinery & Equipment (56)	3.70	2.89	13.97	17.67	6.70	7.95	32	7.5	7.0	28.7	54.2	65.0	62.7	4.7
	2.06	1.82	7.37	10.03	4.98	5.58	45	5.5	12.3	73.7	128.1	102.1	88.9	19.4
	1.64	0.64	3.54	4.74	2.81	3.60	68	3.7	33.0	126.6	181.2	137.1	123.6	50.4
5039 Footwear (58)	2.54	3.31	14.48	15.77	7.27	8.00	41	9.9	1.1	62.9	85.5	57.5	71.9	6.7
	2.03	2.02	9.44	10.32	4.52	5.22	62	5.7	4.2	85.1	142.8	89.9	113.0	13.0
	1.60	1.09	5.09	5.30	3.24	3.42	75	4.0	7.5	149.3	211.6	123.1	157.0	49.2
5048 Fresh Fruits & Vegetables (64)	3.11	2.03	15.32	44.37	12.66	22.34	13	103.0	17.1	23.9	48.9	15.7	138.3	16.9
	2.11	1.14	7.81	14.87	8.67	12.67	21	43.0	37.0	53.3	82.5	38.4	268.1	29.7
	1.12	0.15	1.88	6.54	5.54	7.99	37	19.8	79.0	99.3	162.7	75.5	702.2	71.0
5097 Furniture & Home Furnishings (78)	3.22	3.08	11.15	15.50	7.29	8.25	38	10.4	6.6	39.2	75.2	50.9	72.0	14.2
	1.78	1.50	6.85	7.52	4.68	5.34	51	7.1	15.8	79.0	133.8	81.2	107.3	23.1
	1.58	0.34	2.30	2.50	3.31	4.22	70	5.3	28.5	155.6	201.0	122.2	167.9	51.6
5041 Groceries, General Line (199)	3.26	1.09	12.47	17.90	21.15	26.66	7	17.5	10.6	35.9	89.0	88.6	50.3	16.9
	2.13	0.58	7.70	8.67	13.09	16.39	12	12.6	26.1	76.4	138.1	120.9	77.0	41.3
	1.59	0.24	3.84	4.43	7.52	10.03	17	9.7	63.1	133.0	236.0	168.8	110.5	80.2

¶127

Key Business Ratios (continued)

Wholesaling (continued)

Line of Business (and number of concerns reporting)	Current assets to current debt (Times)	Net profits on net sales (Per cent)	Net profits on tangible net worth (Per cent)	Net profits on working capital (Per cent)	Net sales to tangible net worth (Times)	Net sales to net working capital (Times)	Collection period (Days)	Net Sales to inventory (Times)	Fixed assets to tangible net worth (Per cent)	Current debt to tangible net worth (Per cent)	Total debt to tangible net worth (Per cent)	Inventory to net working capital (Per cent)	Current debt to inventory (Per cent)	Funded debts to net working capital (Per cent)
5072 Hardware (181)	4.30	2.45	8.02	10.09	5.86	6.98	33	6.3	6.2	24.8	45.8	75.7	40.2	6.1
	2.78	1.30	5.16	6.28	3.68	4.40	42	4.7	14.4	48.5	77.1	94.7	60.9	16.9
	1.89	0.59	2.18	2.76	2.60	3.10	52	3.5	26.9	86.9	139.4	121.1	93.9	33.1
5084 Industrial Machinery & Equipment (107)	3.01	2.65	11.84	14.73	7.24	9.45	31	10.1	9.8	39.7	72.4	58.5	65.5	10.3
	2.21	1.26	6.07	6.42	4.66	5.85	44	6.3	23.3	63.8	103.9	89.8	91.2	24.1
	1.61	0.41	1.18	2.11	3.12	3.81	54	4.2	41.5	132.1	203.9	111.6	144.8	77.8
5098 Lumber & Construction Materials (153)	4.30	2.50	9.82	16.10	7.64	9.28	34	10.5	10.6	23.5	61.0	53.4	46.9	16.0
	2.71	1.29	5.89	7.28	4.40	5.14	43	6.9	22.0	51.5	109.7	80.0	95.1	28.4
	1.67	0.44	2.31	2.47	2.77	3.67	58	5.5	39.9	104.5	183.4	107.6	143.9	50.9
5047 Meats & Meat Products (45)	2.67	1.81	21.12	30.13	23.84	35.10	13	100.4	10.2	36.2	78.7	24.2	133.1	19.9
	2.01	0.86	11.58	19.05	16.01	23.45	19	44.9	28.8	76.3	107.4	57.7	210.0	31.7
	1.48	0.27	6.63	9.87	9.80	15.15	24	29.9	50.7	116.0	277.4	79.5	329.2	69.2
5091 Metals & Minerals (76)	3.68	2.74	10.84	16.09	5.69	7.19	37	7.2	11.6	29.7	44.7	65.2	51.5	9.6
	2.37	1.95	7.11	9.17	4.05	5.12	44	5.0	30.1	54.4	84.8	89.5	87.5	22.7
	1.56	1.06	4.09	5.28	2.64	3.72	57	4.1	46.9	110.9	145.1	129.4	114.5	45.0
5028 Paints & Varnishes (41)	4.93	2.36	8.42	11.38	4.91	6.49	28	7.4	9.2	16.6	36.3	58.6	44.1	9.3
	2.60	1.30	5.55	6.47	3.85	4.65	38	6.0	19.8	33.9	67.1	84.5	75.3	26.9
	2.06	0.66	2.36	3.00	2.40	3.01	51	4.6	32.7	68.4	106.8	100.1	102.7	51.5
5096 Paper & Its Products (120)	3.49	2.21	11.22	13.10	7.85	10.19	29	11.8	6.5	31.3	61.6	56.8	59.1	12.5
	2.44	1.21	6.86	7.70	5.33	6.78	40	8.1	14.3	51.6	92.9	80.5	87.8	23.5
	1.77	0.53	3.04	4.03	3.58	4.51	49	6.5	31.5	98.6	161.3	102.7	134.2	52.7
5092 Petroleum & Petroleum Products (85)	3.10	3.13	14.39	41.55	8.01	18.64	23	33.7	35.3	24.2	44.5	28.4	112.1	20.2
	2.01	1.68	8.80	22.34	4.29	10.02	33	21.8	56.6	38.1	76.3	56.7	188.3	46.7
	1.37	0.78	4.58	9.81	3.09	6.07	48	12.4	86.1	85.9	160.1	93.7	327.6	138.6
5033 Piece Goods (117)	3.09	2.46	10.79	11.60	7.41	8.07	31	10.2	2.2	45.2	76.9	58.1	58.6	12.4
	2.11	1.25	5.87	6.54	4.48	5.23	47	6.3	5.8	80.4	111.6	87.8	94.6	22.3
	1.69	0.45	2.48	2.87	3.36	3.51	65	4.3	17.0	117.6	174.9	126.5	165.9	46.0

Key Business Ratios (continued)

Wholesaling (continued)

Line of Business (and number of concerns reporting)	Current assets to current debt	Net profits on net sales	Net profits on tangible net worth	Net profits on net working capital	Net sales to tangible net worth	Net sales to net working capital	Collection period	Net Sales to inventory	Fixed assets to tangible net worth	Current debt to tangible net worth	Total debt to tangible net worth	Inventory to net working capital	Current debt to inventory	Funded debts to net working capital
	Times	Per cent	Per cent	Per cent	Times	Times	Days	Times	Per cent	Per cent	Per cent	Per cent	Per cent	Per cent
5074 Plumbing & Heating Equip. & Sup. (183)	3.91	2.48	10.40	13.05	5.75	6.74	37	7.3	7.1	29.7	51.0	67.6	50.3	7.8
	2.65	1.62	6.71	7.61	4.17	5.14	46	5.6	14.0	53.0	91.6	85.2	77.4	18.1
	1.90	0.78	3.12	3.64	2.90	3.53	58	4.4	27.8	95.1	156.6	112.5	107.3	33.7
5044 Poultry & Poultry Products (42)	3.89	1.58	13.96	23.43	16.51	31.83	15	81.3	16.4	26.1	44.8	24.2	93.6	23.2
	2.31	0.79	8.89	13.82	9.54	19.27	20	54.9	38.0	44.6	60.7	39.0	173.5	42.0
	1.54	0.47	4.64	8.54	5.08	11.07	31	18.8	70.1	103.2	167.9	100.5	379.9	172.9
5093 Scrap & Waste Materials (59)	3.78	3.74	13.87	26.60	6.91	12.78	16	26.4	13.8	18.6	52.8	25.3	75.1	17.9
	2.45	2.10	7.82	13.94	4.46	8.02	27	15.5	37.0	34.4	86.7	50.7	127.6	36.1
	1.71	1.06	4.47	6.36	2.87	5.20	35	10.0	63.1	78.0	176.5	91.3	252.9	74.5
5014 Tires & Tubes (38)	2.77	3.52	14.09	18.35	7.27	8.70	33	7.0	15.8	52.9	134.4	78.6	82.6	15.0
	1.81	1.74	8.36	9.97	5.34	6.22	42	5.3	25.9	111.7	164.9	112.3	118.8	55.5
	1.47	0.66	2.21	3.67	3.59	4.33	62	4.3	57.5	147.2	193.3	157.0	161.7	73.9
5094 Tobacco & Its Products (92)	2.61	1.00	14.07	18.41	22.71	31.14	12	24.2	7.7	53.4	95.5	70.2	81.3	14.5
	1.88	0.72	8.00	10.98	14.08	18.20	16	18.7	14.9	86.2	141.8	95.4	116.6	25.4
	1.43	0.30	4.36	6.22	9.16	11.54	24	14.0	30.3	159.3	225.7	165.2	175.1	46.4

Key Business Ratios (continued)

Manufacturing and Construction

Line of Business (and number of concerns reporting)	Current assets to current debt (Times)	Net profits on net sales (Per cent)	Net profits on tangible net worth (Per cent)	Net profits on net working capital (Per cent)	Net sales to tangible net worth (Times)	Net sales to net working capital (Times)	Collection period (Days)	Net Sales to inventory (Times)	Fixed assets to tangible net worth (Per cent)	Current debt to tangible net worth (Per cent)	Total debt to tangible net worth (Per cent)	Inventory to net working capital (Per cent)	Current debt to inventory (Per cent)	Funded debts to net working capital (Per cent)
2871—72—79 Agricultural Chemicals (43)	2.86	4.53	13.77	29.57	4.04	17.62	37	12.8	29.8	28.8	69.4	35.6	90.6	17.9
	1.84	2.05	8.24	14.00	2.49	5.35	59	8.7	47.5	52.7	135.0	89.6	135.3	118.3
	1.18	0.89	2.75	5.25	2.04	3.33	100	5.3	82.3	91.6	218.2	168.3	243.4	201.1
3722—23—29 Airplane Parts & Accessories (68)	3.43	3.73	9.16	13.59	3.37	5.94	38	8.1	36.7	23.0	51.7	59.0	68.2	27.6
	2.42	1.91	4.46	5.58	2.35	4.08	53	5.2	56.0	36.9	78.9	82.6	100.0	69.0
	1.59	0.07	0.20	0.27	1.89	3.04	74	3.8	81.9	80.0	135.6	135.6	126.7	98.9
2051—52 Bakery Products (65)	2.94	2.85	9.98	43.54	5.73	24.22	17	41.2	57.0	16.6	38.1	33.7	134.7	23.9
	1.94	1.22	5.85	22.02	4.22	13.20	22	30.0	71.6	29.9	52.6	60.3	196.8	63.0
	1.40	0.57	2.62	12.51	3.17	9.06	30	18.4	93.9	44.1	87.9	86.0	303.7	152.5
3312—13—15—16—17 Blast Furnaces, Steel Wks. & Rolling Mills (71)	3.31	5.81	10.54	23.42	2.92	5.56	38	7.5	31.5	14.9	38.8	72.5	57.2	31.8
	2.40	3.07	6.18	13.89	1.96	4.13	44	4.9	59.9	29.7	66.0	90.5	77.6	53.5
	1.50	1.34	3.01	6.19	1.50	3.18	58	4.0	79.8	47.7	99.3	114.0	102.1	132.5
2331 Blouses & Waists, Women's & Misses' (54)	2.13	3.19	23.82	28.84	11.96	16.12	33	22.6	4.4	72.5	86.7	53.3	126.3	7.8
	1.62	1.65	10.50	12.38	9.34	12.33	43	14.3	9.8	136.3	178.4	83.0	174.9	23.1
	1.40	0.64	3.43	4.67	5.85	6.48	56	7.9	18.2	210.5	269.8	120.8	273.2	34.0
2731—32 Books: Publishing, Publishing & Printing (54)	4.21	6.50	13.85	19.68	3.12	4.10	53	6.9	14.1	20.7	37.9	40.6	56.8	7.2
	2.86	3.92	8.22	10.60	2.04	2.86	65	3.9	36.0	43.6	71.7	63.8	81.0	24.2
	1.94	1.50	3.70	4.20	1.43	2.00	100	2.9	48.5	67.6	119.8	86.6	173.8	76.3
2211 Broad Woven Fabrics, Cotton (46)	3.97	3.93	8.68	19.14	2.26	5.79	46	7.7	43.4	12.3	36.7	58.6	50.6	25.6
	3.07	2.75	5.12	11.53	1.92	4.24	58	5.3	58.0	21.2	57.1	74.9	64.6	47.6
	2.13	1.62	2.44	5.32	1.66	3.43	72	3.9	67.1	37.9	97.6	107.3	87.7	71.1
2031—32—33—34—35—36—37 Canned & Preserved Frts, Vegs, Sea Fds (76)	2.93	4.10	14.24	26.51	5.28	12.14	18	7.6	38.1	27.5	73.7	96.8	59.7	21.1
	1.68	2.15	7.73	15.86	3.76	6.74	28	4.3	57.1	70.8	138.9	142.3	90.7	55.2
	1.23	0.84	3.65	4.88	2.20	4.14	37	2.9	88.5	140.9	229.2	294.3	113.2	164.2
2751 Commercial Printing except Lithograph (85)	3.17	3.35	10.46	18.76	4.66	10.82	35	**	41.8	21.5	57.2	**	**	33.8
	2.19	1.37	4.55	8.78	3.04	6.14	45	**	58.5	40.7	104.4	**	**	72.2
	1.62	0.50	1.01	2.41	1.91	4.02	58	**	89.2	69.9	120.9	**	**	159.7

( ) Indicates loss.
**Not computed. Printers carry only current supplies such as paper, ink, and binding materials rather than merchandise inventories for re-sale. As a general rule, such contractors have no customary selling terms, each contract being a special job for which individual terms are arranged. Building Trades contractors have no inventories in the credit sense of the term.

Key Business Ratios (continued)

Manufacturing and Construction (continued)

Line of Business (and number of concerns reporting)	Current assets to current debt (Times)	Net profits on net sales (Per cent)	Net profits on tangible net worth (Per cent)	Net profits on net working capital (Per cent)	Net sales to tangible net worth (Times)	Net sales to net working capital (Times)	Collection period (Days)	Net sales to inventory (Times)	Fixed assets to tangible net worth (Per cent)	Current debt to tangible net worth (Per cent)	Total debt to tangible net worth (Per cent)	Inventory to net working capital (Per cent)	Current debt to inventory (Per cent)	Funded debts to net working capital (Per cent)
3661—62 Communication Equipment (76)	3.86	4.30	12.61	16.31	3.75	4.83	52	7.0	20.6	24.6	43.1	55.0	56.1	17.4
	2.81	2.66	6.33	8.18	2.46	3.04	63	4.7	39.0	35.9	70.9	73.8	84.7	38.6
	1.65	0.19	0.48	0.59	1.88	2.27	96	3.5	66.4	64.1	131.4	102.7	124.3	76.7
3271—72—73—74—75 Concrete, Gypsum & Plaster Products (81)	3.80	4.79	9.68	36.57	2.93	9.12	36	19.7	42.9	12.0	31.2	33.4	68.2	13.1
	2.68	3.13	6.29	16.53	2.10	5.38	53	10.3	60.6	29.0	61.1	56.5	141.4	42.9
	1.56	1.42	3.66	8.63	1.53	3.77	74	6.2	80.2	57.1	106.2	84.8	328.6	136.6
2071—72—73 Confectionery & Related Products (51)	3.53	4.49	13.31	26.12	5.27	11.63	15	10.1	34.4	20.7	41.0	65.1	50.0	24.2
	2.31	1.78	4.96	9.61	3.65	7.32	24	7.7	55.7	35.9	70.7	94.4	69.5	70.6
	1.48	0.06	0.25	0.76	2.26	4.64	40	5.3	90.7	86.0	184.4	175.1	123.2	166.7
3531—32—33—34—35—36—37 Const., Min. & Hdlg. Machy. & Equip. (82)	3.40	5.01	13.09	21.49	4.26	5.13	39	7.6	20.8	24.5	44.9	68.6	51.9	19.2
	2.37	2.96	7.18	11.16	2.36	3.48	61	4.0	35.7	44.4	87.3	84.2	72.7	38.2
	1.68	1.32	3.47	4.96	1.90	2.50	79	2.9	51.6	96.5	155.6	118.9	128.0	70.9
2641—42—43—44—45—46—47—49 Converted Paper & Paperboard Prods. (76)	3.81	4.25	12.25	24.43	4.63	9.63	33	10.8	41.4	19.0	47.0	49.1	64.0	31.8
	2.70	2.48	7.77	12.26	2.87	5.32	42	7.8	61.1	34.3	75.3	74.6	90.9	62.0
	1.72	0.92	3.12	5.64	1.97	3.47	57	6.2	87.0	73.0	109.4	108.6	133.8	112.3
3421—23—25—29 Cutlery, Hand Tools & General Hardware (99)	5.86	5.57	12.32	19.50	3.20	4.51	33	6.7	27.3	13.2	31.8	56.4	35.9	6.9
	3.39	2.92	6.63	11.55	2.21	3.50	46	4.6	38.9	25.4	59.8	77.9	55.3	35.2
	2.15	1.03	2.17	3.47	1.69	2.70	60	3.3	53.9	47.7	105.2	95.4	86.6	58.4
2021—22—23—24—26 Dairy Products (120)	2.06	2.43	13.23	53.27	8.44	34.14	17	51.9	40.2	27.4	56.6	36.6	142.9	34.1
	1.52	1.22	7.32	23.04	6.07	19.59	26	28.7	61.5	58.4	88.6	66.6	257.1	68.0
	1.21	0.50	3.07	8.88	4.06	10.70	35	16.6	89.8	91.7	149.3	131.6	413.4	114.5
2335 Dresses: Women's, Misses' & Junior's (99)	2.14	2.05	15.99	20.90	11.64	15.08	34	19.3	4.5	66.0	83.2	54.7	124.4	11.8
	1.66	1.03	10.20	12.57	8.10	10.59	52	14.4	9.2	122.6	148.5	81.0	182.2	22.6
	1.39	0.37	2.80	3.45	4.70	6.21	60	7.5	20.0	208.3	216.9	124.6	277.1	43.7
2831—33—34 Drugs (66)	3.78	8.96	21.30	38.05	3.30	6.68	40	8.4	23.5	22.9	32.8	53.3	64.7	15.4
	2.36	4.77	12.19	18.59	2.14	4.20	58	6.0	38.6	31.4	56.9	67.1	96.9	34.0
	1.45	2.26	5.08	8.65	1.74	2.89	89	4.2	62.1	50.7	96.6	134.2	112.1	98.8
3641—42—43—44 Electric Lighting & Wiring Equipment (59)	4.32	4.40	12.94	17.54	3.87	6.79	38	7.6	14.9	19.5	37.7	57.0	44.7	11.3
	3.10	2.93	7.98	11.46	2.79	4.15	45	5.4	39.6	36.4	59.9	77.2	68.1	29.2
	1.88	1.02	3.22	5.59	1.95	3.07	55	4.0	58.5	54.8	101.4	107.8	107.1	54.9

¶127

Key Business Ratios (continued)

Manufacturing and Construction (continued)

Line of Business (and number of concerns reporting)	Current assets to current debt	Net profits on net sales	Net profits on tangible net worth	Net profits on net working capital	Net sales to tangible net worth	Net sales to net working capital	Collection period	Net sales to inventory	Fixed assets to tangible net worth	Current debt to tangible net worth	Total debt to tangible net worth	Inventory to net working capital	Current debt to inventory	Funded debts to net working capital
	Times	Per cent	Per cent	Per cent	Times	Times	Days	Times	Per cent	Per cent	Per cent	Per cent	Per cent	Per cent
3611—12—13 Electric Transmission & Distribution Equipment (61)	3.66	4.34	10.68	13.89	3.71	5.81	49	5.5	22.5	24.7	53.6	74.6	44.7	21.6
	2.77	2.81	6.69	9.02	2.52	3.31	61	3.8	40.9	41.6	82.6	90.1	73.2	37.7
	1.82	0.41	1.29	2.21	1.81	2.13	78	2.9	59.8	78.0	151.8	122.4	99.6	68.9
3621—22—23—24—29 Electrical Industrial Apparatus (54)	4.48	4.26	12.51	20.70	4.01	6.84	39	6.8	31.7	21.6	33.2	64.1	42.9	8.6
	2.64	2.78	7.63	10.73	2.80	3.69	51	4.8	45.2	34.5	72.8	84.8	73.8	46.8
	1.78	1.15	2.65	3.85	2.11	2.96	68	3.7	69.4	76.8	134.0	109.1	96.5	87.4
1731 Electrical Work (135)	3.21	3.51	15.17	20.50	8.53	13.30	**	**	11.8	35.0	50.0	**	**	6.7
	2.09	1.74	8.71	11.65	5.14	7.20	**	**	20.3	66.2	108.9	**	**	17.4
	1.44	0.67	3.72	5.60	3.43	4.56	**	**	38.8	134.2	205.0	**	**	41.4
3671—72—73—74—79 Electronic Components & Accessories (96)	3.41	3.41	9.85	16.22	4.29	5.87	42	6.8	27.0	30.8	64.9	62.3	62.0	28.0
	2.41	1.71	4.67	6.65	2.68	4.05	55	4.8	48.4	49.0	102.8	82.1	87.8	60.3
	1.79	(3.64)	(11.64)	(12.56)	1.89	2.87	71	3.4	79.1	77.3	160.4	107.4	128.5	92.6
3811 Engineering, Laboratory & Scientific Instruments (41)	5.18	5.83	11.40	15.43	2.57	3.36	52	4.7	28.0	15.9	47.6	54.6	34.9	16.7
	3.43	2.11	5.74	7.69	2.06	2.50	67	3.8	41.9	34.5	84.9	71.1	60.6	43.3
	2.31	(1.26)	(2.78)	(4.17)	1.78	2.16	86	2.8	61.1	65.2	136.0	90.8	81.7	70.1
3441—42—43—44—46—49 Fabricated Structural Metal Products (128)	4.02	3.52	11.04	17.28	4.82	6.95	42	12.1	24.1	26.3	51.5	46.9	62.5	16.1
	2.44	1.53	4.85	7.49	3.06	4.67	52	6.7	37.0	45.7	87.0	68.1	95.7	38.6
	1.78	0.32	1.06	1.91	2.41	3.40	73	4.8	64.2	79.1	184.6	105.0	171.9	72.2
3522 Farm Machinery & Equipment (80)	4.28	5.52	14.34	21.55	4.07	5.80	25	5.4	21.4	22.1	46.9	68.4	39.0	19.0
	2.59	2.91	8.05	10.65	2.67	3.71	43	3.8	42.6	39.1	80.7	89.8	65.8	36.7
	1.67	1.48	3.07	4.05	1.99	2.61	67	2.9	58.8	79.5	138.6	134.3	92.0	54.5
3141 Footwear (111)	3.75	3.43	12.71	17.21	6.00	8.37	41	9.4	11.8	31.3	66.5	60.0	59.9	8.7
	2.33	1.50	6.27	7.40	4.03	5.30	55	6.0	23.9	59.1	108.4	87.2	90.2	21.9
	1.66	0.42	1.61	2.58	3.07	3.48	67	4.1	38.5	118.0	192.8	139.7	136.8	46.2
2371 Fur Goods (41)	3.29	1.20	8.59	9.32	8.28	9.10	26	17.8	1.7	42.6	196.7	42.9	79.7	17.3
	1.91	0.35	1.62	1.74	5.86	6.24	53	6.0	4.0	97.4	266.8	83.7	135.0	18.3
	1.44	(1.71)	(3.44)	(5.45)	3.07	4.24	87	3.1	7.6	257.7	321.3	122.7	249.9	110.8

( ) Indicates loss.
**Not computed. Printers carry only current supplies such as paper, ink, and binding materials rather than merchandise inventories for re-sale. Building Trades contractors have no inventories in the credit sense of the term. As a general rule, such contractors have no customary selling terms, each contract being a special job for which individual terms are arranged.

Key Business Ratios (continued)

Manufacturing and Construction (continued)

Line of Business (and number of concerns reporting)	Current assets to current debt (Times)	Net profits on net sales (Per cent)	Net profits on tangible net worth (Per cent)	Net profits on net working capital (Per cent)	Net sales to tangible net worth (Times)	Net sales to net working capital (Times)	Collection period (Days)	Net sales to inventory (Times)	Fixed assets to tangible net worth (Per cent)	Current debt to tangible net worth (Per cent)	Total debt to tangible net worth (Per cent)	Inventory to net working capital (Per cent)	Current debt to inventory (Per cent)	Funded debts to net working capital (Per cent)
1511 General Building Contractors (194)	2.04	2.29	17.45	29.94	13.55	20.86	**	**	9.4	67.1	76.5	**	**	8.3
	1.45	1.20	9.10	14.20	8.19	12.54	**	**	24.1	124.8	180.7	**	**	30.8
	1.24	0.53	4.38	7.02	4.27	6.76	**	**	40.1	237.3	315.5	**	**	77.3
3561—62—64—65—66—67—69 General Industrial Machinery & Equip. (113)	4.38	5.50	12.16	19.09	3.46	5.76	44	7.6	27.6	22.6	45.3	52.6	53.5	19.6
	2.76	2.92	8.03	12.36	2.64	3.93	58	4.7	44.8	35.3	68.1	76.6	80.0	37.5
	1.98	0.79	1.98	3.93	1.81	2.91	72	3.6	63.9	62.1	126.0	99.6	128.0	67.9
2041—42—43—44—45—46 Grain Mill Products (71)	4.11	3.24	14.27	31.70	6.39	15.63	21	16.0	30.2	19.5	41.2	40.4	73.4	39.1
	2.11	1.96	7.85	15.44	4.47	9.16	27	11.8	51.0	32.3	78.7	77.3	107.3	65.5
	1.44	0.87	4.14	7.82	3.17	5.62	45	8.1	76.3	65.8	132.3	148.0	145.1	94.5
3431—32—33 Heating Apparatus & Plumbing Fixtures (47)	4.66	3.74	13.22	15.84	4.49	5.57	29	6.9	16.8	21.8	55.8	61.3	49.7	15.3
	2.45	1.84	6.17	8.63	2.81	3.95	46	5.1	37.2	36.8	88.2	88.8	77.1	45.5
	2.00	0.76	2.83	3.93	1.95	3.34	63	3.9	64.0	72.7	190.4	108.9	100.7	62.5
1621 Heavy Construction, except Highway & Street (105)	2.64	3.80	14.62	34.60	5.87	14.53	**	**	33.5	29.8	59.5	**	**	20.4
	1.68	2.08	7.91	19.32	3.96	7.43	**	**	54.5	54.4	107.5	**	**	65.9
	1.22	0.62	2.75	5.97	2.25	4.50	**	**	84.9	111.6	214.5	**	**	121.8
2251—52 Hosiery (57)	3.76	3.75	12.26	24.78	4.87	9.18	29	9.6	34.3	21.3	55.1	58.6	44.2	20.1
	2.10	2.19	6.63	10.95	3.19	5.30	40	5.5	47.7	42.1	85.3	94.3	90.3	45.3
	1.66	1.03	3.44	5.45	2.24	3.32	61	4.0	68.7	78.3	108.0	128.3	128.3	72.0
3631—32—33—34—35—36—39 Household Appliances (48)	3.85	4.71	13.89	31.20	4.44	6.71	42	6.4	18.9	26.5	45.1	71.8	53.9	16.8
	2.38	2.63	9.87	13.33	3.10	4.00	51	4.6	36.0	43.0	84.8	93.4	74.7	35.3
	1.59	1.12	2.72	3.95	2.27	3.19	73	3.5	56.1	86.6	127.9	155.2	100.0	58.6
2812—13—15—16—18—19 Industrial Chemicals (65)	3.04	5.31	10.71	33.97	2.87	10.40	47	10.0	40.9	18.2	46.1	63.4	62.2	44.9
	1.99	2.73	6.83	14.54	2.06	4.91	57	6.0	63.7	32.7	70.6	86.2	100.0	85.8
	1.34	0.88	2.64	3.65	1.65	3.64	100	4.9	81.7	55.8	107.1	138.5	151.3	166.3
3821—22 Instruments, Measuring & Controlling (52)	4.69	4.31	11.25	16.40	3.53	4.61	49	5.1	22.8	17.7	45.1	61.4	38.8	21.3
	3.38	3.21	6.08	7.39	2.34	3.01	61	4.1	37.3	32.3	75.4	77.3	59.8	36.9
	2.00	0.74	2.25	2.49	1.58	2.14	87	3.0	59.6	74.0	160.6	98.3	111.8	77.2

( ) Indicates loss.

**Not computed. Printers carry only current supplies such as paper, ink, and binding materials rather than merchandise inventories for re-sale. Building Trades contractors have no inventories in the credit sense of the term. As a general rule, such contractors have no customary selling terms, each contract being a special job for which individual terms are arranged.

Key Business Ratios (continued)

Manufacturing and Construction (continued)

Line of Business (and number of concerns reporting)	Current assets to current debt (Times)	Net profits on net sales (Per cent)	Net profits on tangible net worth (Per cent)	Net profits on net working capital (Per cent)	Net sales to tangible net worth (Times)	Net sales to net working capital (Times)	Collection period (Days)	Net Sales to inventory (Times)	Fixed assets to tangible net worth (Per cent)	Current debt to tangible net worth (Per cent)	Total debt to tangible net worth (Per cent)	Inventory to net working capital (Per cent)	Current debt to inventory (Per cent)	Funded debts to net working capital (Per cent)
2871—72—79 Iron & Steel Foundries (61)	3.59 / 2.36 / 1.81	3.50 / 2.52 / 0.69	11.18 / 6.14 / 1.82	25.45 / 13.04 / 5.15	3.10 / 2.23 / 1.88	7.91 / 5.67 / 3.77	37 / 45 / 58	15.9 / 10.2 / 6.1	44.3 / 60.5 / 78.0	21.0 / 28.8 / 44.0	37.2 / 57.8 / 88.5	31.9 / 66.2 / 93.1	65.8 / 104.2 / 231.9	19.3 / 41.8 / 77.0
2253 Knit Outerwear Mills (53)	3.06 / 2.45 / 1.72	2.78 / 1.10 / 0.46	11.69 / 4.41 / 1.43	13.95 / 5.33 / 1.54	5.53 / 4.03 / 2.86	8.86 / 5.51 / 3.68	23 / 43 / 66	12.7 / 6.9 / 5.1	5.8 / 23.7 / 39.9	35.8 / 55.3 / 95.4	57.7 / 80.9 / 137.8	47.2 / 68.2 / 103.8	73.2 / 109.7 / 176.4	12.8 / 22.5 / 41.0
2082 Malt Liquors (34)	3.04 / 2.04 / 1.37	4.29 / 1.31 / 0.35	9.86 / 4.48 / 0.56	65.21 / 17.79 / 1.13	3.76 / 2.69 / 2.07	22.44 / 11.03 / 4.99	12 / 17 / 30	22.2 / 14.4 / 10.9	51.9 / 75.0 / 99.4	18.9 / 28.9 / 35.3	50.5 / 58.5 / 79.9	39.1 / 77.0 / 145.5	112.6 / 159.7 / 178.4	81.2 / 99.0 / 270.7
2515 Mattresses & Bedsprings (46)	4.90 / 2.58 / 1.92	3.47 / 1.93 / 0.73	9.57 / 5.31 / 3.31	15.85 / 10.02 / 5.85	5.32 / 3.16 / 2.14	7.68 / 5.30 / 3.77	34 / 50 / 56	10.4 / 8.2 / 5.8	13.2 / 24.1 / 41.4	13.9 / 33.5 / 66.9	55.7 / 75.7 / 136.0	49.7 / 66.9 / 99.5	49.9 / 90.4 / 153.9	25.1 / 39.3 / 64.9
2011 Meat Packing Plants (94)	3.85 / 2.30 / 1.39	1.52 / 0.82 / 0.35	13.75 / 9.45 / 3.72	36.01 / 16.89 / 7.92	15.84 / 10.02 / 7.25	33.69 / 20.79 / 12.65	10 / 14 / 20	52.6 / 34.5 / 22.3	41.0 / 57.0 / 79.6	19.7 / 41.6 / 84.7	53.3 / 84.7 / 153.4	40.6 / 63.4 / 117.0	79.7 / 131.5 / 214.6	19.8 / 49.5 / 121.6
3461 Metal Stampings (103)	3.88 / 2.55 / 1.67	3.50 / 1.64 / 0.39	9.27 / 4.75 / 1.18	18.26 / 9.21 / 3.33	4.66 / 3.04 / 2.07	8.30 / 5.80 / 4.04	29 / 37 / 46	12.6 / 7.7 / 5.9	36.8 / 56.0 / 81.9	17.9 / 35.5 / 63.5	42.6 / 72.4 / 139.5	48.9 / 75.7 / 106.4	57.5 / 99.3 / 165.2	24.8 / 47.5 / 94.4
3541—42—44—45—48 Metalworking Machinery & Equipment (124)	4.10 / 2.45 / 1.61	5.13 / 2.25 / 0.14	11.17 / 5.88 / 0.15	21.30 / 9.52 / 0.61	4.02 / 2.30 / 1.56	7.78 / 3.89 / 2.59	34 / 51 / 67	14.7 / 6.4 / 3.2	38.9 / 49.2 / 71.4	20.4 / 33.6 / 66.7	44.0 / 81.9 / 126.9	38.0 / 81.4 / 106.6	52.8 / 91.2 / 186.2	17.0 / 42.1 / 88.7
2431 Millwork (59)	5.26 / 3.27 / 2.04	4.03 / 1.98 / 1.05	12.39 / 7.92 / 2.93	19.13 / 9.75 / 4.42	5.16 / 3.35 / 2.37	6.87 / 4.30 / 2.99	33 / 45 / 57	9.8 / 6.8 / 5.0	19.9 / 36.0 / 54.6	17.4 / 33.9 / 71.7	46.1 / 75.8 / 184.3	42.4 / 69.4 / 102.0	40.6 / 72.9 / 130.8	16.6 / 26.2 / 42.4
3599 Miscellaneous Machinery, except Electrical (90)	4.09 / 2.76 / 1.99	5.28 / 2.74 / 0.70	13.07 / 7.03 / 1.60	26.51 / 13.49 / 2.68	3.25 / 2.45 / 1.74	7.24 / 4.75 / 3.36	30 / 44 / 55	24.4 / 8.8 / 5.2	29.0 / 46.9 / 69.8	15.2 / 29.7 / 50.3	38.7 / 66.5 / 104.2	21.6 / 53.1 / 91.4	64.9 / 126.2 / 272.3	8.3 / 24.3 / 87.9

Key Business Ratios (continued)

Manufacturing and Construction (continued)

Line of Business (and number of concerns reporting)	Current assets to current debt	Net profits on net sales	Net profits on tangible net worth	Net profits on net working capital	Net sales to tangible net worth	Net sales to net working capital	Collection period	Net sales to inventory	Fixed assets to tangible net worth	Current debt to tangible net worth	Total debt to tangible net worth	Inventory to net working capital	Current debt to inventory	Funded debts to net working capital
	Times	Per cent	Per cent	Per cent	Times	Times	Days	Times	Per cent	Per cent	Per cent	Per cent	Per cent	Per cent
3714 Motor Vehicle Parts & Accessories (88)	3.98	5.21	12.91	20.28	3.81	6.19	35	9.3	29.8	22.7	48.8	57.7	52.6	25.8
	2.87	2.73	7.74	11.67	2.73	3.94	44	5.7	45.8	34.1	77.1	76.8	73.7	44.2
	1.87	1.15	3.80	5.22	2.00	2.80	54	3.8	61.4	63.7	107.4	103.9	115.8	77.0
3361—62—69 Nonferrous Foundries (45)	3.94	4.34	9.44	21.42	3.89	8.54	32	19.7	36.7	18.4	36.0	42.6	82.1	19.5
	2.42	1.41	5.18	9.80	2.82	6.50	42	10.9	51.5	34.8	58.5	61.8	127.2	40.4
	1.73	(1.03)	(2.63)	(3.48)	2.17	3.77	49	7.0	88.7	56.7	114.3	106.4	180.3	129.1
2541—42 Office & Store Fixtures (65)	3.94	5.09	15.12	24.15	5.11	7.38	31	13.2	17.0	24.1	41.4	36.1	74.3	22.0
	2.51	2.06	7.77	12.07	3.16	4.68	47	7.6	35.3	52.5	100.1	67.3	115.0	49.5
	1.75	0.55	1.71	2.70	2.18	2.80	78	4.8	57.9	96.1	149.4	94.9	188.8	83.5
2361—63—69 Outerwear, Children's & Infants' (57)	2.47	2.43	20.40	26.62	10.79	12.42	25	16.3	5.9	56.6	72.7	48.2	81.5	4.8
	1.88	0.99	10.08	12.95	7.21	8.09	38	9.1	15.0	93.6	128.5	76.3	151.5	15.7
	1.36	0.37	2.47	3.19	4.33	5.07	53	5.3	29.5	236.7	280.6	153.9	234.0	53.4
2851 Paints, Varnishes, Lacquers & Enamels (112)	4.83	3.86	10.16	16.58	3.93	6.00	34	8.4	19.7	18.0	39.3	54.2	46.0	20.2
	3.39	2.33	5.86	8.44	2.83	4.43	45	6.8	35.6	25.2	65.1	68.2	67.5	37.2
	2.44	0.79	2.60	4.27	2.13	3.16	60	5.1	56.3	46.6	100.0	87.4	92.2	58.2
2621 Paper Mills, except Building Paper (53)	3.46	4.89	8.99	29.53	2.40	6.18	32	10.4	41.9	17.9	33.5	53.1	65.7	28.9
	2.50	2.66	6.02	15.26	1.89	5.55	42	7.4	78.8	22.0	64.5	66.5	99.2	82.1
	1.55	1.30	2.78	5.10	1.51	3.68	53	5.8	105.7	32.6	88.2	92.9	139.0	179.2
2651—52—53—54—55 Paperboard Containers & Boxes (62)	5.39	4.83	13.13	28.30	4.23	7.76	30	16.8	34.3	12.5	52.4	33.0	53.7	40.3
	2.90	2.61	6.95	13.55	2.94	5.70	39	10.6	68.1	26.1	87.0	54.8	99.3	81.6
	1.86	0.40	2.19	5.62	1.98	3.47	47	5.8	101.2	59.7	128.0	96.8	165.0	142.6
3712—13 Passenger Car, Truck & Bus Bodies (47)	3.79	3.63	11.30	19.98	5.49	8.65	29	9.4	23.9	26.0	40.3	58.4	64.5	10.8
	2.11	1.95	7.07	12.21	3.86	5.60	47	6.6	38.4	54.6	96.6	95.2	102.5	19.8
	1.60	0.32	2.80	5.46	2.40	3.60	56	5.1	69.1	97.0	142.4	129.2	129.1	58.7
2911 Petroleum Refining (53)	1.76	6.80	11.62	68.89	3.39	19.07	44	30.1	7.1	6.6	12.8	40.6	100.0	33.3
	1.09	3.41	7.45	32.35	1.91	8.89	54	14.2	34.1	23.0	35.2	82.6	119.1	112.0
	1.00	1.74	3.62	21.59	1.36	4.70	66	9.0	79.2	62.0	120.1	204.2	184.1	262.9

¶127

Key Business Ratios (continued)

Manufacturing and Construction (continued)

Line of Business (and number of concerns reporting)	Current assets to current debt — Times	Net profits on net sales — Per cent	Net profits on tangible net worth — Per cent	Net profits on net working capital — Per cent	Net sales to tangible net worth — Times	Net sales to net working capital — Times	Collection period — Days	Net sales to inventory — Times	Fixed assets to tangible net worth — Per cent	Current debt to tangible net worth — Per cent	Total debt to tangible net worth — Per cent	Inventory to net working capital — Per cent	Current debt to inventory — Per cent	Funded debts to net working capital — Per cent
2821—22—23—24 Plastics Materials & Synthetics (36)	2.58	5.53	17.13	51.64	4.87	11.40	40	11.7	32.4	26.6	43.5	62.5	79.8	30.6
	1.84	2.34	7.00	17.39	2.96	6.04	60	8.6	54.3	41.4	68.3	86.3	108.8	83.5
	1.27	0.88	3.10	5.50	2.08	4.59	70	6.7	89.7	78.6	135.3	122.2	214.2	137.8
1711 Plumbing, Heating & Air Conditioning (106)	2.82	2.72	18.22	23.30	9.91	13.55	**	**	10.9	41.2	59.9	**	**	10.3
	1.85	1.58	10.11	13.38	6.31	8.45	**	**	20.6	90.1	143.3	**	**	22.1
	1.37	0.57	3.26	4.78	4.31	5.30	**	**	38.0	172.7	250.6	**	**	48.1
2421 Sawmills & Planing Mills (84)	4.87	5.46	11.47	35.48	3.60	7.27	20	9.9	26.1	11.2	26.4	52.3	47.7	15.3
	2.58	2.29	5.32	17.25	2.35	4.97	29	5.6	48.6	26.4	55.3	74.9	73.4	43.0
	1.80	0.61	1.22	3.72	1.43	3.20	46	4.0	79.8	57.4	102.9	117.4	129.8	149.5
3451—52 Screw Machine Products (74)	4.73	5.15	10.37	22.37	3.31	8.04	26	11.6	37.1	14.7	42.5	50.7	54.3	22.8
	2.72	2.70	4.87	10.82	2.45	4.85	36	6.3	55.2	24.8	64.0	74.5	85.7	51.2
	1.95	0.51	1.33	2.50	1.70	3.04	49	4.8	78.3	48.4	84.1	109.5	123.2	89.5
2321—22 Shirts, Underwear & Nightwear, Men's & Boys' (62)	3.09	3.16	11.77	14.42	6.07	6.74	42	7.7	2.6	47.4	78.9	66.1	71.7	6.8
	1.98	1.79	7.04	7.91	4.34	5.02	55	5.3	12.9	93.2	104.3	95.2	102.2	21.4
	1.58	0.45	1.57	2.87	3.43	3.80	68	4.2	30.4	148.1	202.8	137.0	142.0	48.5
2841—42—43—44 Soap, Detergents, Perfumes & Cosmetics (71)	4.31	6.29	15.06	25.87	3.96	6.98	39	10.5	17.3	20.4	38.2	44.3	62.0	12.7
	2.40	3.23	10.04	15.93	2.90	4.40	52	7.3	26.9	34.5	56.2	61.5	95.5	21.8
	1.80	0.67	1.84	2.95	1.79	2.99	83	5.2	39.9	57.6	85.2	95.8	124.7	46.4
2086 Soft Drinks, Bottled & Canned (75)	3.04	6.06	16.35	73.59	4.55	17.68	14	20.3	50.3	19.0	34.8	46.6	97.2	25.3
	1.84	3.74	10.87	39.45	2.88	9.93	20	14.7	79.3	32.2	66.5	83.7	151.7	99.0
	1.42	1.76	7.27	18.92	1.97	5.69	31	10.7	120.6	51.1	119.7	120.6	222.8	276.9
3551—52—53—54—55—59 Special Industry Machinery (94)	5.27	5.42	10.91	16.67	3.04	5.50	42	6.4	25.0	16.3	48.4	51.4	48.4	12.9
	2.70	3.18	7.78	10.55	2.16	2.99	57	4.6	36.4	37.2	75.9	78.7	70.7	32.0
	1.91	0.69	1.28	1.49	1.59	2.12	78	3.1	50.0	63.8	124.6	104.5	111.1	63.2

**Not computed. Printers carry only current supplies such as paper, ink, and binding materials rather than merchandise inventories for re-sale. Building Trades contractors have no inventories in the credit sense of the term. As a general rule, such contractors have no customary selling terms, each contract being a special job for which individual terms are arranged.

Key Business Ratios (continued)

Manufacturing and Construction (continued)

Line of Business (and number of concerns reporting)	Current assets to current debt (Times)	Net profits on net sales (Per cent)	Net profits on tangible net worth (Per cent)	Net profits on net working capital (Per cent)	Net sales to tangible net worth (Times)	Net sales to net working capital (Times)	Collection period (Days)	Net sales to inventory (Times)	Fixed assets to tangible net worth (Per cent)	Current debt to tangible net worth (Per cent)	Total debt to tangible net worth (Per cent)	Inventory to net working capital (Per cent)	Current debt to inventory (Per cent)	Funded debts to net working capital (Per cent)
2337 Suits & Coats, Women's & Misses' (81)	3.30	2.55	11.24	15.47	10.15	11.64	25	22.6	2.7	39.1	60.0	36.0	94.8	8.6
	2.14	1.03	5.33	6.38	5.76	6.58	44	13.4	5.5	69.5	122.0	61.5	139.1	17.2
	1.54	0.33	1.21	1.65	3.66	4.18	62	5.6	12.0	158.0	271.8	85.9	236.9	66.2
2311 Suits, Coats & Overcoats, Men's & Boys' (110)	2.83	2.53	12.10	13.43	6.26	7.50	24	7.9	3.9	47.2	77.0	58.1	67.1	9.0
	2.13	1.29	4.41	5.93	4.26	5.31	53	5.0	8.7	82.4	106.8	90.8	89.5	26.0
	1.68	0.24	1.32	1.59	3.02	3.50	83	3.6	22.1	140.4	172.7	140.4	127.9	40.2
3841—42—43 Surgical, Medical & Dental Instruments (62)	5.29	6.07	14.82	21.16	3.28	4.76	43	7.5	19.2	18.0	32.6	56.0	36.8	10.9
	3.20	4.21	9.97	11.82	2.62	3.52	56	4.8	35.3	30.8	59.4	72.2	62.2	29.5
	2.11	2.20	5.53	6.26	1.89	2.68	67	3.5	51.5	53.5	91.2	98.4	98.5	63.0
3941—42—43—49 Toys, Amusement & Sporting Goods (62)	2.74	4.82	16.66	23.26	4.63	8.79	46	6.9	19.6	41.0	71.7	77.1	77.0	22.6
	1.89	1.80	8.08	9.80	3.74	5.12	60	4.7	39.1	75.1	122.2	106.2	119.1	42.3
	1.35	0.53	2.16	2.92	2.28	3.30	92	3.6	65.5	132.3	176.3	165.7	147.6	61.5
2327 Trousers, Men's & Boys' (49)	3.58	2.95	15.29	15.78	8.22	9.30	36	11.4	4.7	41.0	68.9	50.8	71.9	10.5
	2.05	1.48	8.90	9.98	4.82	5.57	59	5.7	8.3	85.9	102.2	68.4	109.7	19.4
	1.56	0.91	3.04	4.35	3.31	3.45	86	4.1	20.5	158.1	209.5	129.8	190.0	33.9
2341 Underwear & Nightwear, Women's & Children's (77)	3.37	3.40	13.96	19.69	7.45	9.73	32	10.7	8.1	36.4	57.2	60.1	64.2	8.9
	2.21	1.33	7.18	9.58	5.71	6.32	41	7.6	16.0	69.3	114.1	86.7	106.2	21.9
	1.50	0.40	2.20	3.18	4.00	4.67	54	5.2	32.6	132.7	178.6	134.9	162.8	63.1
2511-12 Wood Household Furniture & Upholstered (127)	4.70	3.95	11.89	19.23	4.34	6.77	31	8.8	24.0	19.0	37.7	52.3	44.2	10.0
	2.95	2.15	7.15	9.52	3.01	4.33	42	6.6	37.9	31.3	65.7	82.5	66.8	28.2
	1.94	0.46	2.21	3.78	2.23	3.16	56	4.3	50.2	62.5	112.9	110.3	101.8	58.4
2328 Work Clothing, Men's & Boys' (44)	5.30	4.02	12.89	13.20	4.86	5.24	37	6.6	7.7	23.2	62.0	67.1	43.3	5.3
	2.55	2.28	7.44	8.53	3.61	3.90	53	4.4	15.7	55.5	100.3	84.7	75.7	25.0
	1.95	1.04	3.77	4.82	2.51	2.74	81	3.0	25.4	92.7	129.5	125.2	104.6	39.0

[¶ 128]
Natural Business Year for Various Trades and Industries

Here is a table of suggested closing dates for fiscal years of various industries as prepared by the Natural Business Year Committee of the American Institute of Certified Public Accountants and reproduced with their kind permission:

Trade or Industry	Closing Date *	Trade or Industry	Closing Date *
NONMANUFACTURING		Leather goods	Jan.
		Luggage	Jan.
Agricultural		Lumber & building materials	Nov. to Feb.
Cotton plantations	Mar.	Mail order houses	Jan.
Feed	May to June	Music	Jan.
Hatcheries, chicken	June	Office supplies	May
Orange groves	Sept.	Radio & television	Jan.
Poultry farms	Sept.	Restaurants	June
Poultry supplies	Nov.	Stationery	June
Seeds, wholesale & retail	June	Toys	Jan.
Real Estate and Construction		*Transportation, Service, Misc.*	
Building contractors	Feb.	Advertising, outdoor	Mar.
General contractors	Feb.	Advertising agencies	Dec.
Heating, piping, air		Air transportation companies	Apr.
conditioning contractors	Dec.	Airports	Apr.
Marine contractors	Feb.	Cemeteries	Mar.
Office buildings	May	Cleaning & dyeing	
Paving contractors	Mar.	establishments	Nov.
Real estate, agencies	Sept.	Club, women's	June
Real estate holding companies	Sept.	Colleges	June
		Garages	Sept.
Retail Trade		Hospital (AHA recommendation)	Sept.
Apparel & Clothing—		Hotels, residential	June or July
Corsets & brassieres	July	Hotels, resort (closed part time)	
Men's clothing	Jan.		Last month of season
Ladies' ready-to-wear	Jan.	Hunting & fishing clubs	Feb.
Millinery	June	Laundries	June
Automobiles	Oct.	Photographers	Apr.
Books	June	Schools, private	June or Aug.
Coal	May	Theatres	June
Department stores	Jan.	Warehouses—	
Drugs	Jan.	Cold storage	Mar.
(Drug stores with no		Cotton	June or July
soda or novelty line)	(—possibly July)	Tobacco	May
Dry goods	Jan.	Wharfs	Mar.
Electrical appliances	June		
Filling stations	Sept.	*Wholesale Trade*	
Florists	Sept.	Automotive accessories	Jan.
Furniture	June	Candy	July
General merchandise	Jan.	Coal	Apr.
Gift shops	May	Coffee	May
Groceries	June	Cotton	June or July
Hardware	Jan.	Drugs	June
Jewelry & silverware	Jan.		

---

* Last day of month indicated.

¶128

Natural Business Year for Various Trades and Industries *(continued)*

Trade or Industry	Closing Date *	Trade or Industry	Closing Date *
NONMANUFACTURING *(continued)*		MANUFACTURING *(continued)*	
Wholesale Trade (continued)			
Dry goods	Dec. or Nov.	Brooms & brushes	June
Fruit & vegetable brokers	June	Brush fibres	May
Furs	Jan.	Building materials & supplies—	
Groceries	June	Roofing & waterproof paper	June
Jewelry & silverware	Feb.	Screens, etc.	June
Paper	June	(see also Brick,	
Plumbers' materials	Feb.	Cement, Glass,	
Radio & television	Jan.	Hardware, Lumber, Paints	
Wrapping supplies	Jan.	& Floor coverings)	
		Buttons	June
MANUFACTURING		Candy (see Food)	
		Canning (see Food)	
Agricultural implements		Canvas goods	Nov.
& machinery	Aug. to Oct.	Carpets (see Floor coverings)	
		Cement	Jan.
Apparel—		Chemicals—	
Clothing, men's	Oct.	Insecticides	Oct.
Coats and suits, women's	Nov.	Sulphur—refining	Mar.
Corsets & brassieres	Dec.	(see also Plastics)	
Fur	Mar.	Cigarettes & cigars (see Tobacco)	
Fur coats	Jan. or Feb.	Clocks & watches	Mar.
Garments, silk	Nov.	Clothing (see Apparel)	
Gloves	Nov.	Coal mining	Mar.
Hats	Oct.	Cotton (see Textiles)	
Hats, women's trimmed	Nov.	Cotton compresses	Apr. to July
Hosiery—		Crockery & glassware	Jan.
For those selling		Distillers	Aug.
direct to retailers	Jan.	Elastic webbing	June
For those selling to		Electrical equipment & supplies	
wholesalers & others	Dec.	(including motors, drills, etc.)	Sept.
Underwear	Oct.	Engines, gasoline	Sept.
Work clothing	Nov.	Engines, marine	Sept.
Artificial flower material	Aug.	Fertilizer	June
Automobiles	Sept.	Floor coverings (including	
Automotive accessories	July or Aug.	rugs & carpets)	June
Batteries, automotive			
(see also Tires)	Mar.	Food—	
Aviation—		Bakery products	June
Aeronautical supplies	Sept.	Beverages	Sept.
Aircraft	Nov.	Cereal and grain—	
Awnings and sunshades	Aug.	Flour—milling	Mar. to June
Bags, burlap & buckram	Feb.	Grain dealers	June
Barber shop & beauty parlor supplies	Sept.	Grain, mills & elevators	May to June
Beverages (see Food)		Rice—milling	July
Books (see Publishing)		Confectionery	June
Breweries	Oct.	Dairy & produce companies	Feb. or May
Brick & clay products	Oct. or Mar.		

* Last day of month indicated.

¶128

Natural Business Year for Various Trades and Industries *(continued)*

Trade or Industry	Closing Date *	Trade or Industry	Closing Date *
MANUFACTURING *(continued)*		MANUFACTURING *(continued)*	
Dried fruits—packing	May	Paper	July
Ice cream	Dec.	Paper containers	Apr.
Meat packing	Oct.	Paper novelties	Apr.
Salt	June	Wallpaper	June
Vegetables & fruits—canning	Jan. or Feb.	Petroleum—	
Vegetable oil—		Gasoline—refining	Oct.
cottonseed	June or July	Oil—production	June
in extreme		Oil well supplies	Dec.
South	Apr. or May	Plastics & plastic materials	July
Furniture	Dec. or Nov.	Printing & engraving	July
Glass	June	Printing equipment	Aug.
Hardware	June	Publishing—	
Household appliances—		Books	Jan.
Refrigerators	July	Book, school & college	June
Stoves & furnaces	June	Newspapers	Aug.
(see also Radio)		Radio, television & phonographs	Mar.
Ice, artificial	Oct.	Railroad equipment	Mar.
Jute	Aug.	Rayon (see Textiles)	
Leather & artificial leather	Oct.	Refrigerators (see House-	
Lime	Nov.	hold appliances)	
Liquor (see Distillers)		Rope & cordage	Sept.
Lumber products	Oct.	Rubber (see Tires)	
Machinery and equipment—		Rugs (see Floor coverings)	
Canners' equipment	Sept.	Shipbuilding	June
Laundry equipment	July	Silk (see Textiles)	
Pneumatic machinery	Apr.	Shoes	Nov. or Oct.
Roads machinery	Nov.	Sporting goods	Oct.
Soda fountain equipment	Sept.	Soap	June
Store fixtures	Sept.	Steel & iron products	June
Sugar cane mill equipment	June	Sugar, beet	June
Mattresses	July	Sugar, cane—refining	Dec. or Mar.
Meat (see Food)		Sulphur (see Chemicals)	
Metal products—		Television (see Radio)	
Copper products	June	Textiles—	
Foundries & machine shops	Jan.	Cotton	Sept. or Aug.
Hairpins	May	Cotton gins	Apr. to June
Sheet metal	Mar.	Cotton goods, finished	Sept.
Snapfasteners	June	Linens	Oct.
Wire and fencing	June	Silk & rayon	May
Motion pictures—production	Aug.	Woolens	Oct. or Nov.
Newspapers (see Publishing)		Blankets	Feb.
Office equipment	June	Tires, rubber & rubber goods	Oct.
Oil (see Petroleum)		Tobacco	Feb.
Pads, cotton & sisal	Nov.	Toys	Sept.
Paints, varnish and lacquer	Nov.	Trucks (see Automobiles)	
		Typewriters (see Office equipment)	
		Watches (see Clocks)	
		Woolens (see Textiles)	
		Wiring devices, electrical	Apr.

* Last day of month indicated.

¶128

State-by-State Guide to Maximum Interest
Rates and Usury Laws*

State	Maximum Legal Rate	Maximum Contract Rate	Maximum Judgment Rate	Penalty for Usury	Corporation's Defense of Usury
ALABAMA	6%	8%; 15% corporations $10,000-$100,000; no limit over $100,000.	6%	Forfeit all interest.	No defense over $100,000.
ALASKA	6%	4% above FRB discount rate of 12th Dist.; no limit over $100,000.	8% but in no event over 10%.	Borrower recovers double interest paid; lender forfeits all interest.	No defense.
ARIZONA	6%	10%; 12% over $25,000.	As set out in contract.	Forfeit all interest; borrower recovers interest payments or applies them against principal.	No defense on loan over $5,000 at up to 1½% a month.
ARKANSAS	6%	10%	6%	Contract void.	No provision.
CALIFORNIA	7%	10%	7%	Forfeit all interest; borrower recovers triple interest paid over 10%; violation is felony.	Can defend.
COLORADO	6%	As set out in contract.	6%	No provision.	No defense.
CONNECTICUT	6%	12%; 18% over $10,000.	6%	Fine up to $1,000 or imprisonment up to 6 months or both; loan void.	No provision.
DELAWARE	6%	4% over FRB discount rate; no limit over $100,000 or demand notes over $5,000.	6%	Borrower recovers excess paid; greater of $500 or three times excess paid after maturity.	No defense.
DISTRICT OF COLUMBIA	6%	8%	6% but can set rate to 8%.	Forfeit all interest. Borrower recovers interest or applies it against principal.	No defense.
FLORIDA	6%	10%; 15% over $500,000.	Lesser of 6% or rate agreed on.	Forfeit all interest. When rate 25% or more, forfeit principal and interest.	Usury, if rate over 15%.

*Compilation as of 9/1/74. Most states have special laws governing interest rates on small loans, consumer credit, time sales, bank installment loans, and credit unions. A special check for these laws should be made if the transaction appears to be within possible range.

State-by-State Guide to Maximum Interest Rates and Usury Laws *(continued)*

State	Maximum Legal Rate	Maximum Contract Rate	Maximum Judgment Rate	Penalty for Usury	Corporation's Defense of Usury
GEORGIA	7%	8%; 9% secured by real estate; over 9% VA and FHA loans.	7%	Forfeit all interest; borrower may set off excess interest paid against principal.	No defense on loan over $2,500. No limit on loans of $100,000 or more.
HAWAII	6%	12%	6%	Recover only principal less interest paid.	No defense on loan over $750,000.
IDAHO	8%	10%	8%	Borrower recovers triple interest paid.	12% over $10,000; no defense.
ILLINOIS	5%	8%; 9½% on residential property; no limit on business loans.	6%	Forfeit all interest; borrower recovers double interest.	No defense.
INDIANA	8%	8%	6%, but may contract up to 8%.	Borrower recovers triple interest paid.	No provision.
IOWA	5%	9%; no limit for corporations or REITs.	5%	Forfeit all interest, plus 8% of principal unpaid at time of judgment; additional 8% to school fund.	No defense, including REITs.
KANSAS	6%	10%	8%, but may contract up to 10%.	Forfeit double interest over 10%; excess over 10% applied against principal and interest.	No defense.
KENTUCKY	6%	8½% up to $15,000; no limit over $15,000.	6%	Forfeit interest; debtor can recover twice interest paid, plus reasonable attorney's fees.	No defense unless principal asset is 1- or 2-family house.
LOUISIANA	7%	8%; 10% immovables.	7%	Forfeit all interest.	No defense, including limited and commendam partnerships, endorsers, guarantors and co-makers.
MAINE	6%	No maximum, if in writing; 16% a year simple interest on personal loans over $2,000.	10% after date of judgment.	Personal loans over $2,000, lender forfeits all principal, interest and charges, plus reasonable attorney's fees.	No provision.

State-by-State Guide to Maximum Interest Rates and Usury Laws *(continued)*

State	Maximum Legal Rate	Maximum Contract Rate	Maximum Judgment Rate	Penalty for Usury	Corporation's Defense of Usury
MARYLAND	6%	8%; unsecured loans 12%; no limit on business loans over $5,000. Interest must be stated as a simple annual rate.	6%	Forfeit greater of 3 times excess interest and charges or $500.	No defense.
MASSACHUSETTS	6%	No maximum stated.	No provision.	Criminal penalties.	No provision.
MICHIGAN	5%	7%; no limit on business loans, or realty more than $100,000; 11% other realty.	6%; may contract up to 7%.	Forfeit all interest and charges, plus attorney's fee and court cost. It's criminal usury to charge interest over 25% a year.	No defense. Can agree in writing to pay any rate of interest.
MINNESOTA	6%	8%; no limit over $100,000.	6%	Contract void; borrower gets back all interest paid but one-half goes to school fund.	No defense.
MISSISSIPPI	6%	8%; corporations no limit up to $2,500; 15% over $2,500.	6%, but may contract to 8%.	Forfeit all interest; both interest and principal if rate is over 20%.	No defense if loan under $2,500.
MISSOURI	6%	8%	6%, but may contract to 8%	Forfeit excess and pay costs of action.	No defense.
MONTANA	6%	10%	6%	Double interest charged.	No provision.
NEBRASKA	6%	9%	8%	Forfeit all interest.	No defense.
NEVADA	7%	12%	7%, but may contract to 12%.	Forfeit excess over 12%.	No provision.
NEW HAMPSHIRE	6%	No limit.	6%	No provision.	No provision.
NEW JERSEY	6%	8%; no limit over $50,000; 9½% for residents.	No provision.	Forfeit all interest and costs.	No defense.
NEW MEXICO	6%	10%; 12% if no collateral.	6%, but may contract to 12%.	Forfeit all interest; borrower recovers double interest paid.	No defense.

¶129

State-by-State Guide to Maximum Interest Rates and Usury Laws *(continued)*

State	Maximum Legal Rate	Maximum Contract Rate	Maximum Judgment Rate	Penalty for Usury	Corporation's Defense of Usury
NEW YORK	8%	8½%; no limit on demand notes over $5,000.	6% except where otherwise prescribed by statute.	Contract void. Exception: savings banks and savings and loan assns. forfeit all interest.	No defense unless (1) interest exceeds 25%, or (2) corporation's principal asset is 1- or 2-family house and corporation was formed or control of it acquired within 6 months before loan and mortgage.
NORTH CAROLINA	6%	No limit on loans over $300,000; 12% over $100,000 and up to $300,000; 9% on $100,000 or less, provided nonbusiness and non-real estate loan; 10% over $50,000 and up to $100,000 for business property; 8% on $50,000 or less secured by realty; 10% on $50,000 or less nonresidential realty payable 2 to 10 years; 10% on $7,500 or less secured by realty payable 1 to 10 years. No limit on home mortgage loan of one or more single-family dwelling units, until 6/30/75. Prepayment penalty prohibited.	6%	Forfeit all interest; borrower recovers double interest.	Can defend.
NORTH DAKOTA	4%	Greater of 3% above maximum interest payable on deposits authorized by state banking board, or 7%.	4%	Forfeit all interest plus 25% of principal; borrower recovers double interest paid plus 25% of principal or set off double interest against principal debt; violation is misdemeanor.	No defense. No limit on business loans over $25,000.
OHIO	6%	8%; no limit on loans over $100,000.	6%, but may contract to 8%.	Apply excess interest paid against principal.	No defense.
OKLAHOMA	6%	No limit if not subject to UCC.	10%	Borrower can recover triple interest.	No defense.
OREGON	6%	10%; no limit over $50,000.	6%, but may contract to 10%.	Forfeit entire interest.	No defense on contract not over 12%.
PENNSYLVANIA	6% up to $50,000.	6% on loans up to $35,000 (unsecured) and business loans up to $10,000.	6%	Forfeit excess over 6%; borrower recovers excess paid over 6%.	No defense.
RHODE ISLAND	6%	21%	6%	Contract void; borrower recovers double interest paid.	No provision.

State-by-State Guide to Maximum Interest Rates and Usury Laws *(continued)*

State	Maximum Legal Rate	Maximum Contract Rate	Maximum Judgment Rate	Penalty for Usury	Corporation's Defense of Usury
SOUTH CAROLINA	6%	8%; 10% over $50,000 and up to $100,000; 12% over $100,000 and up to $500,000. Expiring 6/30/75 on realty 9% up to $50,000; 10% over $50,000 and up to $100,000; 12% over 100,000 and up to $500,000; no limit over $500,000; loans up to $100,000 at over 8%, no prepayment penalty.	6%	Forfeit all interest; borrower recovers double interest paid.	No defense capital stock of $40,000 or more has been issued.
SOUTH DAKOTA	6%	10%	8%	Forfeit all interest.	No defense.
TENNESSEE	6%	10%	6%	Forfeit excess interest.	Can plead.
TEXAS	6%	10%	6%, but may contract to 10%.	Forfeit twice amount of interest charged; forfeit both principal and interest if rate is double allowed.	1½% a month; $5,000 or more, no defense.
UTAH	6%	No limit if not subject to UCCC	8%, but may contract to 10%.	Forfeit all interest; borrower recovers triple interest paid and attorney's fees. Violation is misdemeanor.	No provision.
VERMONT	8½%	8½%; no limit for business loans; interest must be stated as an effective annual percentage rate.	No provision.	Lender forfeits all interest and one-half of principal. Borrower recovers excess paid, plus interest costs and attorney's fees. Willful violation $500 fine, 6 mos. jail or both.	No provision.
VIRGINIA	6%	8%; no limit on first mortgage realty loans, if nonagricultural.	No provision.	Borrower can recover twice interest paid if suit brought within two years.	No defense for corporations, partnerships under Title 50, c. 3, professional assn. or real estate investment trusts.

¶129

State-by-State Guide to Maximum Interest Rates and Usury Laws *(continued)*

State	Maximum Legal Rate	Maximum Contract Rate	Maximum Judgment Rate	Penalty for Usury	Corporation's Defense of Usury
WASHINGTON	6%	12%, plus "set up charge" of lesser of 4% or $15 on loans from $100 to $500; minimum of $4 on loans under $100.	8%, but in no case to exceed 10%.	Forfeit all interest; borrower may apply double interest paid against principal; debtor can collect costs and reasonable attorney's fees. Usury laws apply to out-of-state loans made to residents, same as if loan made in state.	No defense for corporations, Massachusetts trusts, associations, limited partnerships, and persons in money lending business or real estate improvement, if transactions over $100,000.
WEST VIRGINIA	6%	8%; 9% for residents.	No provision.	Forfeit all interest; borrower can recover greater of $100 or 4 times interest charged.	No defense.
WISCONSIN	5%	12%; unless consumer transaction; 14% on secured loans.	7%	Forfeit all interest and principal under $2,000; borrower recovers interest paid and $2,000 of principal.	No defense.
WYOMING	7%	UCC limits.	No provision.	Lender forfeits interest. Borrower can recover triple interest.	No provision.

¶129

[¶130]   **State-by-State Guide to Time Sales Acts**

| | | Finance Charge | | |
State	Goods Covered	Limit?	License?	Effective Date
Alaska	All Goods	Yes	No	1/1/63
Arizona	Vehicles	Yes	Yes	7/1/61
California	All Goods	Yes	No	1/1/60
	Vehicles	Yes	No	1/1/62
Colorado	All Goods	Yes	No	5/11/59
	Vehicles	Yes	Yes	1951
Connecticut	All Goods	Yes	Yes	1947
Delaware	All Goods	Yes	No	11/6/60
	Vehicles	Yes	Yes	10/7/60
Dist. of Columbia	Vehicles	Yes	Yes	5/22/60
Florida	All Goods	Yes	Yes	1/1/60
	Vehicles	Yes	Yes	10/1/57
Georgia	All Goods	Yes	No	10/1/67
	Vehicles	Yes	No	10/1/67
Hawaii	All Goods	Yes	No	1941
Illinois	All Goods	Yes	No	1/1/68
	Vehicles	Yes	No	1/1/68
Indiana	All Goods	Yes	Yes	7/1/35
Iowa	Vehicles	Yes	No	7/4/57
Kansas	All Goods	Yes	Yes	5/14/58
Kentucky	Vehicles	Yes	No	1956
	All Goods	No	No	1/1/63
Louisiana	Vehicles	Yes	Yes	1/1/59
Maine	All Goods	No	No	1/1/68
	Vehicles	Yes	Yes	1/1/58
Maryland	All Goods	Yes	No	6/1/41
Massachusetts	All Goods	Yes	Yes	11/1/66
	Vehicles	Yes	Yes	1/19/59
Michigan	All Goods	Yes	No	3/10/67
	Vehicles	Yes	Yes	1051
Minnesota	Vehicles	Yes	Yes	7/1/57
Mississippi	Vehicles	Yes	Yes	7/21/58
Missouri	Vehicles	Yes	Yes	10/11/63
	All Goods	Yes	No	7/1/59
Montana	All Goods	Yes	Yes	5/25/65
Nebraska	All Goods	Yes	Yes	7/1/65
Nevada	All Goods	Yes	No	7/1/65
New Hampshire	Vehicles	Yes	Yes	10/1/61
New Jersey	All Goods	Yes	Yes	9/7/60
New Mexico	Vehicles	Yes	No	6/19/65
	All Goods	Yes	Yes	6/15/59
New York	Vehicles	Yes	Yes	1956
	All Goods	Yes	Yes	7/1/57
North Dakota	All Goods	Yes	No	1949
Ohio	All Goods	Yes	No	1/1/58
Oklahoma	All Goods	Yes	No	9/1/67
Oregon	All Goods	No	No	10/1/63
	Vehicles	Yes	No	6/28/47

¶130

State-by-State Guide to Time Sales Acts *(continued)*

State	Goods Covered	Finance Charge Limit?	License?	Effective Date
Pennsylvania	All Goods	Yes	No	4/1/67
	Vehicles	Yes	Yes	7/1/57
Rhode Island	All Goods	Yes	No	7/1/69
South Carolina	Vehicles	Yes	No	7/31/68
South Dakota	Vehicles	Yes	Yes	3/15/61
Tennessee	All Goods	Yes	No	1953
Texas	All Goods	Yes	No	1/1/68
	Vehicles	Yes	No	1/1/68
Utah	All Goods	Yes	Yes	1/1/64
Vermont	Vehicles	Yes	Yes	1/1/62
	All Goods	Yes	Yes	1/1/64
Virginia	All Goods	No	No	1/1/69
Washington	All Goods	Yes	No	1/1/68
Wisconsin	Vehicles	Yes	Yes	1938

In addition, North Carolina has a statute that requires disclosure of contract terms but not much else.

¶130

[¶131]   State-by-State Guide to Attachments*

The tables set out in this section are designed to give the practitioner quick familiarity with the basic remedies available to the creditor in all jurisdictions and to alert him to possible pitfalls.

State	Can Attachment be Made Prior to Judgment?	Amount of Bond Required
ALABAMA	Yes	Twice amount claimed.
ALASKA	Yes	Amount claimed.
ARIZONA	In specific instances. See statute.	Amount claimed.
ARKANSAS	Yes	Bond that plaintiff shall pay defendant all damages sustained by reason of the attachment.
CALIFORNIA	Yes, when claim is $50 or more.	½ amount claimed, but not under $50.
COLORADO	Fraud only.	Twice amount claimed.
CONNECTICUT	No	None—except cost bond by non-resident plaintiff.
DELAWARE	No	Yes, for costs and damages.
DISTRICT OF COLUMBIA	Yes	Twice amount claimed.
FLORIDA	Yes. See statute.	Twice amount claimed.
GEORGIA	Yes	Twice amount claimed.
HAWAII	Yes. See statute.	Twice amount claimed. If over $50,000, not less than 1½ times amount claimed.
IDAHO	No	Amount claimed.
ILLINOIS	No	Twice amount claimed.
INDIANA	Fraud only.	Yes, for costs and damages.
IOWA	Yes, fraud and other conditions.	Twice amount claimed.
KANSAS	If court permits.	Twice amount claimed.
KENTUCKY	Yes	Limited to twice amount claimed.
LOUISIANA	Yes	Amount claimed.
MAINE	No	Twice amount claimed if personal property.
MARYLAND	Fraud only.	Twice amount claimed.
MASSACHUSETTS	No	Not provided.
MICHIGAN	If court permits.	Tort actions only.
MINNESOTA	Not usually.	At least $250.
MISSISSIPPI	Yes. See statute.	Twice amount claimed.
MISSOURI	Yes	Twice amount claimed.

*Reproduced from PROFITABLE USE OF CREDIT IN SELLING & COLLECTING by Allyn M. Schiffer, published by Fairchild Publications, Inc., New York; price $12 per copy.

State-by-State Guide to Attachments *(continued)*

State	Can Attachment Be Made Prior to Judgment?	Amount of Bond Required
MONTANA	Fraud only.	Twice amount claimed.
NEBRASKA	Yes	Twice amount claimed.
NEVADA	Fraud only.	Not less than $200, nor less than ¼ of claim.
NEW HAMPSHIRE	No	None
NEW JERSEY	If court permits.	Discretion of court.
NEW MEXICO	Yes	Twice amount claimed.
NEW YORK	Fraud only.	For costs. Minimum: $250.
NORTH CAROLINA	No	Not less than $200, at court's discretion.
NORTH DAKOTA	Yes	Not less than $250. $50 in justice-court.
OHIO	Fraud only.	Twice amount claimed.
OKLAHOMA	Yes	Twice amount claimed.
OREGON	No	Amount claimed. Minimum bond $100.
PENNSYLVANIA	Not usually.	Twice amount claimed.
RHODE ISLAND	No	None
SOUTH CAROLINA	Fraud only.	Not less than $250.
SOUTH DAKOTA	Yes	Twice amount claimed. If claim exceeds $1,000, bond is amount claimed, not exceeding $10,000.
TENNESSEE	Yes	Twice amount claimed.
TEXAS	Yes	Twice amount claimed.
UTAH	Fraud only.	Twice amount claimed.
VERMONT	Permitted on partly due running accounts.	Cost bond.
VIRGINIA	Yes	Twice value of property claimed.
WASHINGTON	Yes	Twice amount claimed.
WEST VIRGINIA	Yes	Twice amount claimed.
WISCONSIN	Yes	On unmatured claim, triple amount of claim.
WYOMING	Yes	Twice amount claimed.

¶131

[¶132]   **State-by-State Guide to Garnishments**

The chart below is from Prentice-Hall, Inc.—Installment Sales (looseleaf service current to 9/10/74) and is reproduced by permission of the publisher. It summarizes state laws on garnishment. It shows the limit each state imposes on the amount of wages that can be garnisheed for specific kinds of wage earners (for example, heads of families), and gives the order in which garnishments must be paid (if the state specifies). The penalties employers face for failure to follow garnishment procedure are listed at the end of the chart; the figures in brackets after the name of each state show which penalty applies in that state.

Federal restrictions: Federal law exempts an employee's earnings during a workweek equal to 30 times federal minimum wage or 75% of employee's "disposable earnings," whichever is greater. "Disposable earnings" means earnings minus all deductions required by law. Exemption won't apply, however, to any court order for support of any person; court order of bankruptcy (under Ch. XIII, Bankruptcy Act); or any debt due for any state or federal tax. Union dues, initiation fees, employee's share of health and welfare premiums and repayment of credit union loans aren't considered deductions required by law in determining "disposable income," but amounts withheld for unemployment and workmen's compensation insurance pursuant to state law are considered such deductions.

Law also prohibits employer from firing an employee by reason of the fact his wages have been garnisheed for any one indebtedness (i.e., a single debt regardless of number of garnishment proceedings brought to collect it). Penalty for willful violation: up to $1,000 fine or one year's imprisonment, or both.

If federal and state laws don't agree, law that provides for lesser garnishment or greater restriction on firing will control.

Secretary of Labor, acting through Wage-Hour Division, will enforce federal garnishment provisions.

Alabama: [2] [5] Laborers and resident employees: 75% of wages is exempt.

Alaska: [3] Head of family: $350 of worker's income due him or received by him from any source within 30 days preceding levy of execution, if necessary for his use or use of his family which he supports. Other exemptions are automatic payroll deductions for care of children not in his custody and child support payments ordered paid to court trustee. Single person: $200 is exempt.

Arizona: [2] [5] [8] 50% of earnings for personal services performed within 30 days before service of garnishment papers is exempt, if necessary for the support of the employee's family.

The earnings of a minor child aren't subject to garnishment for parent's debt if the debt wasn't for the special benefit of the child.

Arkansas: [2] Laborers and mechanics: Wages for 60 days are exempt provided employee files affidavit with court stating that 60 days' wages together with other personal property he owns doesn't exceed state constitutional limitation ($200 for single resident not head of a family, $500 for married resident or head of a family). First $25 of net wages of all mechanics and laborers absolutely exempt without need for filing schedule of exemptions. "Net wages" means gross wages less following deductions actually withheld: Ark. income tax, fed. income tax, social security, group retirement, group hospitalization insurance premiums and group life insurance premiums.

Employees of railroads have an exemption of $200 before judgment against them by the creditor.

¶132

California: [1] 100% of earnings is exempt if they are necessary for the support of the employee's family residing within the State. This exemption doesn't apply if the debt is for (1) necessaries or (2) services rendered by an employee of the worker. 50% of the earnings for personal services performed within 30 days before service of garnishment papers is exempt.

Employer may be required to attend court and be examined.

Employee can't be discharged for garnishment for one indebtedness prior to final order or judgment of court.

Colorado: [2] [5] Head of family: 70% of earnings, proceeds of health, accident or disability insurance, and pension or retirement benefits is exempt. Single persons: 35% of such earnings and proceeds is exempt.

The employer can withhold claims that he would have had if he hadn't been garnisheed.

Connecticut: [2] [5] [18] Greater of (a) 75% of disposable earnings for workweek or (b) $65 or 40 times federal minimum hourly wage in effect when earnings are payable is exempt. Tax collector can garnishee wages for local taxes due.

Priorities: Only one execution at a time can be satisfied. They're satisfied in the order they're presented to the employer. Execution of wages for support of wife or minor child (children) takes priority over other executions and two or more can be levied at same time provided total levy doesn't exceed maximum permitted.

Employer can't discipline, suspend, or discharge employee because his wages have been garnisheed, unless more than 7 garnishments in any calendar year.

Delaware: [5] 85% of wages is exempt for all residents, unless debt for state taxes (no wage exemption). In child support cases, 25% of net salary can be attached by court order plus another 5% for each child. Attachment prevails over other exemptions. Employer liable for fine or jail for dismissal of employee due to attachment.

District of Columbia [2] Greater of: (1) 75% of disposable weekly wages or (2) 30 times federal minimum wage. Don't apply to judgments for support; instead, 50% of the employee's gross wages is exempt.

Wages of nonresident (if major portion earned outside D.C.) are exempt to same extent provided by laws of state of his residence. Exemption applies only to contracts or transactions entered into outside D.C.

However, these percentages don't apply to judgments for support of wife, former wife, or children; instead, 50% of the employee's gross wages is exempt.

$200 of earnings (other than wages), insurance, annuities or pension or retirement payments are exempt for each of two months if person is a resident or earns major portion of his livelihood in D.C. and is principal support of a family before service of garnishment papers. If husband and wife live together, their total earnings determine the exemption; $60 of earnings (other than wages), insurance, etc., are exempt for each of 2 months before service of garnishment papers, if person doesn't support a family.

Priorities: Only one attachment is satisfied at one time. Attachments are satisfied in order of priority. However, judgment for support takes priority (in discretion of court).

Florida: [4] Head of family residing in Florida: All wages are exempt.

Earnings of any person or public officer, state or county, (whether the head of a family or not residing in Florida) are subject to garnishment to enforce Florida court order for alimony, suit money, or support. Court, in its discretion, determines the amount to be garnisheed.

Georgia: [4] [5] [8] Greater of 75% of disposable earnings or 30 times federal minimum wage is exempt. $1,000 of wages due deceased employee of railroad company or other corporation is exempt. Salaries of officers of corporations (except municipal corporations) are subject to garnishment if more than $500 a year.

Claim of exemption from garnishment is ineffective against decree for alimony [*Huling v. Huling,* 194 Ga. 819, 22 S.E. (2d) 882, (1942)].

Default judgment against employer for failure to answer can be modified on motion to 125% of amount due employees, less exemption, from time of service to last day on which timely answer could be made but not less than 15% of amount of employee's judgment.

Employee can't be discharged for garnishment for any one indebtedness, whether or not debt is for state taxes [O.A.G., 9-12-69].

Hawaii: [5] The following wages are exempt from garnishment: 95% of the first $100 per month, 90% of the next $100, and 80% of all sums over $200.

The employer can withhold (liquidated) claims that he would have had if he hadn't been garnisheed.

Priorities: Garnishments are paid in order of service of process on employer. If 2 or more are served at the same time, order of issuance from court determines priority.

Employer can't suspend or discharge employee because he was summoned as garnishee in action where employee is debtor or because employee filed petition to pay his debts under wage earners plan of Bankruptcy Act.

Idaho: [4] [5] Greater of 75% of disposable earnings or 30 times federal minimum wage is exempt. Doesn't apply to support orders, debt for state or federal tax.

Illinois: [2] [10] [13] $65 a week if employee is head of family and $50 a week if he isn't, or 85% of gross wages, salary, commissions, bonuses and periodic payments under retirement or pension plan, whichever is greater, or federal limit.

The employer can withhold claims that he would have had if he hadn't been garnisheed.

Employer can deduct greater of $2 or 2% of amount paid because of deduction order on same debt.

Indiana: [4] Resident householder: $15 a week and 90% of the excess wages is exempt.

Single person not a resident householder: 90% of total wages is exempt.

Wages cannot be garnisheed if the employee and creditor are nonresidents and the employer is a resident of Indiana.

Iowa: [2] Greater of 75% of disposable earnings or 30 times federal minimum wage is exempt. Maximum amount that can be garnisheed in one calendar year is $250 for each creditor (except under decrees for support of minors).

Exemptions don't apply to judgments for (1) alimony or (2) support of minors.

Employee can't be discharged for garnishment.

Kansas: [2] [5] [10] Greater of 75% of disposable earnings or 30 times federal minimum wage is exempt. Doesn't apply to court order for child support, order of bankruptcy court or debt for state or federal tax.

No one creditor can issue more than one garnishment during month.

No prejudgment garnishment orders can be issued against wages.

¶132

Employee can't be discharged for garnishment for any one indebtedness.

Kentucky: [3] [5] [12] Greater of 75% of disposable earnings or 30 times federal minimum wage is exempt. Doesn't apply to support orders, bankruptcy orders, debt for state or federal tax. Garnishment is lien on wages during pay period in which served; if less than 2 weeks, or employee paid in advance, during succeeding pay period.

Prejudgment garnishments can be obtained by making demand on defendant in writing, advising him of grounds of suit and his right to hearing.

Priorities: Orders have priority in date of service.

Employee can't be discharged for garnishment for any one indebtedness.

Louisiana: [2] [10] [15] 75% of disposable earnings or excess over 30 times federal minimum wage is exempt, but exemptions can't be less than $70 a week of disposable earnings.

Priorities: A debt between employer and employee is treated as a prior garnishment. However, garnishment of father's wages for child's support always takes priority over all garnishment orders.

Maine: [2] Greater of 75% of disposable earnings or 30 times federal minimum hourly wage is exempt.

Employee can't be discharged because earnings have been subject to garnishment order.

Wages of minor children and wife aren't subject to garnishment for parent's or husband's debt.

Maryland: [2] [9] Greater of $120 times number of weeks wages due at date of attachment were earned, or 75% of such wages due, is exempt.

Caroline, Worcester, Kent, and Queen Anne's Counties: Greater of 75% of wages due or 30 times federal minimum wage is exempt.

Exemption doesn't apply to garnishment for state income tax.

Court can order lien on earnings of defendant in paternity suit. Nonsupport court orders will be lien on notification by Probation Dept.

Employee can't be discharged for garnishment on any one occasion in calendar year.

Massachusetts: [4] $125 a week of wages due is exempt. $100 a week of pensions payable to employee is exempt.

Wages due for personal services of defendant's (employee's) wife or minor child are exempt.

The employer can withhold (liquidated) claims that he would have had if he hadn't been garnisheed.

Michigan: [3] [5] If first garnishment—Householder head of family: 60% of wages is exempt. If wages are for one week's labor (or less), the most that is exempt is $50 and the least is $30. If wages are for more than one week's labor maximum exemption is $90, minimum $60.

Others: 40% of wages is exempt. Most that is exempt is $50 and least is $20.

In all other cases—Householder head of family—60% of wages is exempt. If wages are for one week or less, the most that is exempt is $30 and the least is $12. If wages are for more than one week but no more than 16 days, the most that is exempt is $60 and the least is $24. If wages are for more than 16 days, maximum exemption is $60 and minimum $30.

Others: 30% of wages is exempt. The most that can be exempt is $20 and the least is $10.

Employer can offset claims he has against his employee.

Wages earned by an employee at the time the garnishment papers are served on the employer, but not payable on the next regular payday, are not subject to garnishment.

The amount paid under a judgment for alimony or child support is exempt.

No employer can use garnishment as sole cause of discharge.

Minnesota: [2] [5] Greater of 75% of disposable earnings or an amount of such wages equal to 8 times number of business days and paid holidays (not exceeding 5 per week) times federal minimum wage is exempt. All earnings for preceding 30 days, if needed for use of family supported wholly or partly by his labor, are exempt. If the employee was on relief, all his wages for six months after his return to work are exempt. This exemption can be claimed only once every three years.

The earnings of a minor child cannot be garnisheed for the debt of his parent unless the debt was for the special benefit of the minor.

Priorities: Garnishments are paid in the order of service of the garnishee papers on the employer. More than one garnishment can be paid at a time subject to total nonexempt disposable earnings.

If amount garnisheed is less than $10, garnishment is ineffective and employer is relieved of liability.

Employee can't be discharged for garnishment unless more than 3 garnishments served within 90 days on more than one debt.

Mississippi: [2] [5] 75% of wages, salary or other compensation due resident employees or laborers is exempt.

The proceeds of any trust created by an employer as part of a pension plan, disability or death benefit plan or any trust created under a retirement plan which are exempt from federal income tax aren't subject to garnishment.

Court orders or judgments for payment of alimony, separate maintenance, or child support are exempt from garnishment.

Missouri: [4] Greater of 75% of disposable earnings or 30 times federal minimum wage is exempt. If resident head of family, 90% of disposable earnings is exempt. Doesn't apply to support orders, bankruptcy court orders, or for federal or state taxes due.

Employee can't be discharged for garnishment of any one indebtedness.

Montana: [5] Head of family: All earnings for personal services performed within 45 days before service of garnishment papers are exempt, if necessary for support of employee's family.

If the employee's debts are for the necessaries of life or gasoline, then only 50% of his earnings are exempt.

An unmarried employee, over 60 is entitled to the same exemption as the head of a family.

All earnings for preceding 30 days are exempt in actions for $10 or less.

Employee can't be discharged or laid off for garnishment.

Nebraska: [2] [5] Lesser of (a) 25% of weekly disposable earnings (including pension or retirement program payments); (b) excess of 30 times federal minimum hourly wage; or (c) 15% of disposable earnings if employee head of household, is subject to garnishment. Doesn't apply to support orders of any person, orders of bankruptcy court, or for debt due for state or federal taxes.

No prejudgement action for garnishment may be filed in small claims court.

Priorities: Justice who issued the first garnishment determines the priorities.

Nevada: [1] [5] [8] Lesser of 25% of disposable earnings or excess over 30 times federal minimum wage is subject to garnishment. Exemption doesn't apply to any court orders in bankruptcy for support or debt due for state or federal taxes. Court may allow employer up to $3 for each garnishment levied.

New Hampshire: [2] [5] [10] All wages earned by the employee after the service of garnishment papers on the employer are exempt.

All wages earned before service of garnishment papers are exempt, unless judgment on debt is issued by state court. In such cases, exemption for each week is 50 times federal minimum wage.

Earnings of the wife and minor children of the employee are exempt.

$50 a week of wages earned by the employee before service of garnishment papers is exempt if the main action is based on a small loan contract.

All the wages of a married woman are exempt if the action is based on a small loan contract to which her husband is an obligor.

Exemptions don't apply in an action for taxes by Tax Collector.

New Jersey: [1] If the employee earns at least $48 a week, 10% of his wages may be garnisheed. If the employee earns more than $7,500 a year, a court may increase the percentage to be garnisheed.

Priorities: Only one execution can be satisfied at a time in order in which they are served on employer; except support orders for child or wife have priority over other executions served on same day.

Employer can retain as compensation 5% of amount deducted pursuant to garnishment.

New Mexico: [2] Either 75% of disposable earnings each pay period or 40 times federal minimum hourly wage each week, whichever is greater, is exempt.

Priorities: Liens will be satisfied in order in which they are served.

New York: [1] [5] [7] Income execution: 10% of income is subject to income execution in court of record where judgment debtor (employee) is receiving or will receive more than $85 a week.

Earnings of recipients of public assistance are exempt.

Garnishment proceeding, court not of record: 10% of earnings can be garnisheed by court order if judgment has been recovered in a court not of record against an employee whose income is (1) $30 or more a week if he resides or works in a city of 250,000 or more, or (2) $25 a week in any other case. Only one garnishment can be satisfied at a time. If two or more are issued simultaneously, they are satisfied in the order they are served on the employer.

Priorities: Assignment or court order for support of minor child and/or spouse takes priority over other assignment or garnishment of wages, etc., except as to deductions made mandatory by law.

No employee may be laid off or discharged because one or more executions have been served against his wages.

North Carolina: [2] [10] Head of family: All earnings for 60 days before the service of the garnishment papers are exempt if necessary for the support of the employee's family.

Garnishment for collection of state taxes is limited to 10% of salary or wages paid in any one month, 10% of wages in any one pay period for municipal and county property taxes. Wages due officials or employees of the state, its agencies, instrumentalities and political subdivisions are subject to garnishment for delinquent taxes.

North Dakota: [2] [8] Greater of 75% of disposable earnings or 40 times federal minimum wage is exempt. Doesn't apply to court orders for support of any person or of bankruptcy court or for debt for state or federal taxes.

Ohio: [2] [3] [10] [12] Greater of 175 times federal minimum wage of 82½% of disposable earnings is exempt for services performed within 30 days before issuing garnishment (when exemption claimed in bankruptcy 82½% of gross earnings for 30-day period ending on 10th day prior to filing bankruptcy petition). Garnishment may be granted after judgment. There must be at least 30 days between garnishments. Support orders have priority over all other garnishments and no part of wages is exempt from them. They are paid in order received. Employer may deduct up to 1% as service charge.

Priorities: Priorities may be determined by court order.

Employee can't be discharged solely for attachment for no more than one action in garnishment in any 12-month period.

Oklahoma: [2] [3] [5] Resident head of family: 75% of all wages for services performed within last 90 days is exempt.

Resident employee can apply to have all wages for services performed within 90 days before issuing of garnishment papers exempt if they are necessary for the support of the employee's family, but not if judgment is for child support or maintenance.

Resident not head of family: 75% of all wages for services is exempt.

Wages, bonuses or commissions are exempt from prejudgment garnishment.

Priorities: Priorities may be determined by court order.

Oregon: [3] [4] [5] Greater of 75% of disposable earnings or 30 times federal minimum wage in effect 4-30-69 is exempt. Doesn't apply to support orders (including attorney's fees or court costs), bankruptcy orders, debt for state or federal tax. Protection of law can't be waived.

Any legal process served must indicate whether it's subject to garnishment restrictions.

Pennsylvania: [5] All wages are exempt.

Exceptions: Wages are subject to garnishment for 4 weeks' board and lodging.

A husband's wages can be garnisheed to pay support for wife and children. Likewise, wages can be garnisheed for alimony.

Taxes: On tax collector's written notice and demand, employer must deduct from employee's wages, commissions, or earnings then owing (or that become due within 60 days thereafter), or from any unpaid commissions or earnings in his possession (or that come into his possession within 60 days thereafter), a sum sufficient to pay employee's, or his wife's, delinquent per capita, poll, occupation, occupational privilege, local earned income taxes and costs, and must pay this sum to tax collector within 60 days after receipt of notice. No more than 10% of wages, commissions, or earnings of delinquent taxpayer or husband can be deducted at one time. Employer may keep up to 2% of money collected for extra bookkeeping expenses.

Rhode Island: [2] $50 a week of earnings is exempt; all earnings of wife and minor child of employee are exempt; all wages of seamen are exempt. If employee was on relief, all his wages for one year are exempt.

South Carolina: [3] [10] Head of family: All earnings for personal services performed within 60 days before the garnishment order are exempt, if necessary for a family supported by the employee.

15% of an employee's earnings (but not more than $100) can be garnisheed to pay a judgment for food, fuel, or medicine.

South Dakota: [2] [5] [16] Head of family: All earnings for personal services performed within 60 days before the garnishment order are exempt, if necessary for family's support.

Tennessee: [2] [5] Resident head of family: Greater of 50% of net weekly salary, wages, or income or $20 a week is exempt, subject to a maximum of $50 a week. An additional $2.50 a week for each dependent child under 16 is exempt.

Resident not head of family: Greater of 40% of net weekly salary, wages, or income or $17.50 a week is exempt, subject to a maximum of $40 a week.

Only salary, wages, or income earned at time of service of garnishment papers is affected. Net amount of earnings or income is gross amount earned during pay period to time of service of garnishment papers less social security and withholding taxes.

Exemptions don't apply if judgment against employee is for alimony or child support.

Employee's salary, wages, or income not subject to garnishment by same creditor more than once every other pay period.

Priorities: Priorities are determined by time of service of papers; if served at same time, according to time of filing. If creditor with priority doesn't claim entire amount allowed over exempt amount, next creditor may claim excess; but service of more than one garnishment in a pay period won't reduce exempt amount.

Texas: All current wages for personal services are exempt.

Utah: [2] [5] [10] Head of family or married man: 50% of earnings for personal services performed within 30 days before service of garnishment papers is exempt if necessary for support of family. Minimum exemption, $50.

The earnings of a minor child aren't subject to garnishment for parent's debt if the debt wasn't for the special benefit of the child.

The employer can withhold claims that he would have had if he hadn't been garnisheed.

Vermont: [2] [5] Wages due minor child can't be garnisheed in action against parent; wages due married woman can't be garnisheed in action against husband.

Employer can withhold claims (based on express or implied contract) that he would have had if he hadn't been garnisheed.

Weekly disposable earnings over 30 times federal minimum wage are subject to a lien for delinquent poll tax or old age assistance tax at rate of $4 a week; regardless of any assignment of earnings.

It is unlawful for employer to discharge employee because employee's compensation has been garnisheed unless employer has been previously garnisheed for that employee on 5 or more separate occasions arising from separate actions, or unless employer establishes there were other substantial causes contributing to the discharge.

Virginia: [1] [5] [6] [11] Greater of 75% of disposable earnings or 30 times federal minimum wage is exempt. Doesn't apply to court-ordered support bankruptcy order or debt for state or federal taxes. Earnings include wages, commissions, bonuses, and payments under pension or retirement programs.

Wages of minors are exempt from garnishment for debts of parents.

Employee can't be discharged for garnishment for any indebtedness.

Washington: [2] [5] Greater of 40 times state minimum wage or 75% of disposable earnings is exempt. Deductions as contributions toward pension or retirement plan established pursuant to collective bargaining agreement are not part of disposable earnings.

Exemption doesn't apply to garnishment for child support if (a) based on judgment or court order;

¶132

(b) amount doesn't exceed 2 months, support payments and writ contains such statement. Continuing lien on wages can be obtained.

Priorities: Liens have priority as served. Only 1 garnishment can be satisfied at a time.

Employee can't be discharged for garnishment unless employer is garnisheed on 3 or more separate indebtednesses served within 12 consecutive months.

West Virginia: [1] [5] 80% of earnings in a week is exempt; wages payable to the employee cannot be less than $20 a week.

Priorities: Only one garnishment is satisfied at a time. However, where two or more have been served and the first garnishment has been satisfied, nonexempt wages remaining are applied toward the satisfaction of junior garnishments in the order of their priority.

Wisconsin: [2] Worker with no dependents: Basic exemption—60% of the income of the employee for each 30-day period before service of process in proceeding to collect a debt. The exemption cannot be less than $75 nor more than $100. The employee can elect to have the exemption computed on a 90-day basis.

Worker with dependents: Basic exemption—On the income of the employee for each 30-day period before service of process in proceeding to collect a debt, $120 plus $20 for each dependent. However, the total exemption cannot exceed 75% of total income. The employee can elect to have the exemption computed on a 90-day basis.

"Income" means gross receipts less federal and state withholding and social security taxes.

Subsistence allowance: When earnings are subjected to garnishment, employer pays (on date earnings are payable) subsistence allowance to employee of greater of 75% of disposable earnings or 30 times federal minimum wage. Doesn't apply to support orders, bankruptcy orders, debt for state or federal tax, or orders in voluntary proceedings by wage earners. Garnishment action against earnings can't be started before judgment.

Employer can't discharge employee because his earnings have been garnisheed for any one indebtedness. Willful violation punishable by fine up to $1,000, jail, or both.

Wyoming: [1] [5] Head of resident family: 50% of earnings for personal services performed within 60 days before service of garnishment papers is exempt, if necessary for support of family.

Penalty Footnotes

[1] If employer doesn't follow garnishment procedure, he may be subject to civil action by his employee's creditor.

[2] If employer doesn't follow garnishment procedure, he may be liable for all or part of the claim that the creditor has against his employee.

[3] If employer doesn't follow in garnishment procedure, he may be subject to contempt proceedings; in Mich., he may be liable to arrest.

[4] If employer doesn't follow garnishment procedure, judgment may be taken against him.

[5] Salaries of employees of the state and/or its subdivisions may be subject to garnishment.

[6] Employer is governed by return date of summons but after service or knowledge of issuance of summons employer can only pay employee exempt wages [O.A.G., 6-16-70]. (Va.)

[7] $85 a week floor on garnishable earnings applies to all income executions in effect on or after 9-1-70 even if previously filed [O.A.G., Informal 7-13-70]; applies to gross income [O.A.G., Informal 1-12-71] (N.Y.).

¶132

[8] Based on decision in *Sniadach v. Family Finance Corp. of Bay View* (US S. Ct., 1969) 395 US 337; 89 S. Ct. 1820, laws in following states allowing prejudgment garnishments are invalid: Ariz. [*Termplan Inc. v. Superior Ct. of Maricopa Cty.* Ariz. Sup. Ct., 12-29-69]; Nev. [O.A.G., 4-1-70]; Ga. [*Reeves v. Motor Contract Co.* N.D. Ga., Div. 3-30-71]; N. Dak. [O.A.G., 7-8-69]. (After *Sniadach* decision, Calif. and Wisc. amended their laws to provide for no prejudgment garnishments.)

[9] Under both state and federal law, exemption applies to employee's disposable earnings or net take-home pay after all lawful deductions or withholdings [Ltr., Chief Judge, People's Court of Balt. City to P-H, 9-2-70].

[10] Application denied for exemption from federal restrictions: (Ill., 12-10-70; Kans., 10-30-70; La., 5-25-71; N.H., 8-10-70; N.C., 12-3-70; Ohio, 11-25-70; S.C., 12-3-70; Utah, 2-5-71.

[11] States exempt from federal wage garnishment restrictions: Ky., 12-5-70; Va., 1-12-71. U.S. Labor Dept. retains enforcement responsibility for discharge provisions.

[12] One-garnishment-a-month provision stands (*Hodgson v. Cleveland Munic. Crt.,* 3-19-71; *Hodgson v. Hamilton Municipal Court,* 9-23-71). Resident and nonresident can apply to municipal or county court in jurisdiction of his place of employment for appointment of trustee to avoid garnishment of nonexempt earnings (Ohio).

[13] Eff. 7-1-72, employers must deduct from wages of employees until total amount due on judgment and costs is paid or expiration of employer's payroll period ending immediately prior to 60 days after service of summons, whichever first occurs (Ill.).

[14] When person employed out of state by employer doing "substantial" business in state, particularly when employer has resident agent in state, garnishment for unpaid state income tax can be used and taxpayer needn't be personally served with any papers (Kans.).

[15] State law on garnishment applies to teachers [O.A.G., 9-14-72] (La.).

[16] Eff. 7-1-72, garnishment of earnings is prohibited prior to final judgment in action. Earnings mean compensation paid or payable for personal services (wages), salaries, commission, bonus or otherwise, and includes periodic payments under pension or retirement plans. (S. Dak.).

[17] Reserved.

[18] Eff. 10-1-72, voluntary wage deduction authorization for support will have same force and effect as wage execution ordered and signed by circuit court and have same priority over other executions. Voluntary wage execution will not be allowed unless prior order of support has been filed in circuit court. No voluntary wage deductions authorization will be effective less than 14 days from date of signing authorization (Conn.).

Note: This table does *not* reflect the special garnishment rules enacted by some states for consumer transactions. (For example, see UCC §5.105.)

[¶133]   **State-by-State Guide to**
                            Judgment Notes—Attorneys' Fees

The chart below is from Prentice-Hall, Inc. Installment Sales (looseleaf service current to 9/10/74) and is reproduced by permission of the publisher. It contains state-by-state information showing: whether there is any statutory provision governing judgment notes; whether confessions of judgment are recognized (and any conditions limiting such recognition), as well as indicating conditions and limitations on collection of attorneys' fees for those jurisdictions which permit inclusion of this provision in a contract or note.

Alabama: Judgment notes: No statutory provision. Confessions of judgment: Invalid, if made before suit. In Jefferson County, small loan licensee may not use judgment note. Attorneys' fees: No statutory provision.

Alaska: Judgment notes: No statutory provision. Confessions of judgment: Valid, if made by debtor in person, or by debtor's attorney-in-fact under power of attorney. Attorneys' fees: Retail installment contract may provide for reasonable attorneys' fees.

Arizona: Judgment notes: No statutory provision. Confessions of judgment: Valid only if executed after note matures; cannot be made in connection with small loan. Attorneys' fees: Motor vehicle time sales contract may provide for reasonable fees.

Arkansas: Judgment notes: No statutory provision. Confessions of judgment: Valid if personally made by debtor in court. Attorneys' fees: Enforceable, if doesn't exceed 10% of principal and accrued interest.

California: Judgment notes: No statutory provision. Confessions of judgment: Valid, if taken for money due or to become due, or to secure contingent liability. But prohibited in time sales contract; also, in contracts of industrial loan company or property broker or small loan licensee. Attorneys' fees: May be awarded to prevailing party in any action on time sales contract or installment contract, or on motor vehicle time sales contract.

Colorado: Judgment notes: No statutory provision. Confessions of judgment: Invalid in motor vehicle time sales contracts. Attorneys' fees: Valid in time sales contracts, including motor vehicle time sales contract, if not more than 15% of balance due.

Connecticut: Judgment notes: No statutory provision. Confessions of judgment: May be offered before trial in pending action. Not permitted in time sales contract or installment loan contract, or in small loan contracts. Attorneys' fees: Valid in note or other evidence of indebtedness, but court may modify amount. Also valid in installment contract or installment loan contract, if not more than 15% of balance due.

Delaware: Judgment notes: Valid. Confessions of judgment: Permitted, if made by warrant of attorney. Attorneys' fees: In action on note, contract or other written instrument, court may award reasonable fees (up to 5% of amount awarded for principal and interest), if instrument provides for payment of such fees.

Florida: Judgment notes: No statutory provision. Confessions of judgment: Void if executed before or without an action; not permitted in small loan contracts. Attorneys' fees: Valid in time sales contracts of motor vehicles, or of other goods.

Georgia: Judgment notes: No statutory provision. Confessions of judgment: Permitted only in suit; invalid in small loan contracts. Attorneys' fees: Void, unless debtor given 10 days' notice before suit and fails to pay.

¶133

Hawaii: Judgment notes: No statutory provision. Confessions of judgment: Not enforceable. Attorneys' fees: Court will award in actions on notes or contracts, if fee specified therein or in separate agreement; but not more than 25%.

Idaho: Judgment notes: No statutory provision. Confessions of judgment: Valid, if for money due or to become due, or to secure against contingent liability. Not permitted in small loan contracts or consumer transactions. Attorneys' fees: No statutory provision.

Illinois: Judgment notes: Valid. Confessions of judgment: Enforceable in time sales contracts after buyer's default; valid in small loan contracts. Attorneys' fees: Enforceable in time sales contracts, if fee reasonable.

Indiana: Judgment notes: Void. Confessions of judgment: Void. Use of same or judgment note is misdemeanor. Prohibited in small loan contracts. Attorneys' fees: Valid, if reasonable.

Iowa: Judgment notes: No statutory provision. Confessions of judgment: Valid, if for money due or to become due, or to secure against contingent liability. Not permitted in small loan contracts. Attorneys' fees: If note or contract provides for payment of fees, court will allow 10% of first $200, 5% on excess through $500, 3% on excess through $1,000 and 1% on amounts over $1,000.

Kansas: Judgment notes: No statutory provision. Confessions of judgment: Debtor may offer to confess judgment at any time before trial. Attorneys' fees: No statutory provision.

Kentucky: Judgment notes: No statutory provision. Confessions of judgment: Void. May not be included in motor vehicle time sales contracts, or in small loan contracts. Attorneys' fees: Permitted in motor vehicle time sales contracts, but can't exceed 15% of balance due.

Louisiana: Judgment notes: No statutory provision. Confessions of judgment: May not be made before obligation matures, except for purpose of executory process. Not permitted in small loan contracts. Attorneys' fees: Permitted in motor vehicle time sales contract, if not more than 25% of balance due, with $15 minimum.

Maine: Judgment notes: No statutory provision. Confessions of judgment: Prohibited in home repair agreement, and in small loan contracts. Attorneys' fees: Permitted in home repair agreement and motor vehicle time sales contracts, if reasonable fee.

Maryland: Judgment notes: No statutory provision. Confessions of judgment: Prohibited in all time sales contracts, with a cash sales price of $5,000 or less, if seller takes back security interest; also, in small loan contracts. Attorneys' fees: Permitted in time sales contracts, up to 15% of balance due.

Massachusetts: (General Laws). Judgment notes: Prohibited. Confessions of judgment: Inclusion in note or contract void. Prohibited in time sales contracts of goods and motor vehicles. Attorneys' fees: No statutory provision.

Michigan: Judgment note: No statutory provision. Confessions of judgment: Valid, if made in separate instrument. Prohibited in time sales contracts or retail charge agreements; also, in small loan contracts. Attorneys' fees: Permitted in home repair installment contract, but not to exceed 20% of balance due. Also permitted in retail charge agreement, if fee reasonable.

Minnesota: Judgment notes: No statutory provision. Confessions of judgment: Valid, if made in separate verified statement. Prohibited in motor vehicle time sales contracts, and small loan contracts. Attorneys' fees: Enforceable in motor vehicle time sales contracts up to 15% of balance due.

Mississippi: Judgment notes: Void. Confessions of judgment: Void, including small loan contracts. Attorneys' fees: Enforceable in motor vehicle time sales contracts, up to 15% of balance due.

¶133

Missouri: Judgment notes: No statutory provision. Confessions of judgment: Valid, if taken for money due or to become due, or as security against contingent liability, and made in separate verified statements. Attorneys' fees: Enforceable in time sales contracts of goods, up to 15% of balance due; also, on same conditions, in motor vehicle time sales contracts.

Montana: Judgment notes: Void. Confessions of judgment: Valid, if made by separate verified statement. Attorneys' fees: Enforceable in time sales contracts up to 15% of balance due.

Nebraska: Judgment notes: No statutory provision. Confessions of judgment: May be made by debtor in person, with creditor's consent. Not valid in small loan contracts. Attorneys' fees: Court may grant reasonable fees in action for balance due on purchase of necessaries, up to $1,000, and debtor failed to pay after 90-days' notice.

Nevada: Judgment notes: No statutory provision. Confessions of judgment: Valid, if for sums due or to become due, or to secure contingent liability. Invalid in small loan contracts. Attorneys' fees: Provision for reasonable fee in time sales contract valid. Also valid in small loan contracts, provided fee to be fixed by court in event suit necessary to collect.

New Hampshire: Judgment notes: No statutory provision. Confessions of judgment: Prohibited in motor vehicle time sales contracts; also prohibited in small loan contracts. Attorneys' fees: Provision for reasonable fee in motor vehicle time sales contracts valid; not recognized as permissible delinquency charge under small loan laws.

New Jersey: Judgment notes: Not permitted. Confessions of judgment: Invalid in time sales contracts, or in separate instrument relating thereto; also, invalid in home repair contracts, and small loan contracts. Attorneys' fees: Enforceable in time sales contracts and retail charge accounts if not more than 20% of first $500 and 10% of excess; also in home repair contracts, if "reasonable." If credit union reduces loan to judgment or gives to attorney for collection after default, it may collect attorneys' fees not to exceed 20%.

New Mexico: Judgment notes: Void. Confessions of judgment: Void, if made before cause of action accrues on negotiable instrument or contract to pay money. Not permitted in time sales contracts or retail charge agreements. Attorneys' fees: Provision for reasonable fees permitted in time sales contracts, retail charge agreements or small loan agreements.

New York: Judgment notes: No statutory provision. Confessions of judgment: May be made on debtor's affidavit; but judgment void if entered on affidavit made before debtor's default on installment purchases up to $1,500 of goods for nonbusiness or noncommercial use. Prohibited in time sales contracts, including motor vehicles; revolving credit agreements; small loan contracts. Attorneys' fees: Valid up to 20% of balance due on revolving credit agreements; up to 15% of balance due on motor vehicle time sales contracts; void on retail installment contracts. Credit unions may collect reasonable fees actually spent for necessary court process after debtor's default.

North Carolina: Judgment notes: Not recognized. Confessions of judgment: Enforceable, if made by signed, verified statement for money due or to become due, or to secure against a contingent liability. Prohibited in small loan contracts. Attorneys' fees: Provision not enforceable, but does not affect negotiability of instrument.

North Dakota: Judgment notes: No statutory provision. Confessions of judgment: Enforceable, if entered on debtor's signed, verified statement for a specific sum. Invalid in time sales contracts; also, small loans. Attorneys' fees: Void as against public policy in notes and other evidences of debt.

¶133

Ohio: Judgment notes: Recognized. Confessions of judgment: Invalid, if made in connection with consumer loans or transactions. Attorneys' fees: Not permitted in time sales contracts.

Oklahoma: Judgment notes: No statutory provision. Confessions of judgment: Enforceable, if entered under warrant of attorney acknowledged by debtor, and if debtor first files affidavit as to the facts; void in consumer transactions. Attorneys' fees: Reasonable, if amount financed exceeds $1,000.

Oregon: Judgment notes: No statutory provision. Confessions of judgment: Valid, if for money due or to become due or to secure contingent liability; may be entered, with creditor's consent, if acknowledged by debtor. Unenforceable in motor vehicle time sales contracts; invalid as to small loan contracts. Attorneys' fees: May provide for reasonable fees in time sales contracts of goods and revolving charge accounts; also, in motor vehicle time sales contracts.

Pennsylvania: Judgment notes: Recognized. Confessions of judgment: Invalid in small loan contracts; invalid, if debtor's income less than $10,000. Attorneys' fees: Fees up to 20% of balance due permitted in home improvement contracts. Credit unions may collect fees to public officials and reasonable fees of attorneys and outside collection agencies; but total of such fees can't exceed 20% of outstanding loan balance.

Rhode Island: Judgment notes: No statutory provision. Confessions of judgment: Prohibited in small loan contracts. Attorneys' fees: Reasonable attorney's fees allowed if suit is brought to realize on collateral used to secure loan.

South Carolina: Judgment notes: No statutory provision. Confessions of judgment: Enforceable, if made by verified statement. Prohibited in small loan contracts. Attorneys' fees: May provide in small loan contracts for reasonable fee to be fixed by court.

South Dakota: Judgment notes: No statutory provision. Confessions of judgment: Valid, if made by verified statement. Attorneys' fees: Void as against public policy.

Tennessee: Judgment notes: Invalid. Confessions of judgment: Invalid, if made before action started. Attorneys' fees: No statutory provision.

Texas: Judgment notes: Invalid. Confessions of judgment: Invalid, if made before action started. Prohibited in small loan contracts. Attorneys' fees: No statutory provision.

Utah: Judgment notes: No statutory provision. Confessions of judgment: Valid, if made on debtor's verified statement for money due or to become due or to secure contingent liability. Prohibited in small loan contracts. Attorneys' fees: Not to exceed 15% of unpaid debt in sale; reasonable in small loan contract.

Vermont: Judgment notes: No statutory provision. Confessions of judgment: Valid, if made by debtor in writing, with creditor's consent. Prohibited in small loan contracts and consumer contracts. Attorneys' fees: Enforceable as to time sales contracts of goods, revolving charge accounts, and motor vehicle time sales contracts.

Virginia: Judgement notes: Valid; but warrant of attorney in note must name attorney and court in which judgment may be confessed. Confessions of judgment: May be entered in clerk's office at any time, but debtor has 21 days after notice of entry in which to move to have judgment set aside. Prohibited in small loan contracts. Attorneys' fees: No statutory provision.

Washington: Judgment notes: No statutory provision. Confessions of judgment: May be made on debtor's verified statement. Prohibited in small loan contracts. Attorneys' fees: May provide for reasonable fee in time sales contracts or revolving charge accounts.

¶133

West Virginia: Judgment notes: No statutory provision. Confessions of judgment: May be made in action. Prohibited in small loan contracts. Attorneys' fees: No statutory provision.

Wisconsin: Judgment notes: Void. Confessions of judgment: Void. Attorneys' fees: No statutory provision.

Wyoming: Judgment notes: valid. Confessions of judgment: May be made by debtor in open court, with creditor's consent; but attorney confessing judgment must show warrant of attorney; void in consumer credit transactions. Attorneys' fees: Reasonable.

[¶134]   State Mechanic's Lien Laws

The statutory requirements for mechanic's liens differ widely from state to state in many details. Many states have statutory forms of notice of lien. Here are the main areas to be checked under the law of the particular state you are concerned with:

(1) *Who Is Entitled to the Lien?* Generally, anyone furnishing labor or materials adding to the value of realty, but the mere sale of material without reference to the building in which it is to be used won't give rise to a lien. The labor or material furnished must be on the credit of the building. Some statutes require a written contract describing the building or improvement and fixing the amount.

(2) *Right of Subcontractors*—Some states give the subcontractor a direct lien on the realty regardless of whether there is a debt due the principal contractor; but in other states the subcontractor cannot recover more than is due the contractor.

(3) *Material Furnished Subcontractor*—Some states grant, others deny, liens for materials furnished subcontractors.

(4) *Contractual Provision Against Liens*—In some states a contractual provision against mechanic's liens bars the contractor from claiming a lien; in others it does not bar the contractor but only subcontractors, laborers, and materialmen; but in others, subcontractors, laborers, and materialmen are not barred. Where the contract between the contractor and the subcontractor provides that no lien is to be filed by the subcontractor, or where the subcontract adopts the principal contract containing such a provision, the subcontractor will be barred.

(5) *Time of Filing*—May vary from 30 days to 120 days from completion of work or furnishing materials.

(6) *Contents of Notice of Lien*—Usually must describe claim and property.

(7) *Duration of Lien*—There is usually a short statute of limitations for the commencement of a foreclosure action. Some statutes make provision for renewal of lien.

[¶135]

State-by-State Guide to Age at Which Infants Become Competent to Contract

The following table shows the age at which infants attain majority in the various states. The table also indicates those instances in which females reach majority before males. The effects of marriage upon a minor's disability to contract are shown where such a disability may be removed by appropriate court proceedings.

State	Age	State	Age
Alabama	21 (1)	Montana	*18
Alaska	19 (5)	Nebraska	19 (2)
Arizona	18	Nevada	18
Arkansas	*21	New Hampshire	18
California	18	New Jersey	18
Colorado	18	New Mexico	18 (2)
Connecticut	18	New York	18
Delaware	18	North Carolina	18
District of Columbia	21	North Dakota	18
Florida	18 (2)	Ohio	18
Georgia	18 (1)	Oklahoma	18
Hawaii	18	Oregon	18
Idaho	18 (3)	Pennsylvania	18
Illinois	18	Rhode Island	18
Indiana	18	South Carolina	21
Iowa	18 (2)	South Dakota	18
Kansas	18	Tennessee	18
Kentucky	18	Texas	18
Louisiana	18 (6)	Utah	*21 (2)
Maine	18	Vermont	18
Maryland	18	Virginia	18
Massachusetts	18	Washington	18 (4)
Michigan	18	West Virginia	18
Minnesota	18	Wisconsin	18
Mississippi	21	Wyoming	19
Missouri	18		

*Female attains majority at age 18.
(1) Married person attains majority at age 18.
(2) Married person attains majority upon marriage, regardless of age.
(3) Married male attains majority at age 18, married female upon marriage.
(4) Married person has limited rights, if spouse is over age 21.
(5) Married female attains majority upon marriage.
(6) Married male attains majority upon marriage.

SECTION B
TAX PLANNING AIDS

[¶201]

Federal Income Tax Rates

Single Person				Head of Household			
Taxable Income		Rate on Excess		Taxable Income		Rate on Excess	
$ - $	500	$	14%	$ - $	1,000	$	14%
500 -	1,000	70	15%	1,000 -	2,000	140	16%
1,000 -	1,500	145	16%	2,000 -	4,000	300	18%
1,500 -	2,000	225	17%	4,000 -	6,000	660	19%
2,000 -	4,000	310	19%	6,000 -	8,000	1,040	22%
4,000 -	6,000	690	21%	8,000 -	10,000	1,480	23%
6,000 -	8,000	1,110	24%	10,000 -	12,000	1,940	25%
8,000 -	10,000	1,590	25%	12,000 -	14,000	2,440	27%
10,000 -	12,000	2,090	27%	14,000 -	16,000	2,980	28%
12,000 -	14,000	2,630	29%	16,000 -	18,000	3,540	31%
14,000 -	16,000	3,210	31%	18,000 -	20,000	4,160	32%
16,000 -	18,000	3,830	34%	20,000 -	22,000	4,800	35%
18,000 -	20,000	4,510	36%	22,000 -	24,000	5,500	36%
20,000 -	22,000	5,230	38%	24,000 -	26,000	6,220	38%
22,000 -	26,000	5,990	40%	26,000 -	28,000	6,980	41%
26,000 -	32,000	7,590	45%	28,000 -	32,000	7,800	42%
32,000 -	38,000	10,290	50%	32,000 -	36,000	9,480	45%
38,000 -	44,000	13,290	55%	36,000 -	38,000	11,280	48%
44,000 -	50,000	16,590	60%	38,000 -	40,000	12,240	51%
50,000 -	60,000	20,190	62%	40,000 -	44,000	13,260	52%
60,000 -	70,000	26,390	64%	44,000 -	50,000	15,340	55%
70,000 -	80,000	32,790	66%	50,000 -	52,000	18,640	56%
80,000 -	90,000	39,390	68%	52,000 -	64,000	19,760	58%
90,000 -	100,000	46,190	69%	64,000 -	70,000	26,720	59%
Over $100,000		53,090	70%	70,000 -	76,000	30,260	61%
				76,000 -	80,000	33,920	62%
				80,000 -	88,000	36,400	63%
				88,000 -	100,000	41,440	64%
				100,000 -	120,000	49,120	66%
				120,000 -	140,000	62,320	67%
				140,000 -	160,000	75,720	68%
				160,000 -	180,000	89,320	69%
				Over $180,000		103,120	70%

Federal Income Tax Rates *(continued)*

Married Filing Joint Return; Surviving Spouse		Married Individuals Filing Separately and Estates and Trusts	
Taxable Income	Rate on Excess	Taxable Income	Rate on Excess
$ - $ 1,000	$ 14%	- $ 500	$ 14%
1,000 - 2,000	140 15%	$ 500 - 1,000	70 15%
2,000 - 3,000	290 16%	1,000 - 1,500	145 16%
3,000 - 4,000	450 17%	1,500 - 2,000	225 17%
4,000 - 8,000	620 19%	2,000 - 4,000	310 19%
8,000 - 12,000	1,380 22%	4,000 - 6,000	690 22%
12,000 - 16,000	2,260 25%	6,000 - 8,000	1,130 25%
16,000 - 20,000	3,260 28%	8,000 - 10,000	1,630 28%
20,000 - 24,000	4,380 32%	10,000 - 12,000	2,190 32%
24,000 - 28,000	5,660 36%	12,000 - 14,000	2,830 36%
28,000 - 32,000	7,100 39%	14,000 - 16,000	3,550 39%
32,000 - 36,000	8,660 42%	16,000 - 18,000	4,330 42%
36,000 - 40,000	10,340 45%	18,000 - 20,000	5,170 45%
40,000 - 44,000	12,140 48%	20,000 - 22,000	6,070 48%
44,000 - 52,000	14,060 50%	22,000 - 26,000	7,030 50%
52,000 - 64,000	18,060 53%	26,000 - 32,000	9,030 53%
64,000 - 76,000	24,420 55%	32,000 - 38,000	12,210 55%
76,000 - 88,000	31,020 58%	38,000 - 44,000	15,510 58%
88,000 - 100,000	37,980 60%	44,000 - 50,000	18,990 60%
100,000 - 120,000	45,180 62%	50,000 - 60,000	22,590 62%
120,000 - 140,000	57,580 64%	60,000 - 70,000	28,970 64%
140,000 - 160,000	70,380 66%	70,000 - 80,000	35,190 66%
160,000 - 180,000	83,580 68%	80,000 - 90,000	41,790 68%
180,000 - 200,000	97,180 69%	90,000 - 100,000	48,590 69%
Over $200,000	110,980 70%	Over $100,000	55,490 70%

¶201

[¶202]

After-Tax Income Table

Example of use of this table: Find how much an unmarried man who is not head of household has left from a taxable income of $35,000.

Amount from the line $32,000-$38,000 under column for single persons $21,710
Percentage from that line (50%) times excess of $35,000 over $32,000 ($3,000)   1,500
After-tax income.. 23,210

Single Person			Head of Household		
Taxable Income	After-Tax Income*	Plus This % of Excess	Taxable Income	After-Tax Income*	Plus This % of Excess
$ - $ 500	—	86	$ - $ 1,000	—	86
500 - 1,000	430	85	1,000 - 2,000	860	84
1,000 - 1,500	855	84	2,000 - 4,000	1,700	82
1,500 - 2,000	1,275	83	4,000 - 6,000	3,340	81
2,000 - 4,000	1,690	81	6,000 - 8,000	4,960	78
4,000 - 6,000	3,310	79	8,000 - 10,000	6,520	77
6,000 - 8,000	4,900	76	10,000 - 12,000	8,060	75
8,000 - 10,000	6,410	75	12,000 - 14,000	9,560	73
10,000 - 12,000	7,910	73	14,000 - 16,000	11,020	72
12,000 - 14,000	9,370	71	16,000 - 18,000	12,460	69
14,000 - 16,000	10,790	69	18,000 - 20,000	13,840	68
16,000 - 18,000	12,170	66	20,000 - 22,000	15,200	65
18,000 - 20,000	13,490	64	22,000 - 24,000	16,500	64
20,000 - 22,000	14,770	62	24,000 - 26,000	17,880	63
22,000 - 26,000	16,010	60	26,000 - 28,000	19,020	59
26,000 - 32,000	18,410	55	28,000 - 32,000	20,200	58
32,000 - 38,000	21,710	50	32,000 - 36,000	22,520	55
38,000 - 44,000	24,710	45	36,000 - 38,000	25,720	52
44,000 - 50,000	27,410	40	38,000 - 40,000	25,760	49
50,000 - 60,000	29,810	38	40,000 - 44,000	26,740	48
60,000 - 70,000	33,610	36	44,000 - 50,000	28,660	45
70,000 - 80,000	37,210	34	50,000 - 52,000	31,360	44
80,000 - 90,000	40,610	32	52,000 - 64,000	32,240	42
90,000 - 100,000	43,810	31	64,000 - 70,000	37,280	41
Over $100,000	46,910	30	70,000 - 76,000	39,740	39
			76,000 - 80,000	42,080	38
			80,000 - 88,000	43,600	37
			88,000 - 100,000	46,560	36
			100,000 - 120,000	50,880	34
			120,000 - 140,000	57,680	33
			140,000 - 160,000	64,280	32
			160,000 - 180,000	70,680	31
			Over $180,000	76,880	30

*Lower Amount in First Column

After-Tax Income Table *(continued)*

Married Filing Joint Return; Surviving Spouse			Married Individuals Filing Separately and Estates and Trusts		
Taxable Income	After-Tax Income*	Plus This % of Excess	Taxable Income	After-Tax Income*	Plus This % of Excess
$ - $ 1,000	—	86	$ - $ 500	—	
1,000 - 2,000	860	85	500 - 1,000	430	85
2,000 - 3,000	1,710	84	1,000 - 1,500	855	84
3,000 - 4,000	2,550	83	1,500 - 2,000	1,275	83
4,000 - 8,000	3,380	81	2,000 - 4,000	1,690	81
8,000 - 12,000	6,620	78	4,000 - 6,000	3,310	78
12,000 - 16,000	9,740	75	6,000 - 8,000	4,870	75
16,000 - 20,000	12,740	72	8,000 - 10,000	6,370	72
20,000 - 24,000	15,620	68	10,000 - 12,000	7,810	68
24,000 - 28,000	18,340	64	12,000 - 14,000	9,170	64
28,000 - 32,000	20,900	61	14,000 - 16,000	10,450	61
32,000 - 36,000	23,340	58	16,000 - 18,000	11,670	58
36,000 - 40,000	25,660	55	18,000 - 20,000	12,830	55
40,000 - 44,000	27,860	52	20,000 - 22,000	13,930	52
44,000 - 52,000	29,940	50	22,000 - 26,000	14,970	50
52,000 - 64,000	33,940	47	26,000 - 32,000	16,970	47
64,000 - 76,000	39,580	45	32,000 - 38,000	19,790	45
76,000 - 88,000	44,980	42	38,000 - 44,000	22,490	42
88,000 - 100,000	50,020	40	44,000 - 50,000	25,010	40
100,000 - 120,000	54,820	38	50,000 - 60,000	27,410	38
120,000 - 140,000	62,420	36	60,000 - 70,000	31,210	36
140,000 - 160,000	69,620	34	70,000 - 80,000	34.810	34
160,000 - 180,000	76,420	32	80,000 - 90,000	38,210	32
180,000 - 200,000	82,820	31	90,000 - 100,000	41,410	31
Over $200,000	89,020	30	Over $100,000	44,510	30

*Lower Amount in First Column.

[¶203]

How Much Tax-Exempt Income Is Worth

Example of use of this table:

A single taxpayer with a taxable income of $21,000 purchases a tax-exempt bond yielding 9%. Looking in the table at 9% and under $22,000 (next highest figure), you find 15%, which is the rate he would have to get on a taxable bond to be as well off.

Single Person

Taxable Income up to	Tax %	4%	5%	6%	7%	8%	9%	10%	11%	12%
$ 3,000	19	4.94	6.17	7.41	8.64	9.88	11.11	12.35	13,58	14.81
4,000	21	5.06	6.33	7.59	8.86	10.13	11.39	12.66	13.92	15.20
6,000	24	5.26	6.58	7.89	9.21	10.53	11.84	13.16	14.47	15.79
8,000	25	5.33	6.67	8.00	9.33	10.67	12.00	13.33	14.67	16.00
10,000	27	5.48	6.85	8.22	9.59	10.96	12.33	13.70	15.07	16.44
12,000	29	5.63	7.04	8.45	9.86	11.27	12.68	14.08	15.49	16.90
14,000	31	5.80	7.25	8.70	10.14	11.59	13.04	14.49	15.94	17.39
16,000	34	6.06	7.58	9.09	10.61	12.12	13.64	15.15	16.67	18.18
18,000	36	6.25	7.81	9.38	10.94	12.50	14.06	15.63	17.19	18.75
20,000	38	6.45	8.06	9.68	11.29	12.90	14.52	16.13	17.74	19.35
22,000	40	6.67	8.33	10.00	11.67	13.33	15.00	16.67	18.33	20.00
24,000	40	6.67	8.33	10.00	11.67	13.33	15.00	16.67	18.33	20.00
26,000	45	7.27	9.09	10.91	12.73	14.55	16.36	18.18	20.00	21.82
28,000	45	7.27	9.09	10.91	12.73	14.55	16.36	18.18	20.00	21.82
32,000	50	8.00	10.00	12.00	14.00	16.00	18.00	20.00	22.00	24.00
36,000	50	8.00	10.00	12.00	14.00	16.00	18.00	20.00	22.00	24.00
38,000	55	8.89	11.11	13.33	15.56	17.78	20.00	22.22	24.44	26.67
40,000	55	8.89	11.11	13.33	15.56	17.78	20.00	22.22	24.44	26.67
44,000	60	10.00	12.50	15.00	17.50	20.00	22.50	25.00	27.50	30.00
50,000	62	10.53	13.16	15.79	18.42	21.05	23.68	26.32	28.95	31.58
52,000	62	10.53	13.16	15.79	18.42	21.05	23.68	26.32	28.95	31.58
60,000	64	11.11	13.89	16.67	19.44	22.22	25.00	27.78	30.56	33.33
64,000	64	11.11	13.89	16.67	19.44	22.22	25.00	27.78	30.56	33.33
70,000	66	11.76	14.71	17.65	20.59	23.53	26.47	29.41	32.35	35.29
76,000	66	11.76	14.71	17.65	20.59	23.53	26.47	29.41	32.35	35.29
80,000	68	12.50	15.63	18.75	21.88	25.00	28.13	31.25	34.38	37.50
88,000	68	12.50	15.63	18.75	21.88	25.00	28.13	31.25	34.38	37.50
90,000	69	12.90	16.13	19.35	22.58	25.81	29.03	32.26	35.48	38.71
100,000	70	13.33	16.67	20.00	23.33	26.67	30.00	33.33	36.67	40.00
120,000	70	13.33	16.67	20.00	23.33	26.67	30.00	33.33	36.67	40.00
140,000	70	13.33	16.67	20.00	23.33	26.67	30.00	33.33	36.67	40.00
160,000	70	13.33	16.67	20.00	23.33	26.67	30.00	33.33	36.67	40.00
180,000	70	13.33	16.67	20.00	23.33	26.67	30.00	33.33	36.67	40.00
200,000	70	13.33	16.67	20.00	23.33	26.67	30.00	33.33	36.67	40.00

How Much-Tax-Exempt Income Is Worth *(continued)*

Head of Household

Taxable Income	Tax %	4%	5%	6%	7%	8%	9%	10%	11%	12%
up to $ 3,000	18	4.88	6.10	7.32	8.54	9.76	10.98	12.20	13.41	14.63
4,000	19	4.94	6.17	7.41	8.64	9.88	11.11	12.35	13.58	14.81
6,000	22	5.13	6.41	7.69	8.97	10.26	11.54	12.82	14.10	15.38
8,000	23	5.19	6.49	7.79	9.09	10.39	11.69	12.99	14.29	15.58
10,000	25	5.33	6.67	8.00	9.33	10.67	12.00	13.33	14.67	16.00
12,000	27	5.48	6.85	8.22	9.59	10.96	12.33	13.70	15.07	16.44
14,000	28	5.56	6.94	8.33	9.72	11.11	12.50	13.89	15.28	16.67
16,000	31	5.80	7.25	8.70	10.14	11.59	13.04	14.49	15.94	17.39
18.000	32	5.88	7.35	8.82	10.29	11.76	13.24	14.71	16.18	17.65
20,000	35	6.15	7.69	9.23	10.77	12.31	13.85	15.38	16.92	18.46
22,000	36	6.25	7.81	9.38	10.94	12.50	14.06	15.63	17.19	18.75
24,000	38	6.45	8.06	9.68	11.29	12.90	14.52	16.13	17.74	19.35
26,000	41	6.78	8.47	10.17	11.86	13.56	15.25	16.95	18.64	20.34
28,000	42	6.90	8.62	10.34	12.07	13.79	15.52	17.24	18.97	20.69
32,000	45	7.27	9.09	10.91	12.73	14.55	16.36	18.18	20.00	21.82
36,000	48	7.69	9.62	11.54	13.46	15.38	17.31	19.23	21.15	23.08
38,000	51	8.16	10.20	12.24	14.29	16.33	18.37	20.41	22.45	24.49
40,000	52	8.33	10.42	12.50	14.58	16.67	18.75	20.83	22.92	25.00
44,000	55	8.89	11.11	13.33	15.56	17.78	20.00	22.22	24.44	26.67
50,000	56	9.09	11.36	13.64	15.91	18.18	20.45	22.73	25.00	27.27
52,000	58	9.52	11.90	14.29	16.67	19.05	21.43	23.81	26.19	28.57
60,000	58	9.52	11.90	14.29	16.67	19.05	21.43	23.81	26.19	28.57
64,000	59	9.76	12.20	14.63	17.07	19.51	21.95	24.39	26.83	29.27
70,000	61	10.25	12.82	15.38	17.95	20.51	23.08	25.64	28.21	30.77
76,000	62	10.53	13.16	15.79	18.42	21.05	23.68	26.32	28.95	31.58
80,000	63	10.81	13.51	16.22	18.92	21.62	24.32	27.03	29.73	32.43
88,000	64	11.11	13.89	16.67	19.44	22.22	25.00	27.78	30.56	33.33
90,000	64	11.11	13.89	16.67	19.44	22.22	25.00	27.78	30.56	33.33
100,000	66	11.76	14.71	17.65	20.59	23.53	26.47	29.41	32.35	35.29
120,000	67	12.12	15.15	18.18	21.21	24.24	27.27	30.30	33.33	36.36
140,000	68	12.50	15.63	18.75	21.88	25.00	28.13	31.25	34.38	37.50
160,000	69	12.90	16.13	19.35	22.58	25.81	29.03	32.26	35.48	38.71
180,000	70	13.33	16.67	20.00	23.33	26.67	30.00	33.33	36.67	40.00
200,000	70	13.33	16.67	20.00	23.33	26.67	30.00	33.33	36.67	40.00

How Much Tax-Exempt Income Is Worth *(continued)*

Married Individuals Filing Separately and Estates and Trusts

Taxable Income	Tax %	4%	5%	6%	7%	8%	9%	10%	11%	12%
up to $ 3,000	19	4.94	6.17	7.41	8.64	9.88	11.11	12.35	13.58	14.81
4,000	22	5.13	6.41	7.69	8.97	10.26	11.54	12.82	14.10	15.38
6,000	25	5.33	6.67	8.00	9.33	10.67	12.00	13.33	14.67	16.00
8,000	28	5.56	6.94	8.33	9.72	11.11	12.50	13.89	15.28	16.67
10,000	32	5.88	7.35	8.82	10.29	11.76	13.24	14.71	16.18	17.65
12,000	36	6.25	7.81	9.38	10.94	12.50	14.06	15.63	17.19	18.75
14,000	39	6.56	8.20	9.84	11.48	13.11	14.75	16.39	18.03	19.67
16,000	42	6.90	8.62	10.34	12.07	13.79	15.52	17.24	18.97	20.69
18,000	45	7.27	9.09	10.91	12.73	14.55	16.36	18.18	20.00	21.82
20,000	48	7.69	9.62	11.54	13.46	15.38	17.31	19.23	21.15	23.08
22,000	50	8.00	10.00	12.00	14.00	16.00	18.00	20.00	22.00	24.00
24,000	50	8.00	10.00	12.00	14.00	16.00	18.00	20.00	22.00	24.00
26,000	53	8.51	10.64	12.77	14.89	17.02	19.15	21.28	23.40	25.53
28,000	53	8.51	10.64	12.77	14.89	17.02	19.15	21.28	23.40	25.53
32,000	55	8.89	11.11	13.33	15.56	17.78	20.00	22.22	24.44	26.67
36,000	55	8.89	11.11	13.33	15.56	17.78	20.00	22.22	24.44	26.67
38,000	58	9.52	11.90	14.29	16.67	19.05	21.43	23.81	26.19	28.57
40,000	58	9.52	11.90	14.29	16.67	19.05	21.43	23.81	26.19	28.57
44,000	60	10.00	12.50	15.00	17.50	20.00	22.50	25.00	27.50	30.00
50,000	62	10.53	13.16	15.79	18.42	21.05	23.68	26.32	28.95	31.58
52,000	62	10.53	13.16	15.79	18.42	21.05	23.68	26.32	28.95	31.58
60,000	64	11.11	13.89	16.67	19.44	22.22	25.00	27.78	30.56	33.33
64,000	64	11.11	13.89	16.67	19.44	22.22	25.00	27.78	30.56	33.33
70,000	66	11.76	14.71	17.65	20.59	23.53	26.47	29.41	32.35	35.29
76,000	66	11.76	14.71	17.65	20.59	23.53	26.47	29.41	32.35	35.29
80,000	68	12.50	15.63	18.75	21.88	25.00	28.13	31.25	34.38	37.50
88,000	68	12.50	15.63	18.75	21.88	25.00	28.13	31.25	34.38	37.50
90,000	69	12.90	16.13	19.35	22.58	25.81	29.03	32.26	35.48	38.71
100,000	70	13.33	16.67	20.00	23.33	26.67	30.00	33.33	36.67	40.00
120,000	70	13.33	16.67	20.00	23.33	26.67	30.00	33.33	36.67	40.00
140,000	70	13.33	16.67	20.00	23.33	26.67	30.00	33.33	36.67	40.00
160,000	70	13.33	16.67	20.00	23.33	26.67	30.00	33.33	36.67	40.00
180,000	70	13.33	16.67	20.00	23.33	26.67	30.00	33.33	36.67	40.00
200,000	70	13.33	16.67	20.00	23.33	26.67	30.00	33.33	36.67	40.00

¶203

How Much Tax-Exempt Income Is Worth *(continued)*
Married Filing Joint Return;
Surviving Spouse

Taxable Income	Tax %	4%	5%	6%	7%	8%	9%	10%	11%	12%
up to $ 3,000	17	4.82	6.02	7.23	8.43	9.64	10.84	12.05	13.25	14.46
4,000	19	4.94	6.17	7.41	8.64	9.88	11.11	12.35	13.58	14.81
6,000	19	4.94	6.17	7.41	8.64	9.88	11.11	12.35	13.58	14.81
8,000	22	5.13	6.41	7.69	8.97	10.26	11.54	12.82	14.10	15.38
10,000	22	5.13	6.41	7.69	8.97	10.26	11.54	12.82	14.10	15.38
12,000	25	5.33	6.67	8.00	9.33	10.67	12.00	13.33	14.67	16.00
14,000	25	5.33	6.67	8.00	9.33	10.67	12.00	13.33	14.67	16.00
16,000	28	5.56	6.94	8.33	9.72	11.11	12.50	13.89	15.28	16.67
18,000	28	5.56	6.94	8.33	9.72	11.11	12.50	13.89	15.28	16.67
20,000	32	5.88	7.35	8.82	10.29	11.76	13.24	14.71	16.18	17.65
22,000	32	5.88	7.35	8.82	10.29	11.76	13.24	14.71	16.18	17.65
24,000	36	6.25	7.81	9.38	10.94	12.50	14.06	15.63	17.19	18.75
26,000	36	6.25	7.81	9.38	10.94	12.50	14.06	15.63	17.19	18.75
28,000	39	6.56	8.20	9.84	11.48	13.11	14.75	16.39	18.03	19.67
32,000	42	6.90	8.62	10.34	12.07	13.79	15.52	17.24	18.97	20.69
36,000	45	7.27	9.09	10.91	12.73	14.55	16.36	18.18	20.00	21.82
38,000	45	7.27	9.09	10.91	12.73	14.55	16.36	18.18	20.00	21.82
40,000	48	7.69	9.62	11.54	13.46	15.38	17.31	19.23	21.15	23.08
44,000	50	8.00	10.00	12.00	14.00	16.00	18.00	20.00	22.00	24.00
50,000	50	8.00	10.00	12.00	14.00	16.00	18.00	20.00	22.00	24.00
52,000	53	8.51	10.64	12.77	14.89	17.02	19.15	21.28	23.40	25.53
60,000	53	8.51	10.64	12.77	14.89	17.02	19.15	21.28	23.40	25.53
64,000	55	8.89	11.11	13.33	15.56	17.78	20.00	22.22	24.44	26.67
70,000	55	8.89	11.11	13.33	15.56	17.78	20.00	22.22	24.44	26.67
76,000	58	9.52	11.90	14.29	16.67	19.05	21.43	23,81	26.19	28.57
80,000	58	9.52	11.90	14.29	16.67	19.05	21.43	23.81	26.19	28.57
88,000	60	10.00	12.50	15.00	17.50	20.00	22.50	25.00	27.50	30.00
90,000	60	10.00	12.50	15.00	17.50	20.00	22.50	25.00	27.50	30.00
100,000	62	10.53	13.16	15.79	18.42	21.05	23.68	26.32	28.95	31.58
120,000	64	11.11	13.89	16.67	19.44	22.22	25.00	27.78	30.56	33.33
140,000	66	11.76	14.71	17.65	20.59	23.53	26.47	29.41	32.35	35.29
160,000	68	12.50	15.63	18.75	21.88	25.00	28.13	31.25	34.38	37.50
180,000	69	12.90	16.13	19.35	22.58	25.81	29.03	32.26	35.48	38.71
200,000	70	13.33	16.67	20.00	23.33	26.67	30.00	33.33	36.67	40.00

¶203

[¶204]

Income Tax Reciprocals

Example of use of this table:

To find how much additional income is needed to produce $100 after tax, multiply 100 by the reciprocal. To find the net income from $100, divide by the reciprocal.

Married, Joint Return		*Married, Separate Return*	
Taxable Income (thousands)	*Reciprocal*	*Taxable Income (thousands)*	*Reciprocal*
4-8	1.2346	4.-6	1.2821
8-12	1.2821	6-8	1.3333
12-16	1.3333	8-10	1.3889
16-20	1.3889	10-12	1.4706
20-24	1.4706	12-14	1.5625
24-28	1.5625	14-16	1.6393
28-32	1.6393	16-18	1.7241
32-36	1.7241	18-20	1.8182
36-40	1.8182	20-22	1.9231
40-44	1.9231	22-26	2.0000
44-52	2.0000	26-32	2.1277
52-64	2.1277	32-38	2.2222
64-76	2.2222	38-44	2.3810
76-88	2.3810	44-50	2.5000
88-100	2.5000	50-60	2.6316
100-120	2.6316	60-70	2.7778
120-140	2.7778	70-80	2.9412
140-160	2.9412	80-90	3.1250
160-180	3.1250	90-100	3.2258
180-200	3.2258	over 100	3.3333
over 200	3.3333		

Income Tax Reciprocals *(continued)*

Head of Household			Single Person	
Taxable Income (thousands)	Reciprocal		Taxable Income (thousands)	Reciprocal
4-6	1.2346		4-6	1.2658
6-8	1.2821		6-8	1.3158
8-10	1.2987		8-10	1.3333
10-12	1.3333		10-12	1.3699
12-14	1.3699		12-14	1.4085
14-16	1.3889		14-16	1.4493
16-18	1.4493		16-18	1.5151
18-20	1.4706		18-20	1.5625
20-22	1.5385		20-22	1.6129
22-24	1.5625		22-26	1.6667
24-26	1.6129		26-32	1.8182
26-28	1.6949		32-38	2.0000
28-32	1.7241		38-44	2.2222
32-36	1.8182		44-50	2.5000
36-38	1.9231		50-60	2.6316
38-40	2.0408		60-70	2.7778
40-44	2.0833		70-80	2.9412
44-50	2.2222		80-90	3.1250
50-52	2.2727		90-100	3.2258
52-64	2.3810		over 100	3.3333
64-70	2.4390			
70-76	2.5641			
76-80	2.6316			
80-88	2.7027			
88-100	2.7778			
100-120	2.9412			
120-140	3.0303			
140-160	3.1250			
160-180	3.2258			
over 180	3.3333			

[¶205]

Capital Gains Wealth Multiplier With Tax-Free Injection

The following table is designed to show you two different things:

(1) How much capital gain treatment is worth at different income levels compared to ordinary income.

(2) The buildup possible if you take the tax savings from capital gain treatment and put it into 8% tax-exempt securities or some other tax-sheltered investment.

$12,000–$64,000

If your joint taxable income joint return is *from* to	$12,000 16,000	$16,000 20,000	$20,000 24,000	$24,000 28,000	$28,000 32,000	$32,000 36,000	$36,000 40,000	$40,000 44,000	$44,000 52,000	$52,000 64,000
And you have $1,000 of ordinary income, your federal tax bite is	$250	$280	$320	$360	$390	$420	$450	$480	$500	$530
If you have capital gain income from investments, your federal tax bite is	130	140	160	180	200	210	225	240	250	250
Your tax savings	120	140	160	180	190	210	225	240	250	280
Wealth you can accumulate by investing annual tax savings in 8% tax-exempt securities for 10 years	1,877	2,190	2.503	2,816	2,973	3,286	3,520	3,755	3,911	4,381

$64,000–$300,000

If your taxable income joint return is *from* to	$64,000 76,000	$76,000 88,000	$88,000 100,000	$100,000 120,000	$120,000 140,000	$140,000 160,000	$160,000 180,000	$180,000 200,000	$200,000 300,000
And you have $1,000 of ordinary income, your federal tax bite is	$550	$580	$600	$620	$640	$660	$680	$690	$700
If you have capital gain income from investments, your federal tax bite is	250	250	250	250	250	250	250	250	250
Your tax savings	300	330	350	370	390	410	430	440	450
Wealth you can accumulate by investing annual tax savings in 8% tax-exempt securities for 10 years	4,694	5,163	5,476	5,788	6,102	6,415	6,728	6,884	7,040

Note: This table does not take into account the minimum tax on tax-preference income. See discussion on page 147.

[¶206]

Table of Income Items

The following table is useful in determining which of the following income items are taxable. Sometimes an item may appear in both the "Taxable" column and the "Nontaxable" column. When this happens, it is because an item may be taxable subject to certain exceptions; such as in the case of widow's death payments.

	Taxable	Non-taxable		Taxable	Non-taxable
Accident and disability payments		x	Farm produce consumed by		
Alimony payments, periodic	x		farmer		x
Annuities	x		Foreign income of individual who		
Appreciation of unsold			is abroad for specified period		x
property		x	Foreign income of U.S. citizens		
Awards		x	who are foreign residents		x
Back pay	x		Forgiven indebtedness	x	
Bad debt recoveries (some)		x	Gain on sale of capital assets	x	
Bargain purchases from employer	x		Gain on sale of business or		
Bequests and devises		x	personal property	x	
Board and lodgings (at employer's			Gain on involuntary conversion		x
convenience)		x	Gain on sale of residence		x
Bonuses (including Christmas			Gambling proceeds	x	
bonuses)	x		Gifts		x
Business profits	x		Health and accident insurance		x
Cancellation of indebtedness			Illegal income	x	
(insolvent before and after			Inheritances		x
cancellation)	x	x	Incentive payments	x	
Children's earnings	x	x	Income assigned to another	x	
Citizen abroad (limited amount			Income of decedent	x	
of)		x	Insurance premiums on your life		
Clergymen's fees	x		(proceeds to your beneficiary)		
Commissions, brokers' on in-			paid by your employer but not		
surance or real estate sold to			group insurance	x	
themselves	x		Insurance premiums on "split-		
Commissions, salesmen's	x		dollar" insurance paid by		
Compensation	x		employer		x
Compensation earned without			Interest on bonds of states,		
U.S. by non-resident alien		x	municipalities and U.S. posses-		
Damages for loss of profits	x		sions		x
Damages for injury to business			Interest generally		
property	x		Jury fees	x	
Damages for personal injuries		x	Lease cancellation payments	x	
Death benefits (first $5,000)		x	Legacies		x
Deferred compensation	x		Lessee's improvements		x
Devises		x	Life insurance proceeds		x
Dividends	x		Meals or lodging to employees		x
Dividends received in mer-			Medical expenses reimbursed by		
chandise	x		employer		x
Escrow deposits		x	Merchandise gifts of nominal		
			value to employees		x

Table of Income Items *(continued)*

	Taxable	Non-taxable		Taxable	Non-taxable
Military pay	x	x	Scholarships and fellowships		x
Moving expenses	x	x	Security deposits		x
Net lease payments	x		Social Security Act benefits		x
Pensions	x		Sick pay		x
Preferred Stock Dividend	x		Stock options	x	x
Prizes from contests	x		Stock or notes received in payment	x	
Property or services received in payment	x		Stock dividends	x	x
Recoveries of bad debts, prior taxes, delinquent amounts	x		Strike benefits	x	
Reimbursement of employee expenses		x	Tips	x	
Rent allowance to employee (at convenience of employer)		x	Unemployment or disability insurance		x
			Veterans' benefits		x
Rewards	x		Wages	x	
Royalties	x		Widows, $5,000 death payment		x
Salaries	x		Workmen's compensation		x

¶206

Table of Expense Items

The following table is useful in determining which expenses are deductible and which are not.

	Deduct-ible	Non-deductible		Deduct-ible	Non-deductible
Accident insurance premiums (limitations)	x		Carrying charges (limitation)	x	
Accounting fees:			Cash discounts allowed	x	
paid by investor (in certain cases)	x		Casualty loss (limitation)	x	
preparation of tax returns	x		Chamber of Commerce dues (business)	x	
Admissions tax:			Charitable contributions (limitation)	x	
state excise (but only if paid in business or production of income)	x		Child care expense (limitation)	x	
Advances by stockholder as contribution to capital		x	Chiropodists, fees paid (limitation)	x	
Alimony (certain periodic payments)	x		Chiropractors, fees paid (limitation)	x	
Allowance paid to family members for upkeep		x	Cigarette and cigar tax, federal and state (only if in business or production of income)		x
Ambulance hire (limitation)	x		Circulation expenditures (publishers)	x	
Amortization:			Claims, judgments and damages as business expenses	x	
improvements by lessee	x		Clerical help for investor	x	
premium on bonds (with exception)	x		Club dues		x
Anticipated profits, loss of		x	Commuting expenses		x
Artificial limbs and teeth (limitation)	x		Conservation of soil and water, farmers	x	
Attorney's fees: fees, if business or investor's expense	x		Contested business liabilities, if paid	x	
Auditor's fees in connection with business or production of income	x		Contract:		
			cancellation to prevent loss of earnings	x	
Automobiles:			damages paid for breach (business)	x	
collision, damage (limitation)	x		Contributions (political campaigns)		x
Automobiles (continued):			Convention expenses (business)	x	
interest on purchase loan	x		Cooperative apartment, payments representing taxes and interest; also depreciation, if sublet or used as an office	x	
operating costs, in business and in production of income	x				
inspection fees, license plates, personal operating costs		x	Corporation expense paid by stockholder		x
Bad debts:			Criminal defense, if business-connected, even though unsuccessful	x	
business, completely or partially worthless	x				
nonbusiness, completely worthless (limitation)	x		Custodian fees, investor's expense	x	
Bar examination fees		x	Customers' fees allocated to exempt income		x
Blood donations, value of		x	Damages paid (limitations)	x	
Bonus paid to employees	x		Debts of another, payment of		x
Business: enterprise abandoned	x				
Capital contributions & expenditures		x			

Table of Expense Items *(continued)*

	Deductible	Nondeductible		Deductible	Nondeductible
Demolition loss	x	x	*Expenses (continued)*		
Dental fees and dentures (limitation)	x		uniforms (some)	x	
Depreciation (if in business or production of income)		x	Experimental and research expenses	x	
Development expenses	x		Exploration expenses of mines (limitation)	x	
Diagnosis, fees for (limitation)	x		Explosion damage (limitation)	x	
Discounts to customers	x		Eyeglasses, including examination fee	x	
Doctors' bills (limitation)	x		Farm:		
Drought damage (if unusual)	x		development costs		x
Drugs, cost (limitation)	x		land-clearing expenses	x	
Dues and subscriptions:			soil and water conservation expense	x	
business associations	x		Finance charges	x	
Chamber of Commerce	x		Fines, violation of federal, state and other laws		x
professional societies	x		Fire damage (limitation)	x	
social club membership (business)	x		Flood damage (limitation)	x	
technical journals	x		Freeze damage (limitation)	x	
trade associations	x		Funeral expense		x
unions (for strictly employment matters)	x		Gambling losses (to extent of gambling gains)	x	
Earthquake damage (limitation)	x	x	Gasoline tax:		
Educational expenses		x	federal (if in business or production of income)	x	
Embezzlement loss (limitation)	x		state	x	
Employee's transportation expenses in job (not commuting)	x		Gifts to charity (limitation)	x	
Employment agency fees	x		Gift to individual		x
Entertainment of customers (limits)	x		Guaranteed annual wage plans, contributions to (generally)	x	
Entertainment of employees	x		Health insurance premiums (special rule)	x	
Excise taxes (if in business or production of income)	x		Hearing devices (limitation)	x	
Expenses:			Heirs, payments to deceased employee's	x	x
advertising	x		Home expenses		x
authors (professional)	x		Hospital bills (limitation)	x	
automobile, in business and production of income	x		Hospitalization insurance (special rule)	x	x
experimental (in certain cases)	x		Hurricane damage (limitation)	x	
farmers'	x		Illegal business and payments		x
income, production or collection of	x		Improvements to property	x	x
income tax, determination of	x		Income-producing property, expenses of conserving	x	
investors'	x		Income tax return, fee for preparing	x	
outside salesmen's (deductible in arriving at adjusted gross income)	x		Income tax, state	x	
professional	x		Initiation fees, union	x	
teachers' (some)	x		Installment selling expenses	x	

Table of Expense Items *(continued)*

	Deduct-ible	Non-deductible		Deduct-ible	Non-deductible
Insurance premiums:			Loans, uncollectible	x	x
accident (limitation)	x		Lobbying expense (business-related)	x	x
business property protection	x		Losses:		
credit	x		abandonment (business)	x	
crop	x		automobile collision (limita-tion)	x	
fire (business property)	x				
health (special rule)	x		bad debts (limitation if non-business)	x	
hospitalization (special rule)	x				
liability (business)	x		capital (limitation)	x	
life policy		x	casualty, fire, storm, etc. (limitation)	x	
medical care (special rule)	x				
residential policy		x	damages sustained (limitation)	x	
storm (business)	x		demolition of property (busi-ness generally)	x	
theft (business)	x				
use and occupancy	x	x	embezzlement (limitation)	x	
workmen's compensation	x		foreclosure (business)	x	
Interest:			gambling, to extent of gambling gains	x	
accrued or imputed	x				
to carry tax-exempt insurance		x	indemnity contracts (business)	x	
			net operating loss	x	
to purchase life insurance contracts	x		theft (limitation)	x	
			war	x	
Investment counsel fees	x		wash sales (losses)		x
Investor's expenses, cherical, custodian, office, travel	x		worthless securities	x	
Laboratory fees (limitation)	x		Magazines:		
Laundering and cleaning ex-pense (specialized uniforms; and while on business travel)	x		technical or business	x	
			waiting room of profession-al man	x	
			Maintenance of real property held as investment	x	
Lease (business):					
amortization of improvements by lessee	x		Marital litigation expenses	x	x
			Medical expenses (limitation)	x	
leasehold amortization	x		Medical research organiza-tions, contributions to (limitation)	x	
rentals	x				
restoration of property	x				
taxes paid for landlord	x		Military and naval personnel:		
Lease:			traveling expenses in line of duty to extent not reim-bursed by Gov't	x	
advance payments to secure		x			
advance rentals	x				
buildings removed to obtain rentals		x	uniforms and equipment (some)	x	x
			Mortgage commission paid for obtaining mortgage		x
commissions to obtain im-provements by lessees	x				
			Moving expenses, employees	x	x
Legal expenses, in connection with business, and income-producing or investment property	x		Net operating loss	x	
			Nonbusiness bad debt (limitation)	x	
License fees, automobile (business)	x		Nurses' fees (limitation)	x	
			Obsolescence	x	
Life insurance premiums		x	Patents:		
Litigation settlement expense (some)	x	x	depreciation	x	

Table of Expense Items *(continued)*

Item	Deductible	Non-deductible
Patents (continued):		
infringement litigation:		
damages or settlement payment		x
legal costs		x
Payments in behalf of another		x
Penalties taxes		x
Pensions paid	x	
Personal expenses		x
Premium on bonds, amortization (with exception)	x	x
Production of income, expenses incurred	x	
Profession:		
cost of right to practice, license fee		x
Profit-sharing contributions (employer)	x	
Protection of property (business or income-producing)		x
Proxy fight costs	x	x
Publicity and selling expenses	x	
Real estate tax (except local assessments), state and local	x	
Refunds, expenses connected with	x	
Rentals, fees for collecting	x	
Repairs (business)	x	
Research and experimental work, expenses (business)	x	
Reserve for bad debts, additions	x	
Residence:		
abandoned as home, offered for rent, maintenance expense	x	
casualty damage (limitation)	x	
medically necessary improvements	x	
mortgage interest	x	
real estate taxes	x	x
rented and sold, loss		x
subletting, loss from	x	
Residential improvements, insurance, rent, loss sale		
Returns, income tax, expenses connected with	x	
Safe deposit box:		
rentals by investor	x	
tax on (if in business or production of income)	x	
Sales:		
commissions	x	
expenses	x	
Salesmen, traveling and business expenses	x	
Sales tax, city and state	x	
Seeing-eye dog and maintenance (limitation)	x	
Settlement to avoid litigation expense	x	
Soil and water conservation expenses, farmer's	x	
Stamp taxes:	x	
Federal (in certain cases) state stock transfer	x	
Stock:		
transfers, state taxes on	x	
worthless	x	
Storm damage (plus "clean-up" costs)	x	
Subletting apartment, loss on	x	
Summer school, teacher's professional training (some)	x	
Support payments, divorce or separation (but not child support)	x	x
Surgical fees (limitation)	x	
for medical treatment (limitation)	x	
in connection with business or production of income	x	
Tax returns, cost of preparing		x
Taxes:		
federal (some, business)	x	x
foreign income taxes not taken as credit against tax	x	
gasoline (State)	x	
general sales	x	
property	x	
Social Security (by employer)		x
state income	x	
tenant's payment for landlord	x	
Telephone, telegraph, cable tax:		
federal, if in business or production of income	x	
Theft loss (limitation)	x	
Title:		
expositor performance of property		x

¶207

Table of Expense Items *(continued)*

	Deduct-ible	Non-deductible		Deduct-ible	Non-deductible
Tools used in work (employee)	x		Union dues and initiation fees	x	
Trade association dues	x		Vacation expenses to improve		
Traders in securities, business expense	x		health	x	
			Vandalism damage (limitation)	x	
Training courses	x	x	Veterans organizations,		
Transportation expenses of employee in job (not commuting)	x		contributions	x	
			Wash sales, losses		x
Transportation taxes, federal (if in business or production of income)	x		Wheel chair, cost of (limitation)	x	
			Will contests:		
Traveling expenses:			litigation		x
for charitable work	x		settlement payment		x
Trees and shrubs, storm damage (limitation)	x		Work clothes, specialized only:		
			cleaning	x	
Two-job individual, transportation costs from Job A to Job B	x		cost	x	
			Workmen's compensation fund, payments by employer	x	
Uniforms, if not adaptable to regular wear off duty	x		Worthless securities	x	
			X-rays (limitation)	x	

¶207

[¶208]

Maximum Tax on Earned Income

The Code provides for a 50% ceiling on the tax on earned income. Earned income generally includes wages, salaries, professional fees or compensation for personal services, and, in the case of a taxpayer engaged in a trade or business where both personal services and capital are a material income-producing factor, a reasonable amount but not more than 30% of his share of the net profits of the business.

The 50% limit is applicable to earned income reduced by tax preferences in excess of $30,000 in the current year or the average tax preferences in excess of $30,000 for the current year and the prior four years, whichever is greater. Tax preferences for this purpose are the same as those applicable to individuals under the minimum tax.

Earned taxable income is defined as that proportion of total taxable income which is in the same ratio (but not in excess of 100%) as the ratio of earned income to adjusted gross income. Thus, if 40% of an individual's adjusted gross income is earned income, then 40% of this taxable income is considered earned taxable income.

If, during a taxable year, a taxpayer has earned taxable income which exceeds the 50% tax bracket, he figures his tax as follows:

(1) Takes the lowest amount of taxable income that is taxed over 50%. Computes the tax on this amount.

(2) Takes 50% of the amount by which earned taxable income (as defined above) exceeds the taxable income used in step (1).

(3) Computes the regular tax on the entire taxable income and deducts the tax computed on only the earned taxable income.

(4) Adds the amounts under steps (1), (2), and (3).

How the Tax Is Computed—The following example shows how the tax on earned income is calculated.

In 1972, Jones, married and filing a joint return, has $90,000 salary and bonus, $10,000 dividends on stock, $5,000 unreimbursed travel expenses, and $10,000 itemized deductions and personal exemptions. He has no capital gains, lump-sum pension distribution, or tax-preferred income. His tax is calculated as follows:

Salary and bonus	$ 90,000
Dividends	10,000
Gross Income	$100,000
Unreimbursed travel and entertainment expenses	5,000
Adjusted gross income	$ 95,000
Itemized deductions and personal exemptions	10,000
Taxable income	$ 85,000

The computation is:

(1) Tax on $52,000 (the highest amount on which the tax rate is 50%)..$18,060

(2) Earned taxable income.

$$\frac{\$90,000 \text{ (salary)}}{\$95,000 \text{ (adjusted}} \times \$85,000$$

$95,000 (adjusted)   (taxable
    gross income)   income) = $80,530

    Minus income taxed in step (1)   52,000

    Balance $28,530

    50% of balance = ...$14,265

(3) Tax on $85,000 of total taxable income $36,547

    Tax on $80,530 of earned taxable income......... 33,647

        Balance... 2,850

(4) Total tax (sum of (1), (2), and (3)).. $35,177

The total tax without the special earned-income tax limit would be $36,547. Thus, there is a $1,370 saving by using the special limit.

[¶209]

Minimum Tax on Tax-Preference Income

The Code imposes a minimum tax, in addition to any regular tax, on income defined as "tax-preference income." This income includes:

(1) The excess of investment interests over net investment income. This applies to individuals, Subchapter S corporations, and personal holding companies. It does not apply to regular business corporations because its interest is business rather than investment interest. Beginning in 1972, excess investment interest is not treated as tax-preference income. Instead, the deduction of such interest is limited to 50%.

(2) Accelerated depreciation on real estate.

(3) Accelerated depreciation on personal property subject to a net lease.

(4) Excess depreciation over straight-line for housing rehabilitation expenditures.

(5) Amortization over accelerated depreciation on pollution control equipment.

(6) Amortization over accelerated depreciation on railroad rolling stock.

(7) For executives, the amount by which the fair market value exceeds the option price on the exercise of a qualified stock option.

(8) Bad debt deductions of financial institutions.

(9) The excess of percentage over cost depletion.

(10) Corporate capital gains, but only to the extent of the ratio of the regular corporate tax rate minus the special capital gains rate to the regular corporate rate. In other words, assuming the corporation pays the maximum capital gains rate of 30% (effective beginning in 1972), it would include three-eighths of its long-term capital gains (in excess of net short-term capital losses) as tax-preference income; that is, 48% minus 30% divided by 48%. In the case of individuals, one-half of net long-term capital gain in excess of net short-term capital losses is included as tax-preference income.

The tax is imposed at a rate of 10% on all tax-preference income which exceeds $30,000 plus the amount of regular taxes paid. For example, suppose a corporation has $200,000 of tax-preference income and has a regular tax bill of $50,000. To determine the amount of tax on the tax-preference income, you would subtract the $30,000 exemption plus the $50,000 regular tax. This would leave a balance of $120,000. Applying the 10% rate, the tax on preference income would be $12,000, and the total tax bill would come to $62,000 instead of $50,000.

[¶210]

How Gifts Save Income Taxes

When there is a gift tax to pay, you must deduct the income on the capital consumed to pay the gift tax in order to appraise the income tax saving. The capital represented by the gift tax payment will, itself, be more than washed out in estate tax saving so that the primary concern may be with the loss of future income measured against income tax savings from the switch. Here's how that works out when one-third of family capital is transferred to a child so that, for income tax purposes, income is divided in three equal segments between husband, wife, and child. These figures, based on 1974 rates, assume that other income equals deductions and exemptions and that the exclusions and exemptions total $66,000.

How Gift of One-Third Family Capital Reduces Pre-Tax Income

Family Capital	Gift Tax on 1/3 Property	Income at 8% Before Tax	Loss of 8% Income After Tax	Income After Gift
$ 100,000	$ 0	$ 8,000	$ 0	$ 8,000
300,000	2,790	24,000	223	23,777
600,000	23,175	48,000	1,854	46,146
1,200,000	69,435	96,000	5,555	90,445
2,000,000	135,700	160,000	10,856	149,144
5,000,000	427,860	400,000	34,339	365,771
10,000,000	1,059,680	800,000	84,774	715,226

How Gift of One-Third Family Capital Increases After-Tax Income

Family Capital	After-Tax Income Before Gift	After-Tax Income After Gift	Increase Amount	Increase Percentage
$ 100,000	$ 6,620	$ 6,690	$ 70	1.1%
300,000	18,340	19,150	810	4.4
600,000	31,940	35,510	3,570	11.2
1,200,000	53,220	61,290	8,070	15.2
2,000,000	76,420	88,430	12,010	15.7
5,000,000	149,020	165,930	16,910	11.3
10,000,000	269,020	285,930	16,910	6.3

[¶211]

Income Tax Savings From Family Transfer Of Income-Producing Property

A. Where parent files an individual tax return:

Parent's Taxable Income Before Transfer	Tax Savings if Property Produces Annual Income of		
	$1,000	$5,000	$10,000
$ 20,000	415	1,385	1,900
30,000	495	1,875	3,100
40,000	545	2,065	3,540
50,000	565	2,255	3,940
60,000	585	2,355	4,220
70,000	605	2,455	4,420
80,000	625	2,555	4,620
90,000	645	2,655	4,820
100,000	655	2,705	4,920

B. Where parent files a joint tax return with spouse:

Parent's Taxable Income Before Transfer	Tax Savings if Property Produces Annual Income of		
	$1,000	$5,000	$10,000
$ 20,000	245	625	580
30,000	355	1,115	1,520
40,000	415	1,475	2,280
50,000	465	1,755	2,940
60,000	495	1,905	3,260
70,000	515	2,005	3,440
80,000	545	2,125	3,640
90,000	565	2,195	3,860
100,000	565	2,255	4,020

Note: Both tables disregard the maximum tax rate on earned income and assume the beneficiary receives all of his income from the transferred property.

¶211

[¶212]

Federal Income Tax Rates for Corporations

Taxable Income   Tax Rate

$ 0—25,000   22% (Normal tax)
Over $ 25,000 48% (Combined normal and surtax)

[¶213]

Estimated Tax Payment for Corporations

The former $100,000 exemption from estimated tax is being phased out. By 1977, corporations will be making current payments for all "estimated" tax liability. For taxable years beginning in 1977, a corporation must make payments if its estimated tax is $40 or more.

	Temporary Estimated
Year	*Tax Exemption*
1973	$4,400
1974	3,300
1975	2,200
1976	1,100
1977	0

The following installment payments must be made by corporations required to make installment payments (i.e., tax liability exceeding exemptions shown above).

If the requirements are first met—	The following percentages of the estimated tax shall be paid on the 15th day of the—			
	4th month	6th month	9th month	12th month
Before the 1st day of the 4th month of the taxable year	25	25	25	25
After the last day of the 3rd month and before the 1st day of the 6th month of the taxable year		33-1/3	33-1/3	33-1/3
After the last day of the 5th month and before the 1st day of the 9th month of the taxable year			50	50
After the last day of the 8th month and before the 1st day of the 12th month of the taxable year				100

[¶214]

Best Tax Salary

The best tax salary for an officer-stockholder is the one that will cost the least in taxes when the tax cost to employee and corporation are combined. It strikes a balance: Any increase will cost the employee more in taxes than the extra deduction will save the corporation; any decrease will cost the corporation more because of the reduced deduction than the reduction in tax will save the employee.

If earnings left in the corporation will not be withdrawn in the near future either as dividends or by liquidation and assuming that the stockholder-employee's other income equals his deductions and exemptions, there can be only one best salary taxwise.

Suppose a corporation has earnings of $100,000 before salary. It has a sole stockholder who is married. His best salary is between $40,000 and $44,000. If he were paid more than $44,000, the corporation would be saving taxes at a 48% rate, but sole stockholder would be paying taxes at a 50% rate. If he were paid less than $40,000, the corporation would be paying tax at a 48% rate, but the sole stockholder would be saving tax at a 45% rate.

Assuming the stockholder's other income equals his deductions and exemptions, here are the best salary levels for various corporate incomes.

Corporate Income Before Salary	Single Taxpayer	Married Taxpayer Joint Return	Separate Return
Less than $6,000		Full Corporate Income	
$6,000 to $12,000	$ 6,000	Full Corporate Income	$ 6,000
$12,000 to $31,000	$ 6,000	$12,000	$ 6,000
$31,000 to $47,000 ·		Full Corporate Income Less $25,000	
$47,000 to $57,000		Full Corporate Income Less $25,000	$22,000
$57,000 to $69,000	$32,000	Full Corporate Income Less $25,000	$22,000
Over $69,000	$32,000	$44,000	$22,000

[¶215]

How Much Corporate Dividend Shelter Is Worth

Under Code §243, a special deduction is granted to corporations for dividends received from other corporations. In a normal case, the amount of the deduction is 85% of the dividend received.

The following tables illustrate the tax shelter potential of this deduction. The first table assumes that a businessman in the 50% bracket ($44,000 annual taxable income) incorporates his business, transferring stock in other corporations to his corporation. Because of other deductions, the corporation's taxable income is kept at less than $25,000 a year. The only accumulated earnings are from the dividends, and when they reach $100,000, the businessman liquidates at capital gains rates.

The second table assumes that a corporation has funds to invest, and compares the yield to the corporation if it invests in investments which produce ordinary income, as opposed to investing in stocks which produce dividends.

Table I

Cumulative Dividends Received	After Tax Net to Corporation	Net to Shareholder (After Capital Gains Tax)	After Tax Net to Individual (After 50% Tax)	Savings
$ 5,000	$ 4,835	$ 3,626	$ 2,500	$ 1,126
10,000	9,670	7,553	5,000	2,553
15,000	14,505	10,879	7,500	3,379
20,000	19,340	14,455	10,000	4,455
25,000	24,175	18,009	12,500	5,509
30,000	29,010	21,562	15,000	6,562
40,000	38,680	28,670	20,000	8,670
50,000	48,350	35,694	25,000	10,694
60,000	58,020	42,705	30,000	12,705
70,000	67,690	49,660	35,000	14,660
80,000	77,360	56,526	40,000	16,526
90,000	87,030	63,391	45,000	18,391
100,000	96,700	70,170	50,000	20,170

Note: The alternative tax has not been used in this table. There is no minimum tax at any time on these facts.

Table II

Actual Yield on Investment	Net Yield to Corporation If Received as Dividends at Rate of		Yield Required on Fully Taxable Investment To Get Same Net Result	
	22%	48%	22%	48%
2%	1.934%	1.856%	2.479%	3.569%
3%	2.901	2.784	3.719	5.354
4%	3.868	3.712	4.959	7.138
5%	4.835	4.640	6.199	8.923
6%	5.802	5.568	7.438	10.708
7%	6.769	6.496	8.678	12.492
8%	7.736	7.424	9.918	14.277
9%	8.703	8.352	11.158	16.062
10%	9.670	9.280	12.397	17.846
11%	10.637	10.208	13.637	19.631
12%	11.604	11.136	14.877	21.415
15%	14.505	13.920	18.596	26.769

[¶216]

State Income Taxes—Corporate

State and Rate	When Payable		To Whom
ALA. 5% (32, 40)	Mar. 15	3d, 6th	Dept. Rev.
ALASKA. 18% of Fed tax (11, 31, 37)	Mar. 15	3d	Comr. Rev.
ARIZ. 2½% 1st $M; 4% 2d $M; 5% 3d $M; 6½% 4th $M; 8% 5th $M; 9% 6th $M; 10½% bal. (51)	Apr. 15	2d	Dept. Rev.
ARK. 1% 1st $3M; 2% 2d $3M; 3% next $5M; 5% next $14M; 6% bal. (37)	May 15		Dir. Fin. & Admin.
CALIF. 9% Min. $200 (3, 47)	Mar. 15		Franch. Tax Bd.
COLO. 5% (37)	Apr. 15		Dept. Rev.
CONN. 8% Min. $50 (30)	Apr. 1		Tax Comr.
DEL. 7.2% (31)	Apr. 1		Dept. Fin.
D.C. 8% Min. $25 (49)	Apr. 15		Dept. Fin. & Rev.
FLA. 5% (37, 48)	Apr. 1		Dept. Rev.
GA. 6%	Apr. 15		Comr. Rev.
HAW. 5.85% 1st $25M; 6.435% bal. (37)	Apr. 20		Dir. Tax (12)
IDA. 6.5% plus $10 excise (37)	Apr. 15		Tax Comn.
ILL. 4% (37, 41)	Apr. 15		Dept. Rev.
IND. 3% (10, 31, 37)	Apr. 15		Dept. Rev.
IOWA. 6% 1st $25M; 8% next $75M; $100M or over 10% (43)	Apr. 30		Dept. Rev.
***KAN.** 4½% + 2¼% over $25M (37)	Apr. 15		Dir. Tax.
***KY.** 4% 1st $25M; 5.8% bal.	Apr. 15		Dept. Tax. (4)
LA. 4%	May 15		Coll. Rev.
ME. 4% up to $25,000; $1,000 plus 6% over $25,000 (38, 40)	Mar. 15		Tax Assr.
MD. 7% (18)	Apr. 15		St. Comptr.
MASS. *Excise:* $5.76 per $M on tangible values or net worth (17) +8.55% of net income; $114 Min. (24) *Income:* 5%	Mar. 15		Comr. Corps. & Tax
***MICH.** 7.8% (34, 37)	Apr. 15		Dept. Treas.
MINN. 12% (36) Min. $100	Mar. 15	(28)	Comr. Rev.
MISS. 3% 1st $5M; 4% bal.	Mar. 15	3d, 6th (22)	Comr. Inc. Tax
***MO.** 5% (37)	Apr. 15		Dir. Rev.
MONT. 6¾% Min. $50 (9, 37)	May 15		Dept. Rev.
NEB. 2¾% (25, 26, 37)	Mar. 15		Dept. Rev.
N.H. 7% (6)	May 1		Dept. Rev. Admn.
N.J. 5½%; 7¼% (44)	Apr. 15		Corp. Tax Bur. (4)
N.M. 5% (37)	Mar. 15		Bur. Rev.
***N.Y.** 9% Min. $125 (23, 31, 39)	Mar. 15		Tax Comn.
N.C. 6% (14)	Mar. 15	(5)	Secy. Rev.
N.D. 3% 1st $3M; 4% next $5M; 5% next $7M; 6% bal. (15, 16, 37)	Apr. 15	(27) 3d, 6th, 9th	Tax Comr.
***OHIO.** 4% 1st $25,000; 8% over $25,000 (19). Min. $50	Mar. 30		St. Treas.
OKLA. 4% (33)	Apr. 15		Tax Comn.
***ORE.** 6% Min. $10 (3, 13, 37)	Apr. 15		Dept. Rev.
***PA.** 9.5% (31, 42)	Apr. 15		Dept. Rev.
R.I. 8% (29)	Mar. 15		Tax Admr.
S.C. 6%	Mar 15		Tax Comn.
TENN. 6%	Apr. 1		Dept. Rev.
UTAH. 6% Min. $25 (21, 37)	Apr. 15		Tax Comn.
VT. 5% 1st $10M; 6% next $15M; 7% next $225M; 7½% over $250M. Min. $50 (20)	Mar. 15	(7)	Comr. Taxes
VA. 6%	Apr. 15		Dept. Tax.
W. VA. 6% (35)	Mar. 15		Tax Dept.
WIS. 2.3% 1st $M; 2.8% 2d $M; 3.4% 3d $M; 4.5% 4th $M; 5.6% 5th $M; 6.8% 6th $M; 7.9% bal. (46)	Mar. 15		Dept. Rev.

State Income Taxes—Corporate *(continued)*

NOTES (corresponding to parenthetical numbers in chart):

*Local taxes may also be imposed on corporate income.

(1) Fiscal year returns due on corresponding months for calendar returns.

(2) Later installments are due in these months after return date.

(3) For franchise tax; no min. for income tax.

(4) Payment with return but drawn to order of State.

(5) If $50.01-$400, 2 installments, 3-15 and 9-15; over $400, 4 installments, 3-15, 6-15, 9-15 and next 12-15; 6% interest on deferred payments.

(6) Business profits tax on corporations, partnerships, individuals organizations.

(7) Pay ½ with return and ½ in 90 days after return due date, if tax over $500.

(8) Due in full with return unless installments permitted (see Col. C). Some states require advance payments on estimate.

(9) No minimum for direct income tax. Public contractors get credit against tax for added 1% gross receipts tax.

(10) *Supplemental corporate net income tax:* 2% (eff. 1-1-73); 2½% (1-1-75); 3% (1-1-77). *Gross income tax* (phased out in 20 years): ½%, wholesale-retail sales (0.475% from 4-1-73 to 12-31-73); starting 1974, reduced by 0.025% per year till no tax, 1-1-92; all others 2% (1.9% from 4-1-73 to 12-31-73); starting 1974 reduced by 1/10% till 1-1-92.

(11) Based on Fed rates as of 12-31-63.

(12) Local taxation division, or Honolulu if no local place of business.

(13) Credits: Personalty tax paid (limited); pollution control facility cost.

(14) Credit given for constructing multifamily dwellings for handicapped (for years starting after 12-31-73).

(15) Vietnam bonus surtax—1% of taxable income (min. $10 max. $25).

(16) Added 1% net income tax on privilege of doing business in ND (min. $20). New industry credit: 1% on instate salaries, wages; ½% for 4th and 5th year.

(17) For tax years ending 12-31-73 and after; was $7.98.

(18) Credit for state personalty taxes.

(19) Payable 3-31 based on preceding calendar or fiscal year ended pre-Jan. 1 of payment year. Alternative tax, if higher: 5 mills on net worth.

(20) For years starting after 12-31-73; was 6% (min. $25).

(21) No minimum for direct tax.

(22) In 2 equal installments if tax $50 or more.

(23) Other minimums 9% or 1.6 mills of apportioned income plus compensation, or 1¼ mills on value of apportioned business and investment capital. *Added tax:* 0.8 mills per $1 subsidiary capital.

(24) Credit given for eligible business in depressed areas. 3% (for years ending after 12-30-72 and ending before 12-31-78) investment credit for specified property and credit (elective instead of depreciation or expensed) for pollution control facility. 3% credit on property value of leased tangibles (from regional business development corp.) (for tax yrs. ending after 12-30-73). Also, credit for manufacturing, research & development corporation "full-time" employees increase (for tax yrs. ending after 12-30-73 and before 12-31-78).

(25) For tax years starting 1-1-74; was 3¼%.

(26) Credit for tax on nonhighway fuel less 1⅛% a gal.

¶216

State Income Taxes—Corporate *(continued)*

NOTES *(continued)*

(27) Pay in full $100 or less.

(28) 3-month deferment allowed for ½ tax due after tax credits and estimates.

(29) Alternative tax, if larger, 40¢ per $100 of corporate excess.

(30) Plus added tax equal to amount by which ¼ mills per $1 of corporate excess exceeds 8% rate (max. added tax for companies taking Fed dividends paid deductions: $10,000 or 4/10 mill, if less). Credit for air-water pollution control facilities. Minimum tax $50. Max. $100,000.

(31) Credit against tax allowed for investment in ghetto areas and/or training hard core unemployed. In Alaska, also for industrial incentive. In Del., credit replaced by deduction.

(32) Domestic corporation gets credit for income tax paid outstate.

(33) Credit allowed for pollution control facilities. Manufacturer annual credit for gas used instate: 3-mills per MFC over 25,000 CF.

(34) Limited credit for contributions to public libraries and "educational institutions." Also, inventory (cost over $1500; for years starting after 12-31-73) tax credit for 32% (for years starting after 12-31-73; was 25%) of tax paid or 20% of franchise fee, if no inventory. Starting with period ending after 12-31-74, 39%; 12-31-75, 45%; 12-31-76, 51%; 12-31-77, 57%; 12-31-78, 63%; 12-31-79, 69%; 12-31-80, 75%.

(35) Credit for Business & Occupation tax paid and investment for business expansion.

(36) Credit for occupation tax on copper-nickel mining and pollution control equipment. Also, flat credit of $500.

(37) Multistate Tax Compact member. Compact requires optional tax for small taxpayer (only instate activity sales, not over $100,000 annually, and not owning or renting instate property). Rates on 1st $100,000 of local sales in Haw. ½% (eff. 4-4-74); Ida. 1%; N.M. ¾%; Colo. ½%; Mont. ½%; Ore. ¼% (⅛% if return on sales under 5%); Neb. (under prescribed formula); Mich., prescribed formula or 2/5% of total sales, if less; Utah ½%; N.D. 6/10% to $20,000; 8/10%, over $20,000-$55,000; 1%, $55,000-$100,000. No rates in other compact states: Ark., Ind., Kan., Mo., Fla. disallows this tax.

(38) For 7-1-73 to 12-31-73 (was 4%). Starting 1-1-74, 5% not over $25,000; $1,250 plus 7% over $25,000.

(39) 2% (1% pre-1-1-74) investment credit replaces accelerated depreciation.

(40) Optional tax for small taxpayer: 1% (Me.; repealed for tax years ending after 12-31-72); ¼% (Ala.) disallowed.

(41) $1,000 standard exemption allowed.

(42) For 1974 and fiscals starting in 1974 and after; was 11%.

(43) Optional credit against tax in lieu of fuel tax refund (not watercraft or highway MVs) (after 1974).

(44) Direct income tax (for tax years ending after 12-31-73).

(46) Credit for sales-use tax on fuel-electricity for manufacturing instate.

(47) For tax years ending after 6-30-73; was 7.6%. Prorate for years spanning 7-1-73. Plus added 2.5% minimum tax on specified tax preference items.

(48) Up to $5,000 of net income is exempt.

(49) For tax years starting after 12-31-73; was 7%.

(51) Effective 12-31-73; was 2% 1st $M; 3% 2d $M; 4% 3d $M; 5% 4th $M; 6% 5th $M; 7% 6th $M; 8% bal.

¶216

[¶217]

State Income Taxes—Personal

State and Rate	Exemptions (1)	How Paid	Filing Deadline (28)
*ALA. 1½% 1st $1000; 3% next $2000; 4½% next $2000; 5% bal. (5)	S-$1500; M-$3000; H-$3000; D-$300	Full	4-15 (15th of 4th mo.)
ALASKA 16% of Fed. tax (6, 39, 54)	(45)	Full	4-15 (15th of 4th mo.)
ARIZ. 2% 1st $1000; 3% 2d $1,000; 4% 3rd $1,000; 5% 4th $1,000; 6% 5th $1,000; 7% 6th $1,000; 8% bal. (5, 73)	S-$1000; M-$2000; H-$2000; D-$600; B-$500; A-$1000	Full (34)	4-15 (15th of 4th mo.)
ARK. 1% 1st $2999; 2½% next $3000; 3½% next $3000; 4½% next $6000; 6% next $10,000; 7% bal. (5, 60)	S-$17.50; M-$35; H-$35; D-$6; B-$17.50 (2, 17, 60)	Full	5-15 (15th of 5th mo.)
CALIF. 1% 1st $2,000; $20 + 2% 1st $1,500; $50 + 3% 2d $1,500; $95 + 4% 3d $1,500; $155 + 5% 4th $1,500; $230 + 6% 5th $1,500; $320 + 7% 6th $1,500; $425 + 8% 7th $1,500; $545 + 9% 8th $1,500; $680 + 10% 9th $1500; 11% bal. (5, 22, 54, 74, 78)	S-$25; M-$50; H-$50; D-$8; E-$8; B-$8 (2, 86)	Full	4-15 (15th of 4th mo.)
COLO. 3% 1st $1000; 3½% 2d $1000; 4% 3d $1000; 4½% 4th $1000; 5% 5th $1000; 5½% 6th $1000; 6% 7th $1000; 6½% 8th $1000; 7% 9th $1000; 7½% 10th $1000; 8% bal. (4, 5)	S-$750; M-$1500; D-$750; B-$750; A-$750; E-$750 (44)	Full	4-15 (15th of 4th mo.)
CONN. 6%; (52)	(76)	Full	4-15 (15th of 4th mo.)
*DEL. 1.6% 1st $1000; 2.2% 2nd $1000; 3.3% 3rd $1000; 4.4% 4th $1000; 5.5% 5th $1000; 6.6% 6th $1000; 7.7% next $2000; 8.8% next $12,000; 9.3% next $5000; 9.9% next $5000; 12.1% next $10,000; 13.2% next $10,000; 15.4% next $25,000; 16.5% next $25,000; 19.8% bal. (82)	S-$600; M-$1200; D-$600; B-$600; A-$600; E-$600;	Full	4-30 (30th of 4th mo.)
D.C. 2% 1st $1000; $20 + 3% next $1000; $50 + 4% next $1000; $90 + 5% next $2000; $190 + 6% next $3000 (3, 5)	S-$1000; M-$2000; H-$2000; D-$500; B-$500; A-$500 (65)	Full	4-15 (15th of 4th mo.)
GA. 1% 1st $1000; 2% next $2000; 3% next $2000; 4% next $2000; 5% next $3000; 6% bal. (62)	S-$1500 (36); M-$3000; H-$3000; D-$700 (33); B-$700 (36); E-$700; A-$700 (62)	Full	4-15 (15th of 4th mo.)
HAW. 2.25% 1st $500; $11.25 + 3.25% 2d $500; $27.50 + 4.5% 3rd $500; $50 + 5% 4th $500; $75 + 6.5% next $1000 (5, 46, 54)	(31); B-$5000 (38, 47)	Full	4-20 (20th of 4th mo.)
IDA. 2% 1st $1000; 4% 2d $1000; 4.5% 3rd $1000; 5.5% 4th $1000; 6.5% 5th $1000; 7.5% bal. (5, 29, 54)	(26, 31)	Full	4-15 (15th of 4th mo.)
ILL. 2½%	S-$1000; M-$2000; D-$1000; A-$1000; B-$1000; E-$1000	Full	4-15 (15th of 4th mo.)
*IND. 2%	S-$1000; M-$1000 (30, 35, 63); D-$500; B-$500; A-$500; E-$500	Full	4-15 (15th of 4th mo.)
IOWA ¾% 1st $1000; 1½% 2d $1000; 3% 3rd $1000; 4% 4th $1000; 5% next $3000; 6% next $2000; 7% over $9000 (71)	S-$15 (2); M-$30 (2); H-$30 (2); D-$10 (2); E-$10 (2); B-$15 (2); A-$15 (2)	Full	4-30 (end of 4th mo.)
*KAN. 2% 1st $2000; $40 + 3½% next $1000; $75 + 4% next $2000; $155 + 5% next $2000; $255 + 6½% bal. (5, 54, 61)	S-$600; M-$1200; D-$600; B-$600; E-$600; A-$600	Full	4-15 (15th of 4th mo.)

¶217

State Income Taxes—Personal *(continued)*

State and Rate	Exemptions (1)	How Paid	Filing Deadline (28)
*KY. 2% 1st $3000; 3% 4th $1000; 4% 5th $1000; 5% next $3000; 6% bal. (5)	S-$20; M-$40; D-$20; A-$20; B-$20 (2); E-$20 (2)	Full	4-15 (15th of 4th mo.)
LA. 2% 1st $10,000; 4% next $40,000; 6% bal. (54, 58)	S-$2500 (40); M-$5000 (40); H-$5000 (40); D-$400; B-(27); E-$400 (70)	Full	5-15 (15th of 5th mo.)
ME. 1% 1st $2000; $20 + 2% next $3000; $80 + 3% next $5000; $230 + 4% next $15,000; $830 + 5% next $25,000; $2080 + 6% bal. (54)	S-$1000; M-$2000; D-$1000; A-$1000; B-$1000; E-$1000	Full	4-15 (15th of 4th mo.)
*MD. 2% 1st $1,000; 3% 2d $1000; 4% 3d $1000; 5% bal. (5, 57)	S-$800; M-$1600; D-$800; B-$800 (20); E-$800; A-$800 (20)	Full	4-15 (15th of 4th mo.)
MASS. Part A income (interest, dividends, net capital gains) 9%; Part B income (all other taxable), 5% (5, 88)	(10); S-$2000; M-$2600 (23) D-$600 (55); B-$2000; A-$600; E-$600 (41)	Full (25)	4-15 (15th of 4th mo.)
*MICH. 3.9% (11, 49)	(77)	Full	4-15 (15th of 4th mo.)
MINN. 1.6% 1st $500; 2.2% 2d $500; 3.5% 2d $1000; 5.8% 3d $1000; 7.3% 4th $1000; 8.8% 5th $1000; 10.2% next $2000; 11.5% next $2000; 12.8% next $3500; 14% next $7500; 15% bal. (49)	S-$21 (2); M-$42 (2); M-$42 (2); B-$21 (2, 21); H-$42 (2); D-$21 (2); E-$21 (2); A-$21 (2, 21) (14, 66)	Full (64)	4-15 (15th of 4th mo.)
MISS. 3% 1st $5,000; 4% bal.	S-$4,500; M-$6,500; H-$6,500; D-$750; A-$750; B-$750	2 Instal. (56)	4-15 (15th of 4th mo.)
*MO. (5, 12) 1.5% 1st $1000; 2% next $1000; 2.5% next $1000; 3% next $1000; 3.5% next $1000; 4% next $1000; 4.5% next $1000; 5% next $1000; 5.5% next $1000; 6% over $9000	S-$1200; H-$2000; M-$2400; D-$400	Full	4-15 (105 days)
MONT. (37) 2% 1st $1000; 3% 2d $1000; 4% next $2,000; 5% next $2,000; 6% next $2,000; 7% next $2,000; 8% next $4,000; 9% next $6,000; 10% next $15,000; 11% bal.	S-$650; M-$1300; D-$650 B-$650; A-$650; E-$650 (8)	Full	4-15 (15th of 4th mo.)
NEB. 11% of Fed income tax (SBE sets rate by 11-15) (49, 51, 54)	(45)	Full	4-15
N.H. 4¼% (13)	$600	Full	5-1
4% (9) invalid	$2000	Full	4-15
N.J. None.			
N.M. 0.9% 1st $500; $4.50 + 1.1% 2d $500; $10 + 1.3% 3d $500; $16.50 + 1.5% 4th $500; $24 + 1.6% next $1000; $40 + 1.9% next $1000; $59 + 2.3% next $1000; $82 + 2.4% next $1000; $106 + 3% next $1000; $136 + 3.3% next $1000; (5, 49, 50, 54)	(31)	Full	4-15 (15th of 4th mo.)
*N.Y. 2% 1st $1000; 3% next $2000; 4% next $2000; 5% next $2000; 6% next $2000; 7% next $2000; 8% next $2000; 9% next $2000; 10% next $2000; 11% next $2000; 12% next $2000; 13% next $2000; 14% next $2000; 15% bal. (49, 79)	(48)	Full	4-15 (15th of 4th mo.)
N.C. 3% 1st $2000; 4% next $2000; 5% next $2000; 6% next $4000; 7% bal. (15, 81)	S-$1000; M-$2000; H-$2000; D-$600 (67); B-$1000; E-$600; A-$1000 (75)	Full	4-15 (15th of 4th mo.)

¶217

State Income Taxes—Personal *(continued)*

State and Rate	Exemptions (1)	How Paid	Filing Deadline (28)
N.D. 1% 1st $1000; 2% next $2000; 3% next $2000; 5% next $1000; 7½% next $2000; 10% bal. (42)	(31)	4 Instal. (19)	4-15 (15th of 4th mo.)
***OHIO** ½% 1st $5,000; $25 + 1% next $5,000; $75 + 2% next $5,000; $175 + 2½% next $5,000; $300 + 3% next $5,000; $900 + 3½% bal.	S-$500; M-$1000; D-$500 (83, 85)	Full	4-15 (15th of 4th mo.)
OKLA. ½% 1st $2,000; 1% next $3,000; 2% next $2500; 3% next $2500; 4% next $2500; 5% next $2500; 6% bal. (5, 69)	S-$750; M-$1500; H-$2000; D-$750; E-$750; A-$750	Full	4-15 (15th of 4th mo.)
***ORE.** 4% 1st $500; $20 + 5% next $500; $45 + 6% next $1000; $105 + 7% next $1000; $175 + 8% next $1000; $255 + 9% next $1000; $345 + 10% bal. (5, 54, 84)	(31)	4 Instal. (18, 19)	4-15 (15th of 4th mo.)
***PA.** 2% (7)	None	Full	4-15
R.I. 15% of Fed tax	(45)	Full	4-15
S.C. 2% 1st $2000; 3% next $2000; 4% next $2000; 5% next $2000; 6% next $2000; 7% bal. (5)	S-$800; M-$1600; H-$1600; D-$800; A-$800; B-$800; E-$800	Full	4-15 (15th of 4th mo.)
TENN 6% (16)	None (24)	Full	4-15 (15th 4th mo.)
UTAH 0-$750, 2%; $751-$1500, $15 + 3% over $750; $1501-$2250, $38 + 4% over $1500; $2251-$3000, $68 + 5% over $2250; $3001-$3750, $105 + 6% over $3000; over $3750, $150 + 7¼% over $3750 (59)	(31, 87)	Full	15th of 4th mo.
VT. 25% of Fed income tax (5, 43, 53, 54, 72)	(45)	Full	4-15 (15th of 4th mo.)
VA. 2% 1st $3000; 3% next $2000; 5% next $7000; 5.75% bal.	S-$600; M-$1200; D-$600 B-$600; A-$1000; E-$600	Full	5-1 (15th of 4th mo.)
W. VA. 0-$2000, 2.1%; $2000.01-$4000, $42 + 2.3% over $2000; $4000.01-$6000, $88 + 2.8% over $4000; $6000.01-$8000, $144 + 3.2% over $6000; $8000.01-$10,000, $208 + 3.5% over $8000; $10,000.01-$12,000, $278 + 4% over $10,000; $12,000.01-$14,000, $358 + 4.6% over $12,000; (5, 32, 49, 54)	A-$600; E-$600; S-$600; M-$1200; D-$600; B-$600 A-$600; E-$600	Full	4-15 (15th of 4th mo.)
WIS. 3.1% 1st $1000; 3.4% 2nd $1000; 3.6% 3d $1000; 4.8% 4th $1000; 5.4% 5th $1000; 5.9% 6th $1000; 6.5% 7th $1000; 7.6% 8th $1000; 8.2% 9th $1000; 8.8% 10th $1000; 9.3% 11th $1000; 9.9% 12th $1000; 10.5% 13th $1000; 11.1% 14th $1000; 11.4% over $14,000 (5, 49)	S-$20; M-$40; H-$40; D-$20; A-$25; E-$20 (2, 68)	Full	4-15 (15th of 4th mo.)

NOTES (corresponding to parenthetical numbers in chart):

*Local taxes may also be imposed on income, earnings or payrolls.

(1) S-Single; M-Married; H-Head of family or household; T-Taxpayer; D-Dependents (each); B-Blind (addl.); A-Age 65 min. (addl.); E-Education—student dependent.

(2) Credit against tax.

(3) $370 + 7% next $4000; $650 + 8% next $5000; $1050 + 9% next $8000; $1770 + 10% over $25,000.

¶217

State Income Taxes—Personal *(continued)*

NOTES *(continued)*

(4) Plus 2% surtax on residents' dividends and interest over $5,000. Credit is ½% of net taxable income on the first $9,000.

(5) Optional tax table may be used.

(6) Credit for political campaign contribution (max. $50) (eff 5-10-74).

(7) For tax years starting after 12-31-73; was 2.3%.

(8) For years starting after 12-31-73; was $600 per exemption.

(9) Base: Nonresidents' (commuters) income from NH sources.

(10) Against Part B taxable income.

(11) Credit for sales tax: $10 per dependent (income $5000) to $0 ($15,000) (for tax years starting after 12-31-73).

(12) One rate for total net income, with these credits: 2nd bracket, $5; 3rd $15; 4th, $30; 5th, $55; 6th, $90; 7th, $135.

(13) Base: interest, dividends.

(14) Credit for nonpublic, nonprofit instate elementary-secondary education per pupil "unit"—$100 (max.), 1971-72; after, special formula.

(15) Personal income tax for N.Y. individuals who commute to N.J.

(16) Based on income from stocks and bonds (not deposit certificates); rate is 4% on income from corporations with 75% or more property assessable in state.

(17) Plus $50 credit for mentally retarded child at home. Dependent credit only if his gross under $1,750 and return not jointly with spouse. $17.50 addtl. credit for blind spouse.

(18) 1% interest on installment payments.

(19) Min. total tax installments: N.D., $100; Ore., over $25.

(20) Additional $800 for dependent 65 or older. Added $800 for blind spouse.

(21) Additional credit for spouse: Age $21; Blind $25.

(22) Separate rates for heads of households.

(23) Plus up to $2,000 of earned income (Part B) of spouse with smaller income plus $600 if such spouse's total earned income $2000 (max.).

(24) No return unless income exceeds $25; total income of blind exempt.

(25) Interest is collected at commissioner's discretion.

(26) Also credit: $10 per personal exemption (IRC) claimed (not age or blindness). Residents added $5 credit per IRC exemption, for 1973 only.

(27) Extra $1,000 if deaf, blind, disabled or mentally retarded; similarly, dependents, including deaf dependent.

(28) Dates in parentheses indicate filing date after end of fiscal year.

(29) Plus $10 for each taxpayer (spouses filing jointly, $10 except those blind or receiving public assistance).

(30) Tax credits: Investments in ghetto training centers; contributions to Ind. colleges, etc. "Seniors" and "disableds" get property tax or "rental" credit or refund.

(31) Each IRC (§ 151) exemption: $750. In Ore.: $675.

State Income Taxes—Personal *(continued)*

NOTES *(continued)*

(32) $14,000.01-$16,000, $450 + 4.9% over $14,000; $16,000.01-$18,000, $548 + 5.3% over $16,000; $18,000.01-$20,000, $654 + 5.4% over $18,000; $20,000.01-$22,000, $762 + 6% over $20,000; $22,000.01-$26,000, $882 + 6.1% over $22,000; $26,000.01-$32,000, $1,126 + 6.5% over $26,000; $32,000.01-$38,000, $1,516 + 6.8% over $32,000; $38,000.01-$44,000, $1,924 + 7.2% over $38,000; $44,000.01-$50,000, $2,356 + 7.5% over $44,000; $50,000.01-$60,000, $2,806 + 7.9% over $50,000; $60,000.01-$70,000, $3,596 + 8.2% over $60,000; $70,000.01-$80,000, $4,416 + 8.6% over $70,000; $80,000.01-$90,000, $5,276 + 8.8% over $80,000; $90,000.01-$100,000, $6,156 + 9.1% over $90,000; $100,000.01-$150,000, $7,066 + 9.3% over $100,000; $150,000.01-$200,000, $11,716 + 9.5% over $150,000; over $200,000, $16,466 + 9.6% over $200,000. Separate table for joint returns and surviving spouses.

(33) "Student-dependent" or "disabled", $1400; $700 for student-dependent, qualification for head of household.

(34) If electing installment payments: 1st installment (4-15) 1/3 min.; 2d (8-15) 1/2 of balance; balance (12-15).

(35) Each spouse gets $1,000 or own adjusted gross income if less; min. $500 each.

(36) Allowed for spouse with no gross income and not dependent of another, if taxpayer files separately.

(37) 10% surtax.

(38) Joint return (both blind, deaf or disabled), $10,000; $5000, if only one spouse, instead of personal exemptions.

(39) Retroactive to 1-1-64, based on Fed rates of 12-31-63.

(40) Applied against 1st bracket.

(41) Credits against tax: $4 taxpayer, $4 spouse, $8 each dependent—if income not over $5,000 (spouses combined income $5,000; must file jointly for credits).

(42) Added taxes: (1) 1% (min. $20) on privilege of doing business in ND; (2) 1% Vietnam bonus surtax on taxable income limited from $2.50 to $12.50.

(43) Less % of adjusted gross which is not Vt. income. Plus 9% surcharge on Vt. tax (after 12-31-73; was 12%).

(44) Credit $7 ($21 for 1973 only) times exemptions (not Mentally Retarded, Age or Blind). Also, added $750 for mentally retarded dependent. Senior citizens get credit for property tax or rent paid.

(45) Fed personal exemptions automatically allowed.

(46) $140 + 7.5% next $2000; $290 + 8.5% next $5000; $715 + 9.5% next $4000; $1095 + 10% next $6000; $1695 + 10.5% next $10,000; $2745 + 11% bal. Special rates for Heads of Household.

(47) Credits (based on income): $30-$6 per 9-month resident. (Repealed for tax years starting after 12-31-73; dependent's education, pre-college, $20 to $2; in college $50 to $5. Drug-medical expenses, 4% to 1% low-income household rents, 2% of rent to 1%). To discourage dangerous item sales, 4% per transaction price.

(48) Allows each IRC exemption: $650.

¶217

State Income Taxes—Personal *(continued)*

NOTES *(continued)*

(49) *Minn.* "Seniors" get $2-$720 sliding tax credit for property tax paid. *Mich.:* Credit for Mich. city income tax; property tax (credit for homestead and inventory taxes (32% after 1973 to 75% after 1980, if cost over $1500)) and contributions to public libraries; special credits for paraplegics and disableds (after 1973) and "educational" institutions. *NY:* Credit for industrial investment in ghettos; also, 1% investment credit. *Wis.:* Property tax credit. *W. Va.:* Carrier income tax paid. *Neb.:* Credit for nonhighway gas tax less 1⅛¢ per gal; *N.M.:* Credit for state-local taxes based on income and IRC exemptions.

(50) $169 + 3.6% next $2,000; $241 + 4.3% next $2000; $327 + 6.1% next $8000; $815 + 8% next $30,000; $3215 + 8.5% next $50,000; $7465 + 9% bal: separate rates for heads of households and spouses filing separately or jointly. For tax years starting after 12-31-73; was 1% to 9%.

(51) For tax years starting 1-1-74, 11% (was 13%). Residents get annual $13 (for tax years starting after 12-31-73; was $10) food sales tax credit times exemptions allowed (not blind or aged).

(52) Base: Net capital gains on sales or exchange of capital assets.

(53) Sales tax credit or refund given on basis of income and exemptions (not blind or age). "Seniors" and under "65s" get "credit" for property taxes or rent.

(54) Income splitting allowed. Fed splitting automatic in Alaska, Neb., R.I., Vt.

(55) Includes disabled persons.

(56) In 2 equal installments if tax $50 (min.): 15th of 4th and 5th month after.

(57) Also, not under 20% to 50% (max.) of state income tax imposed in 23 counties and Baltimore City. Also, credit for state, personalty tax paid.

(58) Tax (based on Fed tax; not above current rates) by tax table, for tax years starting after 12-31-74.

(59) Rates for single taxpayers; separate rates for joint or spouses (separately).

(60) Special lower rates for low income. Also, "seniors" get property tax credit or rebate based on income (starting 1973 taxes).

(61) "Elderly" and "disabled" get property tax credit (max. $330).

(62) Separate rates for singles and spouses filing separately. Credit based on income: $15 to $0 (singles, spouses filing separately); $30 to $0 (heads of household, spouses jointly). Also limited credit for educational contributions.

(63) Special exemption per spouse and single taxpayer: Up to $500 adj. gross, none; $500-$1000, excess over $500; over $1000, $500 (max.).

(64) Estate-trusts may defer for 6 months ½ of tax after credits and estimates.

(65) Sales tax credit: $6 for $2000 (max.) income to $2, over $6000 times personal exemptions (not age or blind).

(66) Credit for sales tax paid on electricity for agriculture (retailers certificate; no previous refund.) Also, Special Credits for "low income" taxpayers (starting 1-1-75).

(67) Added $2000 for "severely retardeds".

(68) Starting 1974; was S-$15; M-$30; H-$30; D-$15; A-$20; E-$15.

(69) Seniors and disableds get limited credit for property tax paid based on income (eff 1-1-75).

State Income Taxes—Personal *(continued)*

NOTES *(continued)*

(70) *Credits:* $20 for blind (dependent), deaf, mentally incapacitated, disabled. 0.2% reduction on percentage depletion (old law) for tax years starting after 12-31-74.

(71) Optional credit against tax in lieu of fuel tax refund (not watercraft or highway MVs) (after 1974).

(72) Special capital gains tax applies to land transfers: If land held less than 6 years—tax ranges from 60% (if gain is 200% or more and land held less than 1 year) to 5% (if gain less than 100% and land held 5 to 6 years).

(73) "Seniors" get property tax or rental credit (eff 1974).

(74) Tax credit. *For 1974:* 100%, $4,000 (max.) or $8,000 (max.) for spouses (jointly), heads of households, surviving spouses.

(75) Wife, $2000 (agreement with husband getting only $1000).

(76) Individuals, $100; aged (over 64) added $100; blind, added $100; spouses filing jointly, total (above) for each.

(77) $1500 (for tax years starting after 12-31-73; was $1200) times personal or dependency exemptions on Fed return; plus added $1500 if "paraplegic."

(78) Plus added 2.5% minimum tax on specified tax preference items. Credits: "Renters", $25-$45 (based on income); nonpublic school (nonprofit) costs (grades K thru 12), $125.0 credit (based on income) or educational costs if less, per dependent (ruled unconstitutional). (Student maintenance, $8 per "student" household member.)

(79) 2.5% surcharge; (suspended for 1973 and 1974 in N.Y. and 1973 in N.J.). 6% min. tax on IRC § 57 tax preference items.

(80) Tax applies only to Pa. individuals who commute to NJ.

(81) Credit for constructing dwellings for handicapped.

(82) Eff. 1-1-74; was 1½% 1st $1000; 2% 2nd $1000; 3% 3rd $1000; 4% 4th $1000; 5% 5th $1000; 6% 6th $1000; 7% next $2000; 8% next $12,000; 8½% next $5000; 9% next $5000; 11% next $10,000; 12% next $10,000; 14% next $25,000; 15% next $25,000; 18% bal.

(83) Tax credit (1973; '74) for spouses (jointly)—each with $500 (min.) adjusted gross (not interest, dividends, rents, capital gains): $10,000 (max.) adjusted gross, 20%; $10,001-20,000, 12%; over $20,000, 5%

(84) "Disableds" get $50 credit (for years starting after 12-31-73).

(85) For tax years starting after 12-13-74: S-$650; M-$1,300; D-$650.

(86) Spouses filing separately get $25 credit; surviving spouse or spouses jointly, $50 credit.

(87) Residents get food sales tax credit: $6 times personal exemptions (not age or blindness) claimed for residents (eff 1st tax year *only* starting after 12-31-72).

(88) Part B includes interest on term deposits under $100,000.

[¶218]

Income Apportionment Elements of Multistate Businesses For State Income Tax Purposes

The following chart sets forth the various factors—property, receipts, payroll, and other components—used by the states for fixing taxable net income of businesses operating in more than one state. It also indicates items of income which are separately or specifically allocated (according to situs or tax location), rather than included in the general (or operating) net income that is apportioned by ratios or percentage formulas. The notes to the chart give added details and explain entries where necessary.

State Col. A	Factors of Apportionment Fractions				Items Separately Allocated Col. F	Separate Accounting Allowed? Col. G(2)
	Property Col. B	Receipts Col. C	Payroll Col. D	Others Col. E		
Alabama (1,6)	Yes	Yes	Yes	(6-7)	(1-5)	Yes(2)
Alaska (6,9)	Yes	Yes	Yes	(4)	(1-5)	Yes
Arizona	Yes	Yes	Yes	(1,2)	(2-5,7)	Yes
Arkansas (6,9)	Yes	Yes	Yes	(4,6)	(1-5)	Yes
California (6)	Yes	Yes	Yes	(4)	(1-5)	Yes
Colorado (9)	Yes	Yes	No		(2-4,11,15,17)	Yes
Connecticut	Yes	Yes	Yes	(4,6)	(1,2,4,5)	Yes
Delaware	Yes	Yes	Yes	(7)	(2,3,5,15,16,20)	Yes
Dist. Columbia (5,6)	Yes	Yes	Yes	(4)	(1-5)	Yes
Florida (11)	Yes(1)	Yes(2)	Yes(3)	(4,6)		Yes
Georgia	Yes	Yes	Yes	(4-6)	(2-5,7)	Yes
Hawaii (6,9)	Yes	Yes	Yes	(6)	(1-5)	Yes
Idaho (6,9)	Yes	Yes	Yes		(1-5)	Yes
Illinois (6,9)	Yes	Yes	Yes	(6)	(1-5)	Yes
Indiana (2,6,9)	Yes	Yes	Yes		(1-5)	Yes
Iowa (5)	No	Yes	No	(3,6)	(2-5,8,22)	Yes
Kansas (6,9)	Yes	Yes	Yes	(6)	(1-5)	Yes
Kentucky (6)	Yes	Yes	Yes	(6)	(1-5)	Yes
Louisiana	Yes	Yes	Yes	(4,6)	(1-5,8,14)	Yes
Maine (6,7,8)	Yes	Yes	Yes		(1-5)	Yes
Maryland	Yes	Yes	Yes	(5)	(1-3)	Yes
Massachusetts	Yes	Yes	Yes	(5)	(4)	Yes
Michigan (6,9)	Yes	Yes	Yes	(4,6,13)	(1-5)	Yes
Minnesota (3)	Yes	Yes	Yes	(6)	(1-5,8,19,20)	Yes
Mississippi	Yes	Yes(1)	Yes(1)	(4,6)	(2-5,8,10,12,23)	Yes
Missouri (5,9)	No	Yes	No		(1-5)	Yes
Montana (6,8,9)	Yes	Yes	Yes		(1-5)	Yes
Nebraska (6,7,9)	Yes	Yes	Yes		(1-5)	Yes
New Hampshire	Yes	Yes	Yes	(7)		No
New Jersey	Yes	Yes	Yes	(9)		No
New Mexico (6,9)	Yes	Yes	Yes	(6,12)	(1-5)	Yes
New York	Yes	Yes	Yes	(8)	(1-5,9)	(1,2)
North Carolina (6,8)	Yes	Yes	Yes	(4,6)	(1-5,9)	Yes
North Dakota (6,9)	Yes	Yes	Yes	(6)	(1-5)	Yes
Ohio (6,10)	Yes	Yes	Yes	(6,9)	(1-4,8)	Yes
Oklahoma (6)	Yes	Yes	Yes(2)	(6)	(1-5,6,13,18)	Yes
Oregon (6,8,9)	Yes	Yes	Yes	(4,6)	(1-5)	Yes
Pennsylvania (6)	Yes	Yes	Yes	(14)	(1-5)	Yes
Rhode Island	Yes	Yes	Yes		(1-3,5)	(1)
South Carolina (6)	Yes	Yes	Yes	(10)	(1-5,21)	Yes

Income Apportionment Elements of Multistate Businesses

For State Income Tax Purposes *(continued)*

State Col. A	Property Col. B	Receipts Col. C	Payroll Col. D	Others Col. E	Items Separately Allocated Col. F	Separate Accounting Allowed? Col. G(2)
Tennessee	Yes	Yes	No	(4,6,11)		Yes
Utah (6,8,9)	Yes	Yes	Yes		(1-5)	Yes
Vermont	Yes	Yes	Yes	(7)	(2-3,24)	(1)
Virginia (6)	Yes	Yes	Yes	(4)	(1-5)	Yes
West Virginia (7)	Yes	No	Yes		(2,3,5)	Yes
Wisconsin	Yes	Yes	No(4)	(2-4,15)	(2-5,15,20)	Yes

NOTES (corresponding to number in parentheses in the chart):

Col. A. (1) Domestic corporations and residents can't allocate. (2) Gross income tax allows no allocation. (3) Ratio computed under Columns B, C, and D can't exceed sum of 15% of property percentage (Col. B), 70% of sales percentage (Col. C), and 15% of payroll percentage (Col. D). (5) Single "sales" factor apportionment by seller manufacturing entirely outstate ruled invalid (*Gen. Motors*, US Sup Ct 4-27-65). (6) Uniform Div. of Income for Tax Purposes Act (UDITPA) is substantially adopted (see footnote (9) for states also adopting UDITPA under Multistate Tax Compact). (7) Applies only to corporations. (8) Only for corporations and nonresident individuals. (9) State has adopted Multistate Tax Compact, including 3-factor formula. Taxpayer may elect to allocate by 3-factor formula. (10) For corporation (income) franchise tax (Basis B payable starting 1972); personal income tax (eff. 1-1-72). (11) Disallows equally weighted 3-factor formula though member of Multistate Tax Compact.

Col. B. (1) Factor represents 25% of formula.

Col. C. (1) Sole factor only for manufacturing. (2) Factor is 50% of formula.

Col. D. (1) Manufacturing labor only, direct and indirect. (2) Includes services to extent related to unitary business, but excludes salaries, wages and other compensation as general or administrative expense. (3) Factor is 25% of formula. (4) For tax years starting after 12-31-72, payroll factor replaces cost factor.

Col. E. (1) Purchases factor may be used for purchasing businesses. (2) Manufacturers' costs. (3) For businesses other than manufacture or sale of tangible personalty, tax commission may make rules. (4) Special provisions apply to some or all carriers. (5) Special provisions for construction companies. (6) Special provisions for specialized businesses (e.g., utility, financial and service businesses, etc.). (7) Such other factors as tax administrator deems applicable. (8) 3-factor formula for business income and capital; investment income and capital separately treated. (9) Use assets (in N.J.) and net worth (Ohio) allocation if producing greater percentage than average of Col. B, C and D. (10) Single factor (gross receipts) formula for income other than from manufacturing, producing, collecting, processing, buying, selling, distributing, or dealing in tangible personalty. (11) Wages not a separate factor for sellers but included as part of manufacturer's production cost factor; not applicable to sellers in Tenn. (12) Operating expenses. (13) Special provisions for businesses with small instate sales. (14) Special provisions for transportation, pipeline, natural gas and water transportation companies. (15) Special provisions for financial organizations and public utilities (for tax years starting after 12-31-72).

¶218

Income Apportionment Elements of Multistate Businesses

For State Income Tax Purposes *(continued)*

NOTES *(continued)*

Col. F. (1) Capital gains; in Pa., "gains". (2) Rents. (3) Royalties. (4) Dividends (in Mass. not included in tax base; in Ohio, if not deducted or excluded for corporate net income). (5) Interest. (6) Special rules apply to life insurance companies. (7) Gain on sale of assets not connected with business. (8) Income from personal services. (9) Income from sale of realty. (10) Income from property produced and sold in different places. (11) Interest from intangible personalty not used in business and not inventory. (12) Income from ownership or operation of property. (13) If business not unitary, net income allocated to state where activity conducted. (14) Income from construction repairs and similar services. (15) Gain or loss from capital assets. (16) Gain or loss on realty sale. (17) Gain from sale of corporate stocks. (18) Net income of certain manufacturing or processing enterprises. (19) Farm income. (20) Gain or loss from intangibles. (21) Income from investments of corporations, including investments in subsidiaries, not included in net apportionable income. (22) Special rules apply to gain or loss from sale of assets. (23) Certain instate risks, insurance or reinsurance premiums. (24) Income from any trade, business, all instate.

Col. G. (1) No specific provision. (2) Separate accounting, if permitted, may require official approval.

[¶219]

Receipts in Interstate Apportionment Numerators
For State Income Taxes

Spotting various receipts by situs or tax location is vital in making up overall percentage as applied against total income to be apportioned. Chart below lists determinants that states use to make up the receipts numerator. "X" indicates influential elements that make receipts local, not outstate.

Use this chart to channel your selling operations away from the marked areas as much as possible. For instance, if an "X" falls in Column B for a state where you market your products, arrange to negotiate or execute your sales through offices, agencies or places of business not in that state. If an "X" shows up in Column C, try to store your goods at locations that don't hitch receipts apportionment to the location of property appropriated to an order. And so on.

State Col. A(1)	Office Where Negotiated Col. B	Property at Time of Order Col. C	Receipt or Acceptance of Order Col. D	Nego- tiating Personnel Col. E	Delivery Place Col. F	Shipment Origin Col. G	Other Col. H
Alabama					X	X(4, 5)	X(9)
Alaska					X	X(4, 5)	X(3, 9)
Arizona				X(1, 4)			
Arkansas						X(4, 5)	X(7)
California					X	X(4, 5)	X(3, 9)
Colorado	X(5)			X(4)	X(1)		
Connecticut	X				X		X(2)
Delaware					X(1)	X(5)	
Dist. of Columbia	X(3)				X		X(3, 9)
Florida					X		
Georgia					X(1)		
Hawaii					X	X(4, 5)	X(3, 9)
Idaho					X	X(4, 5)	X(3, 9)
Illinois					X	X(4, 5)	X(3, 9)
Indiana					X	X(4, 5)	X(3, 9)
Iowa					X(1)		
Kansas					X	X(4, 5)	X(3, 9)
Kentucky					X	X(4, 5)	X(3, 9)
Louisiana					X		
Maine					X	X(4, 5)	X(3, 9)
Maryland	X(2, 4)						
Massachusetts	X(3)				X		X(3, 9, 11)
Michigan					X	X(4, 5)	X(3, 9)
Minnesota	X(3)						
Mississippi	X(1)				X(3)	X(3)	
Missouri		X(2)			X	X	X(5)
Montana					X	X(4, 5)	X(3, 9)
Nebraska					X	X(4, 5)	X(3, 9)
New Hampshire	X(3)			X(1, 4)	X(2, 4)	X(3)	X(3, 9)
New Jersey	X				X(5)		X(4)
New Mexico					X	X(4, 5)	X(3, 9)
New York					X		
North Carolina					X		X(2, 6)
North Dakota					X	X(4, 5)	X(3, 9)
Ohio					X(2)		X(1, 3, 9)
Oklahoma					X(4)	X(4, 6, 7)	X(10)

Receipts in Interstate Apportionment Numerators
For State Income Taxes *(continued)*

State Col. A(1)	Office Where Negotiated Col. B	Property at Time of Order Col. C	Receipt or Acceptance of Order Col. D	Nego- tiating Personnel Col. E	Delivery Place Col. F	Shipment Origin Col. G	Other Col. H
Oregon					X	X(4, 5)	X(3, 9)
Pennsylvania					X		X(3, 9)
Rhode Island	X(4)	X(1)	X	X(3)	X	X(2)	X(2)
South Carolina					X	X(4, 5)	X(3, 8, 9)
Tennessee	X(2, 5)				X(2)		
Utah					X		X(3, 9)
Vermont		X(2)	X(1)				X(4)
Virginia					X	X(5)	X(3, 9)
Wisconsin					X	X(4, 5)	X(3, 9)

NOTES (corresponding to parenthetical numbers in chart):

Col. A. (1) For member states of Multistate Tax Compact, see P-H State Tax Guide.

Col. B. (1) Sales assignable to office, agency or place of business where binding sale or agreement first occurs but excluding shipments between outstate points. (2) Sales through or by instate offices, agencies or branches. (3) Sales not negotiated or effected by agents or agencies chiefly at, connected with or sent out from business premises outstate. (4) Includes sales between outstate points if order secured or received by state office or salesman working from it. (5) Public or private warehouse treated like office.

Col. C. (1) Sales crossing state lines included at 50%; also 50% of sales of outstate property not at permanent place of business if orders accepted instate. (2) Property instate on receipt or appropriation to order, or not at taxpayer's outstate business place when order received or accepted instate. (3) If taxpayer not taxed in destination state.

Col. D. (1) See Col. C. (2) One element of sale.

Col. E. (1) Place where selling is performed. (3) Outstate sales included if by local salesman. (4) Promotional acts given weight.

Col. F. (1) Deliveries in state, excluding deliveries for out-shipment. (2) Sales to customers instate. (3) Controlling factor when shipment originated in same state. (4) See Col. H(10). (5) See Col. C.

Col. G. (1) Reserved. (2) One element of local sale. (3) Controlling factor when delivery is in same state. (4) If purchaser is U.S. government. (5) If taxpayer is not taxable in purchaser's state. (6) See Col. H(10). (7) Taxpayer isn't doing business in state of destination.

Col. H. (1) If merely receiving commissions, rents, interest, dividends and fees, allocate by situs. (2) Receipts from instate sources. (3) For sale of intangibles, allocate instate if income-producing activity is instate or Col. H(9) applies. (4) Receipts from instate services and rentals or royalties on property in or used in state. (5) All sales or business transacted in state and half transacted in and outside. (6) Services instate (in N.C., if income producing activities are instate). (7) If income-producing activity is instate for nontangibles; if multistate, Ark. share is same formula percentage used in allocating income of year. (8) Personal property rentals not separately allocated. (9) For sales of intangibles, state with largest share of costs of performance. (10) Though sales of intangibles are included in 3-factor apportionment, there's no provision indicating which go in numerator. (11) If not taxed in purchaser's state (for tax years ending after 12-30-72; was added condition that shipment origin be instate). "Purchaser" includes U.S.

¶219

Interstate Apportionment of Payroll
For State Income Taxes

In allocating and apportioning interstate income, payroll attributed to a particular state carries as much weight as receipts in most states. In other states, payroll factor is diluted (not as much weight) by being included as one element of total costs of production, operations, or business (see Col. C of chart). These are broad categories that the states seize on: (1) office location; (2) time spent instate; and (3) pay earned instate. But, of course, some states have special rules and are indicated in the last column.

When an "X" shows up in a particular column for indicated state, you include pay of employees in numerator of wage factor. For example, in Alaska include in numerator percentage of employee pay based on time spent in the State; in Minn. include pay of all employees working out of Minn. offices.

State Col. A(1)	Office Location Col. B	Time Spent In State(1) Col. C	Compensation Earned in State Col. D	Other Col. E
Alabama	X(7)		X(1, 8, 9)	
Alaska	X(7)	X(2, 3)	X(1, 8, 9)	
Arizona		X		X(2)
Arkansas	X(7)		X(1, 8, 9)	
California	X(7)		X(1, 8, 9)	
Colorado (2)	X(7)		X(1, 8, 9)	
Connecticut	X(7)		X(1, 8, 9)	
Delaware			X(2)	
Dist. of Columbia	X(7)		X(1, 8, 9)	
Florida	X(7)		X(1, 8, 9)	
Georgia			X(3)	
Hawaii	X(7)		X(1, 8, 9)	
Idaho	X(7)		X(1, 8, 9)	
Illinois	X(7)		X(1, 8, 9)	
Indiana	X(7)		X(1, 8, 9)	
Kansas	X(7)		X(1, 8, 9)	
Kentucky	X(7)		X(1, 8, 9)	
Louisiana		X	X	X(2, 4, 6)
Maine	X(7)		X(1, 8, 9)	
Maryland		X		
Massachusetts	X(7)		X(1, 8, 9)	
Michigan	X(7)		X(1, 8, 9)	
Minnesota	X(5)		X(6)	
Mississippi			X(7)	
Missouri (2)	X(7)		X(1, 8, 9)	
Montana	X(7)		X(1, 8, 9)	
Nebraska	X(7)		X(1, 8, 9)	
New Hampshire	X(7)		X(1, 9)	
New Jersey	X(7)		X(1, 8, 9)	X(8)
New Mexico	X(7)		X(1, 8, 9)	
New York	X(1, 8)	X(5)	X(10)	X(1)
North Carolina	X(7)		X(1, 8, 9)	
North Dakota	X(7)		X(1, 8, 9)	
Ohio	X(7)		X(1, 8, 9)	X(2)
Oklahoma		X(6)	X(12)	X(2, 9)
Oregon	X(7)		X(1, 8, 9)	
Pennsylvania	X(7)		X(1, 8, 9)	

Interstate Apportionment of Payroll For State Income Taxes *(continued)*

State Col. A(1)	Office Location Col. B	Time Spent In State(1) Col. C	Compensation Earned in State Col. D	Other Col. E
Rhode Island				X(7)
South Carolina	X(7)		X(1, 8, 9)	
Tennessee			X(11)	
Utah	X(7)		X(1, 8, 9)	
Vermont	X(1, 8)	X(5)	X(10)	X(1)
Virginia	X(7)		X(1, 8, 9)	X(7)
West Virginia	X(7)		X(1, 8, 9)	
Wisconsin	(6)		X(4, 11)	

NOTES (corresponding to numbers in parentheses in chart):

Col. A. (1) For member states of Multistate Tax Compact, see P-H State Tax Guide. (2) As member state of Multistate Tax Compact, 3-factor formula (including payroll) may be elected.

Col. B. (1) N.Y., N.C., Vt.—If chief headquarters. (2) Reserved. (5) Haw., Minn.— If either employed in state or identified with office in state. (6) For tax years starting after 12-31-72, X (7) Col. B applies. (7) If part performance and controlling base instate, or residence instate. (8) N.Y., Vt.—See Col. C, D, E, if inequitable.

Col. C. (1) Applies to those working in and out of state. (2) Alaska.—Seagoing personnel, use port days. (3) Alaska.—Transportation workers, use working days. (4) Reserved. (5) N.Y., Vt.—If office factor inequitable, for employees paid on time basis. (6) Okla.—Itinerant employees' (traveling salesmen) expenses on local-to-total time to further enterprise.

Col. D. (1) If some of the service is performed instate and base of operations or control or individual's residence is instate. (2) Del.—If employee within state. (3) Ga.—Payments to residents deemed incurred (earned) in state. (4) For tax years starting after 12-31-72, X (1,8,9) Col. D applies. (5) Reserved. (6) Haw., Minn.—If either employed in state or identified with office in state. (7) Miss.—Only for manufacturers (use direct or indirect manufacturing labor in state) and noncarrier pipelines (use direct and indirect operating labor). (8) See Col. B. (7). (9) Also service entirely instate, or having only incidentals outside. (10) N.Y., Vt.—If office factor inequitable, see Col. E. (11) Tenn., Included instate costs such as manufacturing, operating, collecting, assembling or processing. (12) Okla.—Compensation is pay for services to extent related to unitary business, but excludes salaries, wages, other pay and administrative expense.

Col. E. (1) N.Y., Vt.—If office factor inequitable: salesmen on commission, on basis of business volume; compensation based on results achieved, on value of instate services. (2) Ariz., Calif., La., Ohio, Okla.—Transportation employees, mileage. (3) Reserved. (4) La.—Salesmen, instate sales. (6) La.—Any method found fair and reasonable by tax administrator. (7) Connected with instate activities or transactions. (8) N.J.—Also if contributions not paid under unemployment compensation law of another state. (9) Okla.—Net income from business of unitary character is separately allocated to state in which activity conducted.

[¶221]
What Pay Is Apportioned in Interstate Apportionment of Payroll For State Income Taxes

Employees and Compensation: The methods the states use to assign payroll to interstate apportionment numerators are detailed in P-H State Tax Guide. However, that is only half the story—the types of employees and pay that are considered in making up the payroll factors are also important.

Class of Employees: The states use different rules for determining what pay works its way into the top of the payroll fraction. The pay of "employees" is generally added to the top of the fraction. But whether the states consider officers, directors, or independent agents in the covered class is indicated by the chart below in Columns B, C, and D.

Type of Compensation: The numerators of most state fractions take in "pay" or "compensation." Some state rules spell out in detail what is in the class, while all of them use a catch-all class of "other" pay. Columns E, F, G, and H detail what the states expect you to consider in figuring out the payroll percentage.

State Col. A(1, 2)	Class of Employee Included				Type of Compensation Included		
	Officers Col. B	Directors Col. C	Agents Col. D	Independent Commissions Col. E	Bonuses Col. F	Board & Lodging Col. G	Other Col. H(1)
Alabama	X			X	X	(3)	X(5)
Alaska	X			X	(2)	(1)	X
Arizona	X			X	X	(1)	X
Arkansas	(1)			X	X	(3)	X(5)
California	X			X	X	(3)	X(5)
Colorado (3)	X			X	X	(3)	X(5)
Connecticut							X(5)
Delaware	(2)						X(5)
Dist. of Columbia	(1)						X
Florida				X	X		X(5)
Georgia	X		X(1)	X		(1)	X
Hawaii	(1)			X	(1)	(1)	X(5)
Idaho	X			X	X	(3)	X(5)
Illinois	(1)			X	X	(3)	X(5)
Indiana	X			X	X	(2, 3)	X(5)
Kansas	X			X	X	(3)	X(5)
Kentucky	X			X	X	(3)	X(5)
Louisiana	X			X	(1)		X
Maine	(1)			X	X	(3)	X(5)
Maryland							X
Massachusetts	(1)			X	(3, 5)	(3)	X(5)
Michigan	(3)	(3)		(1)	(2)	(3)	X(7)
Minnesota	X					(1)	X
Mississippi							X(3)
Missouri (3)	X			X	X	(3)	X(5)
Montana	(1)			X	(2)	(3)	X(5)
Nebraska	(3)	(3)		X	(2)	(3)	X(7)
New Hampshire	X(5)			X	X	X(4)	X(5)
New Jersey	X	(2)		X			X(4)
New Mexico	X			X	X	(3)	X(5)
New York	(2)	(2)		X		(1)	X
North Carolina	(2)	(2)		X	(1)	(2)	X(5)
North Dakota	X			X	X	(3)	X(5)
Ohio (4)							X(8)
Oklahoma				(2)	(4)	(3)	X(6)

What Pay Is Apportioned in Interstate Apportionment of Payroll
For State Income Taxes *(continued)*

State Col. A(1, 2)	Officers Col. B	Directors Col. C	Agents Col. D	Independent Commissions Col. E	Bonuses Col. F	Board & Lodging Col. G	Other Col. H(1)
Oregon X				X	X	(3)	X(5)
Pennsylvania (4)...........							
Rhode Island X			X				X
South Carolina............ (2)		(2)		X	(1)	(3)	X(5)
Tennessee							(2)
Utah .... (2, 4)		(4)	(2)	X	(2)	(3)	X(5)
Vermont .... (2)		(2)		X	(2)		X
Virginia .... (2)		(2)		X	X	(3)	X(5)
West Virginia.............. (1)				X	(3)	(3)	X(5)
Wisconsin (4)							(2)

NOTES (corresponding to numbers in parentheses in chart):

Col. A. (1) See Column H and related comment. (2) For member states of Multistate Tax Compact, see P-H State Tax Guide. (3) As member of Multistate Tax Compact, 3-factor formula (including payroll) may be elected. (4) "Compensation" is only pay apportioned; no other specific provisions noted (for tax years starting after 12-31-72.)

Col. B. (1) Compensation paid to officers is subject to withholding and, by analogy, might be held to be allocable compensation. (2) Pay to executive officers is specifically excluded. (3) Officers might be employees whose wages are included in factor; see Col. B (1). (4) Amounts reportable for Employment Security purposes may be used. (5) No express exclusion for officers' salaries.

Col. C. (1) Reserved. (2) Specifically excluded. (3) See Col. B (3). (4) See Col. B (4).

Col. D. (1) Paid to "agents". (2) Payments to independent contractors specifically excluded. See Col. B (4).

Col. E. (1) Follows federal inclusion in gross income and is allocable compensation. (2) Itinerant employees' (e.g., traveling salesmen) expense.

Col. F. (1) State withholding specifically covers bonuses so, by analogy, probably considered allocable compensation. (2) Follows Fed inclusion in gross income. (3) See Col. H and related comment. (4) See Col. E (1). (5) State "wages" same as IRC § 3401 (a).

Col. G. (1) State income tax may treat board and lodging (subsistence) as taxable compensation, and amounts may be considered allocable compensation. (2) If "income" under IRC. (3) See Col. H. and related comment.

Col. H. (1) Catch-all classification, such as "other compensation", might be extended to embrace some of the preceding types of compensation. (2) Wages and salaries included in cost such as manufacturing, operating, collecting, assembling or processing. No longer applies in Wis. (for tax years starting after 12-31-72). (3) Manufacturing labor. (4) N.J.—Includes all compensation reportable to N.J. Div. of Employment Security, including amounts in excess of $3600. (5) "Compensation" means wages, salaries, commissions and any other form of remuneration paid to exployees for personal services. (6) Compensation is pay for services to extent related to unitary business, but excludes salaries, wages and other pay administrative expense. But see Col. E (2). (7) Compensation same as pay in IRC § 3401 (a). (8) Motor carrier employee pay allocated in ratio of local-to-total mileage.

¶221

[¶222]

Property in Interstate Apportionment Numerators
For State Income Taxes

Property is a primary income producer. Usually, income-spreading formulas "weigh-in" the property factor at 33-1/3% (some at 50%). Two states assign it no role in allocation (Iowa, Mo.). The chart shows numerator elements (values assigned to the taxing state). If an "X" appears in a column, that class or type of property is included.

State(22)	Col. A Owned Realty	Col. B Rented Realty	Col. C Owned Tangible Personalty	Col. D Rented Tangible Personalty	Col. E Goods In Transit	Col. F(7) Migratory Property
Alabama	X	(5)	X	(5)	(1)	(1)
Alaska	X	X(2)	X	X(2)	X(23)	X(13)
Arizona (4)	X	X(2)	X	X(16)	(1)	X(14)
Arkansas	X	X(2)	X	X(2)	(1)	(1)
California	X	X(2)	X	X(2)	X(23)	X(13)
Colorado	X	(1)	X	(1)	(1)	(1)
Connecticut	X	X(2, 17)	X	X(2)	(1)	(14)
Delaware (4)	X	X(2)	X	X(2)	(1)	(1)
Dist. of Columbia	X	X(2)	X	X(2)	(1)	X(14)
Florida	X	X(2)	X	X(2)	(1)	(1)
Georgia	X	X(2)	X	X	(1)	(1)
Hawaii	X	X(2)	X	X(2)	(1)	(1)
Idaho	X	X(15)	X	X(15)	(1)	X(14)
Illinois	X	X	X	X	(1)	X(13)
Indiana (4)	X	X(2)	X	X(2)	X(23)	X(13, 14)
Kansas	X	X(2)	X	X(2)	(1)	X(1)
Kentucky	X	X(2)	X	X(2)	(1)	(1)
Louisiana (4)	X	(1)	X	(1)	(1)	(1)
Maine	X	X	X	X	(1)	(1)
Maryland (4)	X	X(2, 9)	X	X(2, 9)	(1)	(1)
Massachusetts	X	X(2)	X	X(2)	(1)	(1)
Michigan	X	X(2)	X	X(2)	(1)	(1)
Minnesota (4)	X	X(3, 21)	X	X(3, 21)	(1)	(1)
Mississippi (4)	X	X(2)	X	X(2)	(1)	X(14)
Missouri (8)	X	X(2)	X	X(2)	(1)	(1)
Montana	X	X(2)	X	X(2)	(1)	(1)
Nebraska	X	X(2)	X	X(2)	(1)	(1)
New Hampshire	X	(5)	X	(5)	(1)	(1)
New Jersey	X	(6)	X	(6)	(10-12)	(1)
New Mexico	X	X(2)	X	X(2)	(1)	(1)
New York	X	X(2)	X	(6)	(10-12)	X(19)
North Carolina (4)	X	X(2)	X	X(18)	X(23)	X(13)
North Dakota (4)	X	X(2)	X	X(2)	(1)	(1)
Ohio (24)	X	X(2, 20)	X	X(2, 20)	(1)	(1)
Oklahoma (4)	X	X(2)	X	X(2)	(1)	(1)
Oregon (4)	X	X(2)	X	X(2)	(1)	(1)
Pennsylvania (20)	X	X(2)	X	X(2)	(1)	(1)
Rhode Island	X	(1)	X	X(2)	(1)	X(14)
South Carolina (4)	X	X(2)	X	X(2)	(1)	X(14)
Tennessee	X	X	X	(1)	(1)	X(14)
Utah	X	X(2)	X	X(2)	(1)	(1)

Property in Interstate Apportionment Numerators
For State Income Taxes *(continued)*

State(22)	Col. A Owned Realty	Col. B Rented Realty	Col. C Owned Tangible Personalty	Col. D Rented Tangible Personalty	Col. E Goods In Transit	Col. F(7) Migratory Property
Vermont	X	X(2)	X	(6)	(10-12)	(1)
Virginia	X	X(2)	X	X(2)	(1)	X(14)
West Virginia	X	X(2)	X	X(2)	(1)	(1)
Wisconsin (4)	X	X(2)	X	X(2)	(1)	(1)

NOTES (corresponding to parenthetical numbers in chart):

(1) No specific provision.

(2) 8 times net annual rent.

(3) 7½ times annual rent.

(4) If income specifically allocated, non-allocable or exempt, exclude property.

(5) Property used is included.

(6) Specifically excluded.

(7) Carrier rules for rolling stock listed in absence of general rules.

(8) As member of Multistate Tax Compact, 3-factor formula (including property) may be elected.

(9) Leased storage facilities for tangibles excluded.

(10) Goods in transit between outstate points included in denominator only.

(11) Goods in transit between instate points included in numerator and denominator.

(12) Goods in transit between instate and outstate points excluded from numerator and denominator.

(13) In numerator on percentage of time spent basis.

(14) In numerator on percentage of mileage basis.

(15) Annual rent capitalized at 6%.

(16) 4 times annual rent.

(17) For minimum tax base, exclude value of improvements to realty rented to taxpayer (made by or for him), which revert to owner-lessor.

(18) Multiply annual rent of manufacturing and processing machinery and equipment by 3; furniture and fixtures by 3; office machinery and equipment by 2; delivery and mobile equipment by 1.

(19) Rolling equipment on time spent, mileage or other basis.

(20) Property owned or rented must be used (1) instate (numerator); (2) everywhere in tax period (denominator).

(21) Rented *and* used instate during tax period.

(22) For member states of Multistate Tax Compact, see P-H State Tax Guide.

(23) In numerator (destination state) if regularly included in denominator.

(24) For income tax payable 1972 and after. Instate pollution control facilities are excluded.

Ownership and leaseholds. In building up the property fraction, some states concentrate on ownership (Cols. A and C). Others make rental property an integral part of the property fractions (Cols. B and D). Specialized types of property have a part in many property fractions, but it is not uncommon for state rules to ignore or eliminate such items as Columns E and F show.

¶222

[¶223]

State Sales and Use Taxes

Alabama:	4%	Missouri:	3%
Arizona:	4%	Nebraska:	2½% NT
Arkansas:	3%	Nevada:	3%
California:	4¾% NT	New Jersey:	5%
Colorado:	3%	New Mexico:	4%
Connecticut:	6%	New York:	4%
District of Columbia:	5%	North Carolina:	3%
Florida:	4%	North Dakota:	4%
Georgia:	3%	Ohio:	4%
Hawaii:	4%	Oklahoma:	2%
Idaho:	3%	Pennsylvania:	6%
Illinois:	4%	Rhode Island:	5%
Indiana:	4%	South Carolina:	4%
Iowa:	3%	South Dakota:	4%
Kansas:	3%	Tennessee:	3½% NT
Kentucky:	5%	Texas:	4%
Louisiana:	3%	Utah:	4%
Maine:	5%	Vermont:	3%
Maryland:	4%	Virginia	4%
Massachusetts:	3%	Washington:	4½% NT
Michigan:	4%	West Virginia:	3%
Minnesota:	4%	Wisconsin:	4%
Mississippi:	5%	Wyoming:	3%

NOTE: These figures are for the state rates; additional local (county, city, town, school district, etc.) taxes ranging from ⅛% to 4% are imposed in many states including: Ala., Alaska, Ariz., Ark., Calif., Colo., Ill., Kan., La., Minn., Mo., Neb., Nev., N.M., N.Y., N.C., Ohio, Okla., S.D., Tenn., Tex., Utah, Va., Wash., and Wis.

[¶224]

State Corporate Costs and Fees

Corporate costs and fees, including initial organizational costs, annual maintenance, and the costs of corporate qualification outside the home state, may play a vital part in corporate planning and operation. The tables in this section are designed as planning aids in this area.

Incorporating Fees (Principal Incorporating States)

The tables below are designed as a quick and ready reference for counsel to determine the costs incident to organizing corporations in the principal states used for incorporating outside the home state.

California

Filing fee—based on authorized capital stock—nonpar valued at $10 per sh.

To $75,000	$ 25
$75,001—500,000	75
$500,001—1,000,000	100
Over $1,000,000	100 + 50 per 500,000 or part

Incidental fees $2—$25
Minimum franchise tax $200

State Corporate Costs and Fees *(continued)*

Incorporating Fees (Principal Incorporating States) *(continued)*

Connecticut

Organization tax, based on authorized capital stock as follows: Under 10,000—1 cent; 10,000 to 100,000—½ cent; 100,000 to 1,000,000—¼ cent; over 1,000,000—1/5 cent; (minimum $50)
Filing fee—$20.00
Incidental fees—$15.00

Delaware

Tax based on authorized capital stock ($10 minimum)

Par Value Stock–per $100		No-Par Value Stock–per share	
To $2,000,000	1¢	To 20,000 shares	½¢
Next $18,000,000	$200 + 1/2¢	Next 1,980,000 shares	$100 + 1/4¢
Over $20,000,000	$1,100 + 1/5¢	Over 2,000,000 shares	$5,050 + 1/5¢

Minimum tax—$10.00
Other fees—About $50.00

Illinois

License tax based on value, expressed in dollars, of entire consideration received for issued shares, at rate of 1/20 of 1%.
Franchise tax on that proportion of total stated capital and paid-in surplus which sum of value property instate and gross business transacted in or from Illinois bears to sum of value of all property and gross amount of business; rate is 1/10% of proportion. Pay tax in advance: initially on qualification and on the anniversary thereafter. Min. $25; max. $1 million.
Filing fee—$75.00
Recording—$8.00 (approx.)

Massachusetts

Filing fee based on total authorized capital stock at rate of 1/20 of 1% on par value stock (minimum par value allowed for filing fee purposes is $1.00 per share even if actual par value is lower than $1.00 per share) and 1¢ per share on nonpar stock with a $125.00 minimum.

New Jersey

Tax based on authorized capital stock at rate of 1¢ per share up to 10,000 shares of par value stock and 1/10 of a cent per share over 10,000 shares with a minimum fee of $25.00; maximum fee $1,000.

New York

Tax based on authorized capital stock at rate of 50¢ per $1,000 (1/20%) on par value stock and 5¢ per share on no-par shares with a $10.00 minimum.
Filing fee—$50.00

Pennsylvania

Excise tax based on authorized capital stock at rate of ½ of 1% on the total stated capital and authorized capital stock.
Filing—$75.00
Publications costs—$40.00 (approx.)

¶224

[¶225]

Annual Corporate Taxes and Fees
(Principal Incorporating States)

The tables below show the annual costs, taxes, and fees incident to operation of a corporation in the principal states used for incorporating.

California

Franchise–income tax-9% of net income allocated to the state based on income for the preceding calendar of fiscal year with a minimum of $200 on the franchise tax.

Connecticut

Annual report filing fee—$16.00
Business income tax—8% of net income from instate business plus added tax equal to amount by which ¼ mill per $1.00 corporate excise allocable to Connecticut exceeds 8% tax on net income. Minimum $45.00.

Delaware

Annual report filing fee—$10.00
Franchise tax lesser of (1) or (2) with *minimum* of $20.00 and *maximum* of $110,000.
(1) Based on number of authorized shares, par or nonpar:

To 1,000 shares	$20.00	3,001 to 5,000 shares	$30.25
1,001 to 3,000 shares	$24.20	5,001 to 10,000 shares	$60.50
		Each added 10,000 shares or fraction	$30.25

(2) $121 for $1,000,000 of assumed par value capital. If the assumed par value is less than $1,000,000, the tax is proportionately reduced.
Income tax—7.2% of taxable income from business carried on and property located in Delaware. Corporations merely maintaining statutory corporate office in Delaware are exempt.

Illinois

Franchise tax—Based on that proportion of total stated capital and paid-in surplus which sum of value property instate and gross business transacted in or from Illinois bears to sum of value of all property and gross amount of business; rate is 1/10% of proportion. Min. $25.00; max. $1 million.
Corporation income tax—4% on the taxable income received in Illinois.

Massachusetts

Excise tax—Tax is total of:
   (a) 8.55% of allocated net income; and
   (b) $5.76 per $1,000 of the corporation's taxable tangible property, if a tangible property corporation, or, of the corporation's taxable net worth if an intangible property corporation.
   (c) Minimum tax is $114.

Income tax—5%
Filing certificate of condition—$35.00.

¶225

Annual Corporate Taxes and Fees
(Principal Incorporating States) *(continued)*

New Jersey

Filing fee for annual report—$10.00.

Franchise tax—the greatest of following:

(a) Tax on proportion of entire net worth allocable to State, at rates of 2 mills per dollar on first $100,000,000; 4/10ths mill on second; 3/10ths mill on third; and 2/10ths mill on excess.

(b) Tax on 5/10ths mill per dollar on first $100,000,000 of total assets allocated to State; 2/10ths mill per dollar thereafter.

(c) Tax which is least of (a) amount based on authorized capital stock if not over 5000 shares; $25.00, 5001 to 10,000; $55.00 over 10,000 shares; $55.00 plus $27.50 per added 10,000 shares or (b) 11/100 mill per $1.00 on total assets or (c) $100,000.

(d) Minimum tax of $25.00.

In lieu of all the foregoing, corporations having total assets everywhere (less reasonable reserves for depreciation) of less than $150,000 may elect to pay tax based upon total assets, according to a special tax table.

Income tax—5½% of net income allocable to the State.

New York

Franchise tax—the greatest one of the following:

1. (a) 9% of the entire net income or the portion thereof allocated to New York, or

   (b) 1.6 mills per dollar of the total business and investment capital or portion thereof allocated to New York, or

   (c) 9% of the following or the proportion thereof allocated to New York: 30% of the entire net income plus salaries and other compensation paid to elected or appointed officers and to every stockholder owning more than 5% of the issued capital stock, less an exemption of $15,000 and any net loss for the reported year, or

   (d) $125

And in addition:

2. A tax of 8/10 mill for each dollar of the portion of the subsidiary capital allocated to New York.

Pennsylvania

Capital stock tax—10 mills per dollar of taxable portion of the actual value of the whole capital stock at the end of the tax year. Exemption is allowed for capital which is invested in manufacturing, research or development, the distillation or other production of alcoholic liquors, and processing.

Income tax—9.5% of net income allocated to Pennsylvania.

Corporate loans tax—Tax of 4 mills per dollar on corporate loans held by Pennsylvania residents, if interest is paid thereon. Tax is on the holders of corporate loans but must be withheld by the corporation from the interest when paid, and remitted to the Commonwealth.

¶225

[¶226]
Excise Taxes (Federal)

Item	Rate
Retailer's excise taxes: Diesel fuel and special motor fuel	4¢ per gal
Manufacturers' excise taxes:	
Trucks, buses and trailers with a gross weight over 10,000 lbs.	10% of price, 5% as of 10-1-77
Truck and bus parts and	8% of price, 5% as of 10-1-77
Tires, etc.:	
Highway type	10¢ per lb., 5¢ as of 10-1-77
Other	5¢ per lb.
Inner tubes	10¢ per lb., 9¢ as of 10-1-77
Tread rubber	5¢ per lb. until 10-1-77
Laminated tires (nonhighway)	1¢ per lb.
Gasoline	4¢ per gal[1]
Lubricating oil (highway use)	6¢ per gal.
Fishing equipment	10% of price
Bows and arrows (hunting) and parts and accessories	11% of price as of 1-1-75
Pistols and revolvers	10% of price
Other firearms, shells and cartridges	11% of price
Airport and airway user taxes:	
Fuel and gasoline used in general noncommercial aviation	7¢ per gallon[2] as of 7-1-70.[3]
Domestic transportation of persons by air	8% of amount paid as of 7-1-70, 5% after 6-30-80
International transportation of persons by air (flights beginning in U.S.)	$3 added to amount paid as of 7-1-70[3]
Domestic transportation of property by air	5% of amount paid as of 7-1-70[3]
Annual registration tax on civil aircraft:	
Turbine engine powered aircraft	$25 plus 3½¢ per lb. as of 7-1-70[3]
Other aircraft	$25 plus 2¢ per lb. of total weight if it exceeds 2500 lbs., as of 7-1-70 through 6-30-71. $25 plus 2¢ per lb. of weight in excess of 2500 lbs. as of 7-1-71.
Miscellaneous excise taxes:	
Communications:	
Local and toll telephone service and teletype- writer service	7% as of Jan. 1, 1975; reduce to 6% Jan. 1, 1976; to 5% Jan. 1, 1977; to 4% Jan. 1, 1978; to 3% Jan. 1, 1979; to 2% Jan. 1, 1980; to 1% Jan. 1, 1981; repeal Jan. 1, 1982.
Foreign insurance policies	4¢ or 1¢ per dollar of premium
Wagering:	
Wagers	2% of amount of wager as of 12-1-74
Occupation of accepting wagers	$500 per year as of 12-1-75[5]
Coin-operated gaming devices	$250 per device per year
Use tax on certain highway vehicles	$3 per 1,000 lbs. per year[4]
Sugar	0.53¢ per lb.[6]
Import tax on oleomargarine	15¢ per lb.

¶226

Excise Taxes (Federal) *(continued)*

Item	Rate
Regulatory taxes:	
White phosphorus matches	2¢ per hundred
Adulterated butter	10¢ or 15¢ per lb.
Process butter	¼¢ per lb.
Occupational taxes: adulterated or process butter	$48 to $600 per year
Cotton futures	2¢ per lb.
Bank circulation tax (other than national banks)	1/12 of 1%; 1/6 of 1%; 10%
Alcohol taxes:	
Distilled spirits	$10.50 per proof gal.
Beer	$9 per barrel

[1] Reduce to 1½ cents per gallon on and after 10-1-77. See also footnote [2].

[2] The 7¢ per gallon tax on gasoline used in general noncommercial aviation is reached by adding a 3¢ (5½¢ on and after 10-1-77) retailers' tax to the 4¢ (1½¢ on and after 10-1-77) manufacturers' tax on gasoline listed above.

[3] The tax will not apply on and after 7-1-80.

[4] The tax is 75¢ per 1,000 lbs. for the taxable period 7-1-77 through 9-30-77. The tax will not apply on and after 10-1-77.

[5] Only $50 for fiscal year ending 6-30-75 if taxpayer became liable for and paid such tax before 12-1-74.

[6] This tax is set to phase out as of 6/30/75.

¶226

[¶227]

Depreciation Methods Available
For Different Types of Property

Under Code §167, any method of depreciation which is reasonable and consistently applied may be used for any type of property. However, no method may be used which would permit depreciation to exceed amounts necessary to recover cost or other basis, less salvage value, during the remaining useful life of the property (Reg. §1.167(b)-0(a)). Also, depreciation can be deducted only if it is taken on property used in the trade or business or to produce income.

The methods which the Code explicity recognizes as reasonable are the straight-line method, 200%-declining-balance method, 150%-declining-balance method, sum-of-the-years-digits method, and, finally, any other method which does not exceed, during the first two-thirds of the useful life of the property, the amount of depreciation which would be allowable under the 200%-declining-balance method. However, all the above methods, except straight-line and 150%-declining-balance, are limited to tangible assets with useful lives of more than three years whose original use commenced with the taxpayer after December 31, 1953.

All the above methods depend, of course, on the definition of the useful life of property. The useful life for any property begins and ends with the taxpayer's use for the property and depends upon the facts and circumstances of his case. As an aid to taxpayers (and to IRS), Bulletin F was promulgated, which listed Treasury estimates of useful lives for various types of property. In 1962 this Bulletin was superseded by *Rev. Proc. 62-21*. The effect of this was to create "class lives" as a method of estimating useful lives. Under the class life system, all items in a particular class or group were given "guideline lives," regardless of a taxpayer's actual experience for the particular items. Then, they could be depreciated individually or as a class. A taxpayer could also use individual useful lives if he chose, but Bulletin F would no longer be the basis of automatic treasury approval.

This system applied until 1971, when the "Class Life ADR" system was promulgated. ADR is discussed in subsequent tables and applies to all property placed in service after 1970. For property placed in service before 1970 but depreciated through later years, either system may be used. The guideline lives of *Rev. Proc. 62-21* may not be used for any property placed in service after 1970. However, the useful life system may be used for any property, whenever placed in service.

Regardless of the system used for estimating useful life, either some or all depreciation may be recaptured under §1250, which applies to real estate, or §1245, which applies to all other depreciable property. These sections come into play when an asset which has been depreciated is sold for more than its adjusted basis. If it is §1245 property, all the gain attributable to depreciation is recaptured as ordinary income, rather than treated as capital gain. If it is §1259 property, less depreciation may be recaptured, depending on the type of real estate and how long it has been held. However, this favorable treatment for real estate is offset by the fact that most new real estate is limited to either straight-line or 150%-declining-balance depreciation, and most used real estate is limited to straight-line under §167(j).

[¶228]

Comparative Depreciation Tables

The following tables show the annual and cumulative depreciation for various useful lives under the straight-line, 200%-declining-balance, 150%-declining-balance, 125%-declining-balance, and sum-of-the-years-digits methods. *All amounts are expressed as percentages of the basis of the property at the time the useful life begins.*

Year	Straight-Line Annual %	Straight-Line Cum. %	200%-Declining-Balance Annual %	200%-Declining-Balance Cum. %	150%-Declining-Balance Annual %	150%-Declining-Balance Cum. %	Sum-of-Digits Annual %	Sum-of-Digits Cum. %
3-Year Life								
1	33.33	33.33	66.66	66.66	50.00	50.00	50.00	50.00
2	33.33	66.66	22.22	88.88	25.00	75.00	33.33	83.33
3	33.34	100.00	7.41	96.29	12.50	87.50	16.67	100.00
4-Year Life								
1	25.00	25.00	50.00	50.00	37.50	37.50	40.00	40.00
2	25.00	50.00	25.00	75.00	23.44	60.94	30.00	70.00
3	25.00	75.00	12.50	87.50	14.65	75.59	20.00	90.00
4	25.00	100.00	6.25	93.75	9.15	84.74	10.00	100.00
5-Year Life								
1	20.00	20.00	40.00	40.00	30.00	30.00	33.33	33.33
2	20.00	40.00	24.00	64.00	21.00	51.00	26.67	60.00
3	20.00	60.00	14.40	78.40	14.70	65.70	20.00	80.00
4	20.00	80.00	8.64	87.04	10.29	75.99	13.33	93.33
5	20.00	100.00	5.18	92.22	7.20	83.19	6.67	100.00
6-Year Life								
1	16.67	16.67	33.34	33.34	25.00	25.00	28.57	28.57
2	16.67	33.34	22.22	55.56	18.75	43.75	23.81	52.38
3	16.66	50.00	14.81	70.37	14.06	57.81	19.05	71.43
4	16.67	66.67	9.87	80.24	10.55	68.36	14.29	85.72
5	16.67	83.34	6.58	86.82	7.91	76.27	9.52	95.24
6	16.66	100.00	4.39	91.21	5.93	82.20	4.76	100.00
7-Year Life								
1	14.28	14.28	28.57	28.57	21.43	21.43	25.00	25.00
2	14.28	28.56	20.41	48.98	16.83	38.26	21.43	46.43
3	14.29	42.85	14.58	63.56	13.23	51.49	17.86	64.29
4	14.29	57.14	10.41	73.97	10.40	61.89	14.29	78.58
5	14.29	71.43	7.44	81.41	8.17	70.06	10.71	89.29
6	14.29	85.72	5.31	86.72	6.42	76.48	7.14	96.43
7	14.28	100.00	3.79	90.51	5.04	81.52	3.57	100.00
8-Year Life								
1	12.50	12.50	25.00	25.00	18.75	18.75	22.22	22.22
2	12.50	25.00	18.75	43.75	15.23	33.98	19.44	41.66
3	12.50	37.50	14.06	57.81	12.38	46.36	16.67	58.33
4	12.50	50.00	10.55	68.36	10.06	56.42	13.89	72.22
5	12.50	62.50	7.91	76.27	8.17	64.59	11.11	83.33
6	12.50	75.00	5.93	82.20	6.64	71.23	8.33	91.66
7	12.50	87.50	4.45	86.65	5.39	76.62	5.56	97.22
8	12.50	100.00	3.34	89.99	4.38	81.00	2.78	100.00

Comparative Depreciation Tables *(continued)*

Year	Straight-Line		200%-Declining-Balance		150%-Declining-Balance		125%-Declining-Balance *		Sum-of-Digits	
	Annual %	Cum. %	Annual %	Cum. %	Annual %	Cum. %	Annual %	Cum. %	Annual %	Cum. %
9-Year Life										
1	11.11	11.11	22.22	22.22	16.67	16.67			20.00	20.00
2	11.11	22.22	17.28	39.50	13.89	30.56			17.78	37.78
3	11.11	33.33	13.44	52.94	11.57	42.13			15.56	53.34
4	11.11	44.44	10.45	63.39	9.65	51.78			13.33	66.67
5	11.11	55.55	8.13	71.52	8.04	59.82			11.11	77.78
6	11.11	66.66	6.32	77.84	6.70	66.52			8.89	86.67
7	11.11	77.77	4.92	82.76	5.58	72.10			6.67	93.34
8	11.11	88.88	3.83	86.59	4.65	76.75			4.44	97.78
9	11.12	100.00	2.98	89.57	3.88	80.63			2.22	100.00
10-Year Life										
1	10.00	10.00	20.00	20.00	15.00	15.00			18.18	18.18
2	10.00	20.00	16.00	36.00	12.75	27.75			16.37	34.55
3	10.00	30.00	12.80	48.80	10.84	38.59			14.56	49.09
4	10.00	40.00	10.24	59.04	9.21	47.80			12.73	61.82
5	10.00	50.00	8.19	67.23	7.83	55.63			10.91	72.73
6	10.00	60.00	6.56	73.79	6.66	62.29			9.09	81.82
7	10.00	70.00	5.24	79.03	5.66	67.95			7.27	89.09
8	10.00	80.00	4.19	83.22	4.81	72.76			5.46	94.55
9	10.00	90.00	3.36	86.58	4.09	76.85			3.63	98.18
10	10.00	100.00	2.68	89.26	3.47	80.32			1.82	100.00
15-Year Life										
1	6.67	6.67	13.33	13.33	10.00	10.00			12.50	12.50
2	6.66	13.33	11.56	24.89	9.00	19.00			11.67	24.17
3	6.67	20.00	10.01	34.90	8.10	27.10			10.83	35.00
4	6.67	26.67	8.68	43.58	7.29	34.39			10.00	45.00
5	6.66	33.33	7.53	51.11	6.56	40.95			9.17	54.17
6	6.67	40.00	6.51	57.62	5.90	46.85			8.33	62.50
7	6.67	46.67	5.65	63.27	5.32	52.17			7.50	70.00
8	6.66	53.33	4.90	68.17	4.78	56.95			6.67	76.67
9	6.67	60.00	4.25	72.42	4.30	61.25			5.83	82.50
10	6.67	66.67	3.67	76.09	3.88	65.13			5.00	87.50
11	6.66	73.33	3.19	79.28	3.49	68.62			4.17	91.67
12	6.67	80.00	2.76	82.04	3.14	71.76			3.33	95.00
13	6.67	86.67	2.40	84.44	2.82	74.58			2.50	97.50
14	6.66	93.33	2.07	86.51	2.54	77.12			1.67	99.17
15	6.67	100.00	1.80	88.31	2.29	79.41			.83	100.00

*Available for used residential real estate acquired after 7/24/69 and having useful life of 20 years or more when acquired.

¶228

Comparative Depreciation Tables *(continued)*

Year	Straight-Line		200%-Declining-Balance		150%-Declining-Balance		125%-Declining-Balance		Sum-of-Digits	
	Annual %	Cum. %	Annual %	Cum. %	Annual %	Cum. %	Annual %	Cum. %	Annual %	Cum. %
20-Year Life										
1	5.00	5.00	10.00	10.00	7.50	7.50	6.25	6.25	9.52	9.52
2	5.00	10.00	9.00	19.00	6.94	14.44	5.86	12.11	9.05	18.57
3	5.00	15.00	8.10	27.10	6.42	20.86	5.49	17.60	8.57	27.14
4	5.00	20.00	7.29	34.39	5.94	26.80	5.15	22.75	8.10	35.24
5	5.00	25.00	6.56	40.95	5.49	32.29	4.83	27.58	7.62	42.86
6	5.00	30.00	5.91	46.86	5.08	37.37	4.53	32.11	7.14	50.00
7	5.00	35.00	5.31	52.17	4.70	42.07	4.24	36.35	6.67	56.67
8	5.00	40.00	4.78	56.95	4.35	46.42	3.98	40.33	6.19	62.86
9	5.00	45.00	4.31	61.26	4.02	50.44	3.73	44.06	5.71	68.57
10	5.00	50.00	3.87	65.13	3.71	54.15	3.50	47.55	5.24	73.81
11	5.00	55.00	3.49	68.62	3.44	57.59	3.28	50.83	4.76	78.57
12	5.00	60.00	3.14	71.76	3.18	60.77	3.07	53.90	4.29	82.86
13	5.00	65.00	2.82	74.58	2.94	63.71	2.88	56.79	3.81	86.67
14	5.00	70.00	2.54	77.12	2.72	66.43	2.70	59.49	3.33	90.00
15	5.00	75.00	2.29	79.41	2.52	68.95	2.53	62.02	2.86	92.86
16	5.00	80.00	2.06	81.47	2.33	71.28	2.37	64.39	2.38	95.24
17	5.00	85.00	1.85	83.32	2.15	73.43	2.23	66.62	1.90	97.14
18	5.00	90.00	1.67	84.99	1.99	75.42	2.09	68.70	1.43	98.57
19	5.00	95.00	1.50	86.49	1.84	77.26	1.96	70.66	.95	99.52
20	5.00	100.00	1.35	87.84	1.70	78.96	1.83	72.49	.48	100.00
25-Year Life										
1	4.00	4.00	8.00	8.00	6.00	6.00	5.00	5.00	7.69	7.69
2	4.00	8.00	7.36	15.36	5.64	11.64	4.75	9.75	7.39	15.08
3	4.00	12.00	6.77	22.13	5.30	16.94	4.51	14.26	7.07	22.15
4	4.00	16.00	6.23	28.36	4.98	21.92	4.29	18.55	6.77	28.92
5	4.00	20.00	5.73	34.09	4.68	26.60	4.07	22.62	6.47	35.39
6	4.00	24.00	5.27	39.36	4.40	31.00	3.87	26.49	6.15	41.54
7	4.00	28.00	4.86	44.22	4.14	35.14	3.68	30.17	5.85	47.39
8	4.00	32.00	4.46	48.68	3.89	39.03	3.49	33.66	5.53	52.92
9	4.00	36.00	4.10	52.78	3.66	42.69	3.32	36.98	5.23	58.15
10	4.00	40.00	3.78	56.56	3.43	46.12	3.15	40.13	4.93	63.08
11	4.00	44.00	3.48	60.04	3.23	49.35	2.99	43.12	4.61	67.69
12	4.00	48.00	3.19	63.23	3.03	52.38	2.84	45.96	4.31	72.00
13	4.00	52.00	2.94	66.17	2.86	55.24	2.70	48.67	4.00	76.00
14	4.00	56.00	2.71	68.88	2.68	57.92	2.57	51.23	3.69	79.69
15	4.00	60.00	2.49	71.37	2.52	60.44	2.44	53.67	3.39	83.08
16	4.00	64.00	2.29	73.66	2.37	62.81	2.32	55.99	3.07	86.15
17	4.00	68.00	2.11	75.77	2.23	65.04	2.20	58.19	2.77	88.92
18	4.00	72.00	1.94	77.71	2.10	67.14	2.09	60.28	2.47	91.39
19	4.00	76.00	1.78	79.49	1.97	69.11	1.99	62.26	2.15	93.54
20	4.00	80.00	1.64	81.13	1.85	70.96	1.89	64.15	1.85	95.39

Comparative Depreciation Tables (continued)

Year	Straight-Line		200%-Declining-Balance		150%-Declining-Balance		125%-Declining-Balance		Sum-of-Digits	
	Annual %	Cum. %	Annual %	Cum. %	Annual %	Cum. %	Annual %	Cum. %	Annual %	Cum. %
25-Year Life (continued)										
21	4.00	84.00	1.51	82.64	1.74	72.70	1.79	65.94	1.53	96.92
22	4.00	88.00	1.39	84.03	1.64	74.34	1.70	67.65	1.23	98.15
23	4.00	92.00	1.28	85.31	1.54	75.88	1.62	69.26	.93	99.08
24	4.00	96.00	1.17	86.48	1.45	77.33	1.54	70.80	.61	99.69
25	4.00	100.00	1.08	87.56	1.36	78.69	1.46	72.26	.31	100.00
30-Year Life										
1	3.33	3.33	6.67	6.67	5.00	5.00	4.16	4.16	6.45	6.45
2	3.34	6.67	6.22	12.89	4.75	9.75	3.99	8.16	6.24	12.69
3	3.33	10.00	5.81	18.70	4.51	14.26	3.83	11.99	6.02	18.71
4	3.33	13.33	5.42	24.12	4.29	18.55	3.67	15.65	5.81	24.52
5	3.34	16.67	5.06	29.18	4.07	22.62	3.51	19.17	5.59	30.11
6	3.33	20.00	4.72	33.90	3.87	26.49	3.37	22.54	5.37	35.48
7	3.33	23.33	4.40	38.30	3.68	30.17	3.23	25.76	5.17	40.65
8	3.34	26.67	4.12	42.42	3.49	33.66	3.09	28.86	4.94	45.59
9	3.33	30.00	3.84	46.26	3.32	36.98	2.96	31.82	4.73	50.32
10	3.33	33.33	3.58	49.84	3.15	40.13	2.84	34.66	4.52	54.84
11	3.34	36.67	3.34	53.18	2.99	43.12	2.72	37.38	4.30	59.14
12	3.33	40.00	3.12	56.30	2.84	45.96	2.61	39.99	4.09	63.23
13	3.33	43.33	2.92	59.22	2.70	48.66	2.50	42.49	3.87	67.10
14	3.34	46.67	2.72	61.94	2.57	51.23	2.40	44.89	3.65	70.75
15	3.33	50.00	2.53	64.47	2.44	53.67	2.30	47.19	3.44	74.19
16	3.33	53.33	2.37	66.84	2.32	55.99	2.20	49.39	3.23	77.42
17	3.34	56.67	2.21	69.05	2.20	58.19	2.11	51.50	3.01	80.43
18	3.33	60.00	2.07	71.12	2.09	60.28	2.02	53.52	2.80	83.23
19	3.33	63.33	1.92	73.04	1.99	62.27	1.94	55.45	2.58	85.81
20	3.34	66.67	1.80	74.84	1.89	64.16	1.86	57.31	2.36	88.17
21	3.33	70.00	1.68	76.52	1.80	65.96	1.78	59.09	2.15	90.32
22	3.33	73.33	1.56	78.08	1.70	67.66	1.70	60.79	1.94	92.26
23	3.34	76.67	1.46	79.54	1.62	69.28	1.63	62.43	1.72	93.98
24	3.33	80.00	1.37	80.91	1.54	70.82	1.57	63.99	1.61	95.49
25	3.33	83.33	1.27	82.18	1.46	72.28	1.50	65.49	1.29	96.78
26	3.34	86.67	1.19	83.37	1.39	73.67	1.44	66.93	1.07	97.85
27	3.33	90.00	1.11	84.48	1.32	74.99	1.38	68.31	.86	98.71
28	3.33	93.33	1.03	85.51	1.25	76.24	1.32	69.63	.65	99.36
29	3.34	96.67	.97	86.48	1.19	77.43	1.27	70.89	.43	99.79
30	3.33	100.00	.90	87.38	1.13	78.56	1.21	72.11	.21	100.00
33-1/3 Year Life										
1	3.00	3.00	6.00	6.00	4.50	4.50	3.75	3.75	5.82	5.82
2	3.00	6.00	5.64	11.64	4.30	8.80	3.61	7.36	5.65	11.47
3	3.00	9.00	5.30	16.94	4.10	12.90	3.47	10.83	5.47	16.95
4	3.00	12.00	4.98	21.93	3.92	16.82	3.34	14.18	5.30	22.25
5	3.00	15.00	4.68	26.61	3.74	20.56	3.22	17.40	5.17	27.37

Comparative Depreciation Tables *(continued)*

Year	Straight-Line Annual %	Cum. %	200%-Declining-Balance Annual %	Cum. %	150%-Declining-Balance Annual %	Cum. %	125%-Declining-Balance Annual %	Cum. %	Sum-of-Digits Annual %	Cum. %
33-1/3 Year Life *(continued)*										
6	3.00	18.00	4.40	31.00	3.57	24.14	3.10	20.49	4.95	32.32
7	3.00	21.00	4.14	35.15	3.41	27.55	2.98	23.47	4.76	37.10
8	3.00	24.00	3.89	39.03	3.16	30.81	2.87	26.34	4.60	41.70
9	3.00	27.00	3.66	42.70	3.11	33.93	2.76	29.11	4.43	46.13
10	3.00	30.00	3.44	46.14	2.97	36.90	2.66	31.77	4.25	50.38
11	3.00	33.00	3.23	49.37	2.84	39.74	2.56	34.32	4.08	54.46
12	3.00	36.00	3.04	52.41	2.71	42.45	2.46	36.79	3.90	58.36
13	3.00	39.00	2.86	55.26	2.59	45.04	2.37	39.16	3.73	62.10
14	3.00	42.00	2.68	57.95	2.47	47.51	2.28	41.44	3.55	65.64
15	3.00	45.00	2.52	60.47	2.36	49.88	2.20	43.63	3.38	69.02
16	3.00	48.00	2.37	62.84	2.26	52.13	2.11	45.75	3.20	72.22
17	3.00	51.00	2.23	65.07	2.15	54.29	2.03	47.78	3.03	75.25
18	3.00	54.00	2.10	67.17	2.06	56.34	1.95	49.74	2.85	78.10
19	3.00	57.00	1.97	69.14	1.96	58.31	1.88	51.63	2.68	80.78
20	3.00	60.00	1.85	71.00	1.88	60.18	1.81	53.44	2.50	83.28
21	3.00	63.00	1.74	72.73	1.79	61.97	1.75	55.19	2.33	85.61
22	3.00	66.00	1.64	74.37	1.71	63.69	1.68	56.87	2.15	87.77
23	3.00	69.00	1.54	75.90	1.63	65.32	1.62	58.48	1.98	89.75
24	3.00	72.00	1.45	77.35	1.56	66.88	1.56	60.04	1.81	91.56
25	3.00	75.00	1.36	78.71	1.49	68.37	1.50	61.54	1.63	93.19
26	3.00	78.00	1.28	79.99	1.42	69.79	1.44	62.98	1.46	94.64
27	3.00	81.00	1.20	81.19	1.36	71.15	1.39	64.37	1.28	95.92
28	3.00	84.00	1.13	82.32	1.30	72.45	1.34	65.71	1.11	97.03
29	3.00	87.00	1.06	83.38	1.24	73.69	1.29	66.99	.93	97.96
30	3.00	90.00	1.00	84.37	1.18	74.88	1.24	68.23	.76	98.72
31	3.00	93.00	.94	85.31	1.13	76.01	1.19	69.42	.58	99.30
32	3.00	96.00	.88	86.19	1.08	77.09	1.15	70.57	.41	99.71
33	3.00	99.00	.93	87.02	1.03	78.12	1.10	71.67	.23	99.94
33⅓	1.00	100.00	.26	87.28	.33	78.45	.35	71.95	.06	100.00
35-Year Life										
1	2.86	2.86	5.71	5.71	4.29	4.29	3.57	3.57	5.56	5.56
2	2.86	5.72	5.38	11.09	4.10	8.39	3.44	7.02	5.40	10.96
3	2.85	8.57	5.07	16.16	3.93	12.32	3.32	10.34	5.24	16.20
4	2.86	11.43	4.78	20.94	3.76	16.08	3.20	13.54	5.08	21.28
5	2.86	14.29	4.51	25.45	3.60	19.68	3.09	16.63	4.92	26.20
6	2.85	17.14	4.25	29.70	3.44	23.12	2.98	19.60	4.76	30.96
7	2.86	20.00	4.01	33.71	3.29	26.41	2.87	22.48	4.60	35.56
8	2.86	22.86	3.78	37.49	3.15	29.56	2.77	25.24	4.44	40.00
9	2.85	25.71	3.56	41.05	3.02	32.58	2.67	27.91	4.29	44.29
10	2.86	28.57	3.36	44.41	2.89	35.47	2.57	30.49	4.13	48.42

¶228

Comparative Depreciation Tables *(continued)*

Year	Straight-Line		200%-Declining-Balance		150%-Declining-Balance		125%-Declining-Balance		Sum-of-Digits	
	Annual %	Cum. %	Annual %	Cum. %	Annual %	Cum. %	Annual %	Cum. %	Annual %	Cum. %
35-Year Life *(continued)*										
11	2.86	31.43	3.17	47.58	2.77	38.24	2.48	32.97	3.97	52.39
12	2.85	34.28	2.99	50.57	2.65	40.89	2.39	35.36	3.81	56.20
13	2.86	37.14	2.82	53.39	2.53	43.42	2.31	37.67	3.65	59.85
14	2.86	40.00	2.66	56.05	2.42	45.84	2.23	39.90	3.49	63.34
15	2.85	42.85	2.51	58.56	2.32	48.16	2.15	42.05	3.33	66.67
16	2.86	45.71	2.37	60.93	2.22	50.38	2.07	44.12	3.18	69.85
17	2.86	48.57	2.23	63.16	2.13	52.51	2.00	46.11	3.02	72.87
18	2.85	51.42	2.10	65.26	2.03	54.54	1.92	48.04	2.86	75.73
19	2.86	54.28	1.98	67.24	1.95	56.49	1.86	49.89	2.70	78.43
20	2.86	57.14	1.87	69.11	1.86	58.35	1.79	51.68	2.54	80.97
21	2.86	60.00	1.76	70.87	1.79	60.14	1.73	53.41	2.38	83.35
22	2.86	62.86	1.66	72.53	1.71	61.85	1.66	55.07	2.22	85.57
23	2.86	65.72	1.57	74.10	1.64	63.49	1.60	56.68	2.06	87.63
24	2.85	68.57	1.48	75.58	1.56	65.05	1.55	58.22	1.90	89.53
25	2.86	71.43	1.40	76.98	1.50	66.55	1.49	59.72	1.75	91.28
26	2.86	74.29	1.32	78.30	1.43	67.98	1.44	61.15	1.59	92.87
27	2.85	77.14	1.24	79.54	1.37	69.35	1.39	62.54	1.43	94.30
28	2.86	80.00	1.17	80.71	1.31	70.66	1.34	63.88	1.27	95.57
29	2.86	82.86	1.10	81.81	1.26	71.92	1.29	65.17	1.11	96.68
30	2.85	85.71	1.04	82.85	1.20	73.12	1.24	66.41	.95	97.63
31	2.86	88.57	.98	83.83	1.15	74.27	1.20	67.61	.79	98.42
32	2.86	91.43	.92	84.75	1.10	75.37	1.16	68.77	.63	99.05
33	2.85	94.28	.87	85.62	1.06	76.43	1.12	69.88	.47	99.52
34	2.86	97.14	.82	86.44	1.01	77.44	1.08	70.96	.32	99.84
35	2.86	100.00	.77	87.21	.97	78.41	1.04	72.00	.16	100.00
40-Year Life										
1	2.50	2.50	5.00	5.00	3.75	3.75	3.13	3.13	4.88	4.88
2	2.50	5.00	4.75	9.75	3.61	7.36	3.03	6.15	4.75	9.63
3	2.50	7.50	4.51	14.26	3.47	10.83	2.93	9.09	4.64	14.27
4	2.50	10.00	4.29	18.55	3.34	14.17	2.84	11.93	4.51	18.78
5	2.50	12.50	4.07	22.62	3.22	17.39	2.75	14.68	4.39	23.17
6	2.50	15.00	3.87	26.49	3.10	20.49	2.67	17.34	4.27	27.44
7	2.50	17.50	3.68	30.17	2.98	23.47	2.58	19.93	4.14	31.58
8	2.50	20.00	3.49	33.66	2.87	26.34	2.50	22.43	4.03	35.61
9	2.50	22.50	3.32	36.98	2.76	29.10	2.42	24.85	3.90	39.51
10	2.50	25.00	3.15	40.13	2.66	31.76	2.35	27.20	3.78	43.29
11	2.50	27.50	2.99	43.12	2.56	34.32	2.27	29.48	3.66	46.95
12	2.50	30.00	2.84	45.96	2.46	36.78	2.20	31.68	3.54	50.49
13	2.50	32.50	2.71	48.67	2.37	39.15	2.13	33.82	3.41	53.90
14	2.50	35.00	2.56	51.23	2.28	41.43	2.07	35.88	3.29	57.19
15	2.50	37.50	2.44	53.67	2.20	43.63	2.00	37.89	3.18	60.37

Comparative Depreciation Tables (continued)

Year	Straight-Line		200%-Declining-Balance		150%-Declining-Balance		125%-Declining-Balance		Sum-of-Digits	
	Annual %	Cum. %	Annual %	Cum. %	Annual %	Cum. %	Annual %	Cum. %	Annual %	Cum. %
40-Year Life *(continued)*										
16	2.50	40.00	2.32	55.99	2.11	45.74	1.94	39.83	3.04	63.41
17	2.50	42.50	2.20	58.19	2.03	47.77	1.88	41.71	2.93	66.34
18	2.50	45.00	2.09	60.28	1.96	49.73	1.82	43.53	2.81	69.15
19	2.50	47.50	1.99	62.27	1.88	51.61	1.76	45.30	2.68	71.83
20	2.50	50.00	1.88	64.15	1.81	53.42	1.71	47.01	2.56	74.39
21	2.50	52.50	1.79	65.94	1.75	55.17	1.66	48.66	2.44	76.83
22	2.50	55.00	1.71	67.65	1.68	56.85	1.60	50.27	2.32	79.15
23	2.50	57.50	1.62	69.27	1.62	58.47	1.55	51.82	2.19	81.34
24	2.50	60.00	1.53	70.80	1.56	60.03	1.51	53.33	2.07	83.41
25	2.50	62.50	1.46	72.26	1.50	61.53	1.46	54.78	1.94	85.37
26	2.50	65.00	1.39	73.65	1.44	62.97	1.41	56.20	1.82	87.19
27	2.50	67.50	1.32	74.97	1.39	64.36	1.37	57.57	1.71	88.90
28	2.50	70.00	1.25	76.22	1.34	65.70	1.33	58.89	1.59	90.49
29	2.50	72.50	1.19	77.41	1.29	66.99	1.28	60.18	1.46	91.95
30	2.50	75.00	1.13	78.54	1.24	68.23	1.24	61.42	1.34	93.29
31	2.50	77.50	1.07	79.61	1.19	69.42	1.21	62.63	1.22	94.51
32	2.50	80.00	1.02	80.63	1.15	70.57	1.17	63.79	1.10	95.61
33	2.50	82.50	.97	81.60	1.10	71.67	1.13	64.93	.98	96.58
34	2.50	85.00	.92	85.52	1.06	72.73	1.10	66.02	.86	97.44
35	2.50	87.50	.87	83.39	1.02	73.75	1.06	67.08	.73	98.17
36	2.50	90.00	.83	84.22	.98	74.73	1.03	68.11	.61	98.78
37	2.50	92.50	.79	85.01	.95	75.68	1.00	69.11	.49	99.27
38	2.50	95.00	.75	85.76	.91	76.59	.97	70.07	.36	99.63
39	2.50	97.50	.71	86.47	.88	77.47	.94	71.01	.25	99.88
40	2.50	100.00	.68	87.15	.85	78.32	.91	71.92	.12	100.00
45-Year Life										
1	2.22	2.22	4.44	4.44	3.33	3.33	2.78	2.78	4.35	4.35
2	2.22	4.44	4.24	8.68	3.22	6.55	2.70	5.48	4.25	8.60
3	2.22	6.66	4.05	12.73	3.12	9.67	2.63	8.10	4.15	12.75
4	2.23	8.89	3.87	16.60	3.01	12.68	2.55	10.66	4.06	16.81
5	2.22	11.11	3.70	20.30	2.91	15.59	2.48	13.14	3.96	20.77
6	2.22	13.33	3.54	23.84	2.81	18.40	2.41	15.55	3.86	24.63
7	2.22	15.55	3.38	27.22	2.72	21.12	2.35	17.90	3.77	28.40
8	2.23	17.78	3.23	30.45	2.63	23.75	2.28	20.18	3.67	32.07
9	2.22	20.00	3.09	33.54	2.54	26.29	2.22	22.40	3.57	35.64
10	2.22	22.22	2.95	36.49	2.46	28.75	2.16	24.55	3.48	39.12
11	2.22	24.44	2.82	39.31	2.38	31.13	2.10	26.65	3.38	42.50
12	2.23	26.67	2.69	42.00	2.30	33.43	2.04	28.68	3.28	45.78
13	2.22	28.89	2.57	44.57	2.22	35.65	1.98	30.67	3.19	48.97
14	2.22	31.11	2.46	47.03	2.15	37.80	1.93	32.59	3.09	52.06
15	2.22	33.33	2.35	49.38	2.07	39.87	1.87	34.46	3.00	55.06

¶228

Comparative Depreciation Tables *(continued)*

Year	Straight-Line		200%-Declining-Balance		150%-Declining-Balance		125%-Declining-Balance		Sum-of-Digits	
	Annual %	Cum. %	Annual %	Cum. %	Annual %	Cum. %	Annual %	Cum. %	Annual %	Cum. %
45-Year Life *(continued)*										
16	2.23	35.56	2.25	51.63	2.00	41.87	1.82	36.28	2.90	57.96
17	2.22	37.78	2.15	53.78	1.94	43.81	1.77	38.05	2.80	60.76
18	2.22	40.00	2.05	55.83	1.87	45.68	1.72	39.77	2.70	63.46
19	2.22	42.22	1.96	57.79	1.81	47.49	1.67	41.45	2.61	66.07
20	2.23	44.45	1.87	59.66	1.75	49.24	1.63	43.07	2.51	68.58
21	2.22	46.67	1.79	61.45	1.69	50.93	1.58	44.66	2.42	71.00
22	2.22	48.89	1.71	63.16	1.64	52.57	1.54	46.19	2.32	73.32
23	2.22	51.11	1.63	64.79	1.58	54.15	1.49	47.69	2.22	75.54
24	2.23	53.34	1.56	66.35	1.53	55.68	1.45	49.14	2.13	77.67
25	2.22	55.56	1.49	67.84	1.48	57.16	1.41	50.55	2.03	79.70
26	2.22	57.78	1.42	69.26	1.43	58.59	1.37	51.93	1.93	81.63
27	2.22	60.00	1.36	70.62	1.38	59.97	1.34	53.26	1.84	83.47
28	2.23	62.23	1.30	71.92	1.33	61.30	1.30	54.56	1.74	85.21
29	2.22	64.45	1.24	73.16	1.29	62.59	1.26	55.82	1.64	86.85
30	2.22	66.67	1.18	74.34	1.25	63.84	1.23	57.05	1.55	88.40
31	2.22	68.89	1.13	75.47	1.21	65.05	1.19	58.24	1.45	89.85
32	2.23	71.12	1.08	76.55	1.17	66.22	1.16	59.40	1.35	91.20
33	2.22	73.34	1.03	77.58	1.13	67.35	1.13	60.53	1.26	92.46
34	2.22	75.56	.98	78.56	1.09	68.44	1.10	61.63	1.16	93.62
35	2.22	77.78	.94	79.50	1.05	69.49	1.07	62.69	1.06	94.68
36	2.22	80.00	.90	80.40	1.02	70.51	1.04	63.73	.97	95.65
37	2.23	82.23	.86	81.26	.98	71.49	1.01	64.74	.87	96.52
38	2.22	84.45	.82	82.08	.95	72.44	.98	65.72	.77	97.29
39	2.22	86.67	.78	82.86	.92	73.36	.95	66.67	.68	97.97
40	2.22	88.89	.75	83.61	.89	74.25	.93	67.59	.58	98.55
41	2.23	91.12	.72	84.33	.86	75.11	.90	68.49	.48	99.03
42	2.22	93.34	.69	85.02	.83	75.94	.88	69.37	.39	99.42
43	2.22	95.56	.66	85.68	.80	76.74	.85	70.22	.29	99.71
44	2.22	97.78	.63	86.31	.78	77.52	.83	71.05	.19	99.90
45	2.22	100.00	.60	86.91	.75	78.27	.80	71.85	.10	100.00
50-Year Life										
1	2.00	2.00	4.00	4.00	3.00	3.00	2.50	2.50	3.92	3.92
2	2.00	4.00	3.84	7.84	2.91	5.91	2.44	4.94	3.85	7.77
3	2.00	6.00	3.69	11.53	2.82	8.73	2.38	7.31	3.76	11.53
4	2.00	8.00	3.54	15.07	2.74	11.47	2.32	9.63	3.69	15.22
5	2.00	10.00	3.39	18.46	2.66	14.13	2.26	11.89	3.60	18.82
6	2.00	12.00	3.26	21.72	2.58	16.71	2.20	14.09	3.53	22.35
7	2.00	14.00	3.14	24.86	2.50	19.21	2.15	16.24	3.45	25.80
8	2.00	16.00	3.00	27.86	2.42	21.63	2.09	18.33	3.38	29.18
9	2.00	18.00	2.89	30.75	2.35	23.98	2.04	20.38	3.29	32.47
10	2.00	20.00	2.77	33.52	2.28	26.26	1.99	22.37	3.22	35.69

¶228

Comparative Depreciation Tables *(continued)*

Year	Straight-Line		200%-Declining-Balance		150%-Declining-Balance		125%-Declining-Balance		Sum-of-Digits	
	Annual %	Cum. %	Annual %	Cum. %	Annual %	Cum. %	Annual %	Cum. %	Annual %	Cum. %
50-Year Life *(continued)*										
11	2.00	22.00	2.66	36.18	2.21	28.47	1.94	24.31	3.15	38.82
12	2.00	24.00	2.55	38.73	2.15	30.62	1.89	26.20	3.06	41.88
13	2.00	26.00	2.45	41.18	2.08	32.70	1.84	28.05	2.98	44.86
14	2.00	28.00	2.35	43.53	2.02	34.72	1.80	29.84	2.91	47.77
15	2.00	30.00	2.26	45.79	1.96	36.68	1.75	31.60	2.82	50.59
16	2.00	32.00	2.17	47.96	1.90	38.58	1.71	33.31	2.74	53.33
17	2.00	34.00	2.08	50.04	1.84	40.42	1.67	34.98	2.67	56.00
18	2.00	36.00	2.00	52.04	1.79	42.21	1.63	36.60	2.59	58.59
19	2.00	38.00	1.92	53.96	1.73	43.94	1.58	38.19	2.51	61.10
20	2.00	40.00	1.84	55.80	1.68	45.62	1.55	39.73	2.43	63.53
21	2.00	42.00	1.77	57.57	1.63	47.25	1.51	41.24	2.35	65.88
22	2.00	44.00	1.70	59.27	1.58	48.83	1.47	42.71	2.28	68.16
23	2.00	46.00	1.62	60.89	1.53	50.36	1.43	44.14	2.19	70.35
24	2.00	48.00	1.57	62.46	1.49	51.85	1.40	45.54	2.12	72.47
25	2.00	50.00	1.50	63.96	1.44	53.29	1.36	46.90	2.04	74.51
26	2.00	52.00	1.44	65.40	1.40	54.69	1.33	48.23	1.96	76.47
27	2.00	54.00	1.39	66.79	1.36	56.05	1.29	49.52	1.87	78.35
28	2.00	56.00	1.33	68.12	1.32	57.37	1.26	50.78	1.81	80.16
29	2.00	58.00	1.27	69.39	1.28	58.64	1.23	52.01	1.72	81.88
30	2.00	60.00	1.22	70.61	1.24	59.89	1.20	53.21	1.65	83.53
31	2.00	62.00	1.18	71.79	1.20	61.09	1.17	54.38	1.57	85.10
32	2.00	64.00	1.13	72.92	1.17	62.26	1.14	55.52	1.49	86.59
33	2.00	66.00	1.08	74.00	1.13	63.39	1.11	56.63	1.41	88.00
34	2.00	68.00	1.04	75.04	1.10	64.49	1.08	57.72	1.33	89.33
35	2.00	70.00	1.00	76.04	1.07	65.56	1.06	58.78	1.26	90.59
36	2.00	72.00	.96	77.00	1.03	66.59	1.03	59.81	1.18	91.77
37	2.00	74.00	.92	77.92	1.00	67.59	1.00	60.81	1.09	92.86
38	2.00	76.00	.88	78.80	.97	68.56	.98	61.79	1.02	93.88
39	2.00	78.00	.85	79.65	.94	69.50	.96	62.75	.94	94.82
40	2.00	80.00	.81	80.46	.92	70.42	.93	63.68	.87	95.69
41	2.00	82.00	.78	81.24	.89	71.31	.91	64.58	.78	96.47
42	2.00	84.00	.75	81.99	.86	72.17	.89	65.47	.71	97.18
43	2.00	86.00	.72	82.71	.84	73.01	.86	66.33	.62	97.80
44	2.00	88.00	.69	83.40	.81	73.82	.84	67.18	.55	98.35
45	2.00	90.00	.67	84.07	.79	74.61	.82	68.00	.47	98.82
46	2.00	92.00	.64	84.71	.76	75.37	.80	68.80	.40	99.22
47	2.00	94.00	.61	85.32	.74	76.11	.78	69.58	.31	99.53
48	2.00	96.00	.59	85.90	.72	76.83	.76	70.34	.24	99.77
49	2.00	98.00	.57	86.47	.70	77.53	.74	71.08	.15	99.92
50	2.00	100.00	.54	87.01	.67	78.20	.72	71.80	.08	100.00

¶228

Composite Rates *(Rev. Proc. 62-21)*

(These rates may be used for property placed in service before 1970)

Group One: Guidelines for Depreciable Assets
Used by Business in General

1. Office Furniture, Fixtures, Machines, and Equipment .... 10 years

Includes furniture and fixtures which are not a structural component of the building, and machines and equipment used in the preparation of papers or data. Includes such assets as desks; files; safes; typewriters; accounting, calculating, and data processing machines; communications, duplicating and copying equipment.

2. Transportation Equipment

Includes the following types of transportation equipment:

    (a) Aircraft (air frames and engines, except aircraft of air transport companies) .... 6 years

    (b) Automobiles, including taxis .... 3 years

    (c) Buses .... 9 years

    (d) General-purpose trucks:

        Light (actual unloaded weight less than 13,000 pounds) .... 4 years

        Heavy (actual unloaded weight 13,000 pounds or more) .... 6 years

    (e) Railroad cars (except cars of railroad companies) .... 15 years

    (f) Tractor units (over-the-road) .... 4 years

    (g) Trailers and trailer-mounted containers .... 6 years

    (h) Vessels, barges, tugs, and similar water transportation equipment .... 18 years

3. Land Improvements .... 20 years

Includes land improvements such as paved surfaces, sidewalks, canals, waterways, drainage facilities and sewers, wharves, bridges, all fences except farm fences, landscaping, shrubbery, and similar improvements. Includes agricultural land improvements not classified as soil and water conservation expenditures under the Internal Revenue Code of 1954.

Excludes land improvements which are the major asset of a business, such as cemeteries or golf courses. The depreciable life of such land improvements shall be determined according to the particular facts and circumstances.

Excludes land improvements of electric, gas, steam, and water utilities; telephone and telegraph companies; and pipeline, water, and rail carriers. (These improvements are covered under Group Four.)

4. Buildings

Includes the structural shell of the building and all integral parts thereof. Includes equipment which services normal heating, plumbing, air conditioning, fire prevention, and power requirements, and equipment such as elevators and escalators.

Excludes special-purpose structures which are an integral part of the production process and which, under normal practice, are replaced contemporaneously with the equipment which they house, support, or serve. Nonindustrial and general-purpose industrial buildings, such as warehouses, storage facilities, general factory buildings, and commercial buildings, are not special-purpose structures. Special-purpose structures shall be classified with the equipment which they house, support, or serve, and their depreciable lives determined by reference to the appropriate guidelines for the particular industries.

Composite Rates *(Rev. Proc. 62-21) (continued)*

Group One: Guidelines for Depreciable Assets Used by Business in General *(continued)*

Type of Building
Apartments ....40 years
Banks ....50 years
Dwellings ....45 years
Factories ....45 years
Garages ....45 years
Grain Elevators ....60 years
Hotels ....40 years
Loft Buildings..50 years
Machine Shops...45 years
Office Buildings ....45 years
Stores ....50 years
Theaters ....40 years
Warehouses ....60 years

5. Subsidiary Assets[1]

Includes equipment such as jigs, dies, molds, and patterns; returnable containers and pallets; crockery, glassware, linens, and silverware; and other subsidiary assets which are commonly and properly accounted for separately from those assets falling within the guideline classes in Group Two, Three, or Four.

Where assets in this class are accounted for under a method of depreciation using a life expressed in terms of years, [a] the life shall be determined according to the facts and circumstances.

Group Two: Guidelines for Nonmanufacturing Activities, Excluding Transportation, Communications, and Public Utilities

In general, a single guideline class is specified for each industry included in this group. This single guideline class includes all depreciable property that is not covered by another guideline class. Thus, a single industry guideline class includes production machinery and equipment; power plant machinery and equipment; special equipment; and special-purpose structures (as defined in guideline class 4 under Group One).

Where more than one guideline class is specified for a particular industry, each guideline class covers that portion of the total depreciable property appropriate to the class.

The guideline classes in this group exclude depreciable assets covered under Group One.

---

[1]Amendment I, published in I.R.B. 1962-46, 23, making various changes in Part 1 has been incorporated herein; this amendment was released as Technical Information Release 405, dated October 19, 1962.

[a]These items are more usually and properly accounted for under a method of accounting other than a method of depreciation using a life expressed in terms of years. The method used by the taxpayer may be continued if it is consistently used and clearly reflects income. It should be noted that the cost (or other basis) of any asset used in a trade or business and having a useful life of one year or less may be deducted currently and is not subject to depreciation.

¶229

Composite Rates *(Rev. Proc. 62-21) (continued)*

Group Two: Guidelines for Nonmanufacturing Activities, Excluding Transportation, Communications, and Public Utilities *(continued)*

1. Contract Construction

Includes general building, special trade, heavy construction and marine contractors.

(a) General Contract Construction .... 5 years

Excludes assets used only in marine contract construction.

(b) Marine Contract Construction..12 years

Includes assets used only in marine contract construction.

2. Recreation and Amusement ....10 years

Includes recreation, entertainment and amusement establishments, such as bowling alleys, billiard and pool establishments, theaters, concert halls, and amusement parks.

Excludes facilities which consist primarily of specialized land improvements or structures, such as golf courses, swimming pools, tennis courts, sports stadia, and race tracks. The depreciable life of such facilities shall be determined according to the particular facts and circumstances.

3. Services ....10 years

Includes the providing of personal services such as those offered by hotels and motels, laundry and dry cleaning establishments, beauty and barber shops, photographic studios and mortuaries. Includes the providing of professional services such as those offered by doctors, dentists, lawyers, accountants, architects, engineers, and veterinarians. Includes the providing of repair and maintenance services.

4. Wholesale and Retail Trade ....10 years

Includes purchasing, selling, and brokerage activities at both the wholesale and retail level and related assembling, sorting and grading of goods.

Note: For specific lives in Manufacturing (Group Three) and Utilities (Group Four), see *Rev. Proc. 62-21.*

[¶230]

Class Life System

This system supersedes the class lives of *Rev. Proc. 62-21* and cannot be used for pre-1971 assets. It is based on vintage accounts which are depreciation accounts including all new or used tangible property placed in service for the year listed. Normally there would be a separate vintage account for each year for all items falling within a "guideline class" as defined by *Rev. Proc. 72-10*, but there may be more than one vintage account for similar assets. Then, the "depreciation range" (which is similar to useful life) for all the assets in the vintage account is from 80 to 120% (rounded to the nearest half-year) of the guideline period for that class of assets, and the account as a whole is depreciated annually using one of the recognized depreciation methods. These are straight-line, 200%-declining-balance, or sum-of-the-years-digits where all property in a vintage account is new, and the selected depreciation range is more than three years; straight-line or 150%-declining-balance where some of the property is used, but the selected depreciation range is more than three years; and straight-line where the selected depreciation range is less than three years.

[¶231]

Checklist of Basic Class Life Rules

1. You can't use ADR for assets put into service before 1971.
2. You have a choice of averaging conventions to determine when an asset is first put into service.
3. You must complete a Form 4832 in order to elect ADR.
4. You can't combine §1245 and §1250 property in a single vintage account.
5. If you amortize lessee's improvements over the term of a lease, the improvements are excluded from election under ADR.
6. Property is placed in service when it is in a "state of readiness and availability" and not before.
7. If you trade in ADR property for new ADR property the adjusted basis of the old property (plus any additional consideration) is the basis of the new property.
8. If a trade-in of an ADR asset is tax free, the unadjusted basis of the old asset is removed from the unadjusted basis of the vintage account.
9. If you elect ADR for any eligible property, you must elect for all eligible property (except in special circumstances).
10. You may exclude used property which is otherwise eligible from your ADR election if its value exceeds 10% of the unadjusted basis of all eligible property placed in service in your election year.
11. Regardless of the method of depreciation used, salvage value is not taken into account under ADR. However, salvage value does set a limit on the amount that may be depreciated.

[¶232]

First-Year Convention Under Class Life Rules

If you elect to use ADR, you must also specify the "first-year convention" you intend to use. This is simply the method you choose for averaging the depreciation period for the first year in which assets are placed in service. This is necessary because not all the assets put in service in a given year are put into service at the same time.

There are two basic conventions you can use: the half-year convention and the modified half-year convention. There is also a short-year convention, but this is used only for election years shorter than 12 full months. Regardless of the convention adopted, all vintage accounts in a given year must use the same convention. However, accounts in subsequent years may use other conventions.

Depreciation of the vintage accounts for the first year under the two conventions is computed as follows:

Half-Year

1. All property in the account is treated as having been placed in service on the first day of the second half of the taxable year.

2. All extraordinary retirements are treated as occurring on the first day of the second half of the taxable year.

Modified Half-Year

1. All property in the account is treated as having been placed in service according to the table below.

2. All extraordinary retirements are treated as occurring in accordance with the table below.

Modified Half-Year Table

Actual Date Placed in Service or Retired	Treated as Placed in Service or Retired
1(a) Placed in service during first half of taxable year.	1(a) First day of taxable year.
(b) Retired during first half of year.	(b) First day of year.
(c) Retired during second half of year.	(c) First day of second half of year.
2(a) Placed in service during second half of taxable year.	2(a) First day of the next taxable year.
(b) Retired during first half of year.	(b) First day of second half of year.
(c) Retired during second half of year.	(c) First day of the next year.

As can be seen, the half-year convention is better for you if you have actually placed most of the assets in service in the second half of the taxable year, and the modified half-year convention is better if you have actually placed most of the assets in service in the first half of the year.

ADR Classes and Guidelines

Asset guideline class	Description of assets included	Asset depreciation range (in years)			Annual asset guideline repair allowance percentage
		Lower limit	Asset guideline period	Upper limit	
00.0	Depreciable Assets Used in All Business Activities, Except as Noted:				
00.2	Transportation Equipment:				
00.21	Aircraft (airframes and engines) except aircraft of air transportation companies	5	6	7	14.0
00.22	Automobiles, taxis	2.5	3	3.5	16.5
00.23	Buses	7	9	11.0	11.5
00.24	General purpose trucks, including concrete ready-mix trucks and ore trucks for use over-the-road:				
00.241	Light (actual unloaded weight less than 13,000 pounds)	3	4	5	16.5
00.242	Heavy (actual unloaded weight 13,000 pounds or more)	5	6	7	10.0
00.25	Railroad cars and locomotives, except those owned by railroad transportation companies	12	15	18	8.0
00.26	Tractor units used over-the-road	3	4	5	16.5
00.27	Trailers and trailer-mounted containers	5	6	7	10.0
00.28	Vessels, barges, tugs, and similar water transportation equipment, except those used in marine contract construction	14.5	18	21.5	6.0
00.3	Land Improvements:[1]				

Improvements directly to or added to land that are more often than not directly related to one or another of the specific classes of economic activity specified below. Includes only those depreciable land improvements which have a limited period of use in the trade or business, the length of which can be reasonably estimated for the particular improvement. That is, general grading of land, such as in the case of cemeteries, golf courses and general site grading and leveling costs not directly related to buildings or other structural improvements to be added, are not depreciable or included in this class, but such costs are added to the cost basis of the land.

Includes paved surfaces such as sidewalks and roads, canals, waterways, drainage facilities and sewers; wharves and docks; bridges; all fences except those included in specific classes described below (i.e., farm and railroad fences); landscaping, shrubbery and similar improvements; radio and television transmitting towers, and other inherently permanent physical structures added to land except buildings and their structural components.

[1]This class is established for a three-year transition period in accordance with Section 109(e)(1) of the Revenue Act of 1971 (P.O. 92-178, I.R.B. 1972-3, 14) and will be in effect for the period beginning January 1, 1971 and ending January 1, 1974 or at such earlier date as of which asset classes incorporating the assets herein described are represcribed or modified.

ADR Classes and Guidelines *(continued)*

Asset guideline class	Description of assets included	Asset depreciation range (in years)			Annual asset guideline repair allowance percentage
		Lower limit	Asset guideline period	Upper limit	
	Excludes land improvements of electric, gas, steam and water utilities; telephone and telegraph companies; and pipeline, water and rail carriers which are assets covered by asset guideline classes specific to their respective classes of economic activity		20		
01.0 to 79.0	Depreciable Assets Used in the Following Activities:[2]				
01.0	Agriculture:				
	Includes only such assets as are identified below and that are used in the production of crops or plants, vines and trees (including forestry); the keeping, grazing, or feeding of livestock for animal products (including serums), for animals increase, or value increase; the operation of dry lot or farm dairies, nurseries, greenhouses, sod farms, mushroom cellars, cranberry bogs, apiaries, and fur farms; the production of bulb, flower, and vegetable seed crops; and the performance of agricultural, animal husbandry and horticultural services.				
01.1	Machinery and equipment, including grain bins and fences but no other land improvements	8	10	12	11.0
01.2	Animals:				
01.21	Cattle, breeding or dairy	5.5	7	8.5	
01.22	Horses, breeding or work	8	10	12	
01.23	Hogs, breeding	2.5	3	3.5	
01.24	Sheep and goats, breeding	4	5	6	
01.3	Farm buildings	20	25	30.0	5.0
10.0	Mining:				
	Includes assets used in the mining and quarrying of metallic and non-metallic minerals (including sand, gravel, stone, and clay) and the milling beneficiation and other primary preparation of such materials	8	10	12	6.5
13.0	Petroleum and natural gas production and related activities:				
13.1	Drilling of oil and gas wells:				
	Includes assets used in the drilling of onshore oil and gas wells on a contract, fee or other basis and the provision of geophysical and other exploration services; and the provision of such oil and gas field services as				

---

[2]All asset classes defined below include subsidiary assets within the meaning of Section 109(e)(2) of the Revenue Act of 1971 whenever such assets are used in the economic activities specified. However, in accordance with the provisions of that section of the Act, during the period beginning on January 1, 1971 and ending January 1, 1974 or such earlier date as of which asset classes incorporating the subsidiary assets are represcribed or modified, taxpayers may exclude from an election all subsidiary assets in a specified class provided that at least 3 percent of all the assets placed in service in the class during the taxable year subsidiary assets. See Section 1.167(a)-11(b)(5)(vii) for application of 3 percent test.

¶233

ADR Classes and Guidelines *(continued)*

Asset guide-line class	Description of assets included	Asset depreciation range (in years)			Annual asset guide-line repair allow-ance percent-age
		Lower limit	Asset guide-line period	Upper limit	
	chemical treatment, plugging and abandoning of wells and cementing or perforating well casings; but not including assets used in the performance of any of these activities and services by integrated petroleum and natural gas producers for their own account	5	6	7	10.0
13.2	Exploration for petroleum and natural gas deposits: Includes assets used for drilling of wells and production of petroleum and natural gas, including gathering pipelines and related storage facilities, when these are related activities undertaken by petroleum and natural gas producers	11	14	17	4.5
13.3	Petroleum refining: Includes assets used for the distillation, fractionation, and catalytic cracking of crude petroleum into gasoline and its other components	13	16	19	7.0
13.4	Marketing of petroleum and petroleum products: Includes assets used in marketing, such as related storage facilities and complete service stations, but not including any of these facilities related to petroleum and natural gas trunk pipelines	13	16	19	4.0
15.0	Contract construction: Includes such assets used by general building, special trade, heavy construction and marine contractors; does not include assets used by companies in performing construction services on their own account.				
15.1	Contract construction other than marine	4	5	6	12.5
15.2	Marine contract construction .... Includes floating, self-propelled and other drilling platforms used in offshore drilling for oil and gas.	9.5	12	14.5	5.0
20.0	Manufacture of foods and beverages for human consumption, and certain related products, such as manufactured ice, chewing gum, vegetable and animal fats and oils, and prepared feeds for animals and fowls:				
20.1	Grain and grain mill products: Includes assets used in the production of flours, cereals, livestock feeds, and other grain and grain mill products	13.5	17	20.5	6.0
20.2	Sugar and sugar products: Includes assets used in the production of raw sugar, syrup or finished sugar from sugar cane or sugar beets	14.5	18	21.5	4.5
20.3	Vegetable oils and vegetable oil products: Includes assets used in the production of oil from vegetable materials and the manufacture of related vegetable oil products	14.5	18	21.5	3.5

¶233

ADR Classes and Guidelines *(continued)*

Asset guide-line class	Description of assets included	Asset depreciation range (in years)			Annual asset guide-line repair allow-ance percent-age
		Lower limit	Asset guide-line period	Upper limit	
20.4	All other food and kindred products: Includes assets used in the production of foods, beverages and related production not included in classes 20.1, 20.2 and 20.3	9.5	12	14.5	5.5
20.5	Manufacture of food and beverages—special handling devices: Includes assets defined as specialized materials handling devices such as returnable pellets, pelletized containers, and fish processing equipment including boxes, baskets, carts, and flaking trays used in activities as defined in classes 20.1, 20.2, 20.3, 20.4. Special handling devices are specifically designed for the handling of particular products and have no significant utilitarian value and cannot be adapted to further or different use after changes or improvements are made in the design of the particular product handled by the special devices. Does not include general purpose small tools such as wrenches and drills, both hand and power-driven, and other general purpose equipment such as conveyors, transfer equipment, and materials handling devices	3	4	5	20.0
21.0	Manufacture of tobacco and tobacco products: Includes assets used in the production of cigarettes, cigars, smoking and chewing tobacco, snuff and other tobacco products	12	15	18	5.0
22.0	Manufacture of textile mill products:				
22.1	Knitwear and knit products: Includes assets used in the production of knit fabrics, knit apparel, and yarns processed for knitting, such as boarding machines, dryers, knitting machines, loopers, warpers, winders, seaming machines, twisting machines, twist setting machines, texturizing machines, and collection system equipment	7	9	11	7.0
22.2	Textile mill products, except knitwear: Includes assets used in the production of spun yarn and woven or non-woven fabrics, mattresses, carpets, rugs, pads, sheets, and of other products of natural or synthetic fibers, such as preparatory equipment for fibers, and machinery for carding, combing, drawing, roving, spinning, twisting, warping, winding, slashing, and weaving	11	14	17	4.5
22.3	Finishing and dyeing:				

¶233

ADR Classes and Guidelines *(continued)*

Asset guide-line class	Description of assets included	Asset depreciation range (in years)			Annual asset guide-line repair allow-ance percent-age
		Lower limit	Asset guide-line period	Upper limit	
	Includes assets used in the finishing and dyeing of natural and synthetic fibers, yarns, fabrics including knit materials, and knit apparel, such as assets used for washing, bleaching, finishing, printing and dye-ing, and drying	9.5	12	14.5	5.5
23.0	Manufacture of apparel and other finished products:				
	Includes assets used in the production of clothing and fabricated textile products by the cutting and sewing of woven fabrics, other textile products and furs; but does not include assets used in the manufacture of apparel from rubber and leather	7	9	11	7.0
24.0	Manufacture of lumber and wood products:				
24.1	Cutting of timber:				
	Includes logging machinery and equipment and road building equipment used by logging and sawmill operators and pulp manufacturers on their own ac-count	5	6	7	10.0
24.2	Sawing of dimensional stock from logs:				
	Includes machinery and equipment installed in perma-nent or well-established sawmills	8	10	12	6.5
24.3	Sawing of dimensional stock from logs:				
	Includes machinery and equipment installed in sawmills characterized by temporary foundations and a lack, or minimum amount, of lumber-handling, drying, and residue disposal equipment and facilities	5	6	7	10.0
24.4	Manufacture of lumber, wood products, and furniture:				
	Includes assets used in the production of plywood, hardboard, flooring, veneers, furniture and other wood products, including the treatment of poles and timber	8	10	12	6.5
26.0	Manufacture of paper and allied products:				
26.1	Manufacture of pulps from wood and other cellulose fibers and rags:				
	Includes assets used in the manufacture of paper and paperboard, but does not include the assets used in pulpwood logging nor the manufacture of hardboard .	13	16	19	4.5
26.2	Manufacture of paper and paperboard:				
	Includes assets used in the production of converted products such as paper coated off the paper machines, paper bags, paper boxes, and envelopes	9.5	12	14.5	5.5
27.0	Printing, publishing and allied industries:				
	Includes assets used in printing by one or more of the com-mon processes, such as letterpress, lithography, gravure, or screen; the performance of services for the printing				

¶233

ADR Classes and Guidelines *(continued)*

Asset guide-line class	Description of assets included	Asset depreciation range (in years)			Annual asset guide-line repair allow-ance percent-age
		Lower limit	Asset guide-line period	Upper limit	
	trade, such as bookbinding, typesetting, engraving, photoengraving, and electrotyping; and the publication of newspapers, books, and periodicals, whether or not carried out in conjunction with printing	9	11	13	5.5
28.0	Manufacture of chemicals and allied products:				
	Includes assets used in the manufacture of basis chemicals such as acids, alkalies, salts, and organic and inorganic chemicals; chemical products to be used in further manufacture, such as synthetic fibers and plastics materials, including petro-chemical processing beyond that which is ordinarily a part of petroleum refining; and finished chemical products, such as pharmaceuticals, cosmetics, soaps, fertilizers, paints and varnishes, explosives, and compressed and liquified gases. Does not include assets used in the manufacture of finished rubber and plastic products or in the production of natural gas products, butane, propane, and byproducts of natural gas production plants ..	9	11	13	5.5
30.0	Manufacture of rubber and plastics products:				
30.1	Manufacture of rubber products:				
	Includes assets used for the production of products from natural, synthetic, or reclaimed rubber, gutta percha, balata, or gutta siak, such as tires, tubes, rubber footwear, mechanical rubber goods, heels and soles, flooring, and rubber sundries; and in the recapping, retreading, and rebuilding of tires	11	14	17	5.0
30.11	Manufacture of rubber products—special tools and devices:				
	Includes assets defined as special tools, such as jigs, dies, mandrels, molds, lasts, patterns, specialty containers, pallets, shells and tire molds and accessory parts such as rings and insert plates used in activities as defined in Class 30.1.				
	Does not include tire building drums and accessory parts and general purpose small tools such as wrenches and drills, both power and hand-driven, and other general purpose equipment such as conveyors and transfer equipment	3	4	5	
30.2	Manufacture of miscellaneous finished plastics products:				
	Includes assets used in the manufacture of plastics products and the molding of primary plastics for the trade. Does not include assets used in the manufacture of basic plastics materials nor the manufacture of phonograph records	9	11	13	5.5

¶233

ADR Classes and Guidelines *(continued)*

Asset guide-line class	Description of assets included	Asset depreciation range (in years)			Annual asset guide-line repair allow-ance percent-age
		Lower limit	Asset guide-line period	Upper limit	
30.21	Manufacture of miscellaneous finished plastic products —special tools: Includes assets defined as special tools such as jigs, dies, fixtures, molds, patterns, gauges, and specialty transfer and shipping devices, used in activities as defined in Class 30.2. Special tools are specifically designed for the production or processing of particular parts and have no significant utilitarian value and cannot be adapted to further or different use after changes or improvements are made in the model design of the particular part produced by the special tools. Does not include general purpose small tools, such as wrenches and drills, both hand and power-driven, and other general purpose equipment such as conveyors, transfer equipment, and materials handling devices	3	3.5	4	5.5
31.0	Manufacture of leather: Includes assets used in the tanning, currying, and finishing of hides and skins; the processing of fur pelts; and the manufacture of finished leather products, such as footwear, belting, apparel, luggage and similar leather goods .	9	11	13	5.5
32.0	Manufacture of stone, clay, glass, and concrete products:				
32.1	Manufacture of glass products: Includes assets used in the production of flat, blown, or pressed products of glass, such as float and window and window glass, glass containers, glassware, and fiberglass. Does not include assets used in the manufacture of lenses	11	14	17	12.0[6]
32.11	Manufacture of glass products—special tools: Includes assets defined as special tools such as molds, patterns, pallets, and specialty transfer and shipping devices such as steel racks to transport automotive glass, used in activities as defined in Class 32.1. Special tools are specifically designed for the production or processing of particular parts and have no significant utilitarian value and cannot be adapted to further or different use after changes or improvements are made in the model design of the particular part produced by the special tools. Does not include general purpose small tools such as wrenches and drills, both hand power-driven, and other general purpose equipment such as conveyors, transfer equipment, and materials handling devices	2	2.5	3	10.0

¶233

ADR Classes and Guidelines *(continued)*

Asset guide-line class	Description of assets included	Asset depreciation range (in years)			Annual asset guide-line repair allow-ance percent-age
		Lower limit	Asset guide-line period	Upper limit	
32.2	Manufacture of cement: Includes assets used in the production of cement, but does not include any assets used in the manufacture of concrete and concrete products nor in any mining or extraction process	16	20	24	3.0
32.3	Manufacture of other stone and clay products: Includes assets used in the manufacture of products from materials in the form of clay and stone, such as brick, tile and pipe; pottery and related products, such as vitreous-china, plumbing fixtures, earthenware and ceramic insulating materials; and also includes assets used in manufacture of concrete and concrete products. Does not include assets used in any mining or extraction processes. Includes assets used in the smelting and refining of ferrous and nonferrous metals from ore, pig, or scrap, the rolling, drawing, and alloying of ferrous and nonferrous metals; the manufacture of castings, forgings, and other basic products of ferrous and nonferrous metals; and the manufacture of nails, spikes, structural shapes, tubing, and wire and cable.	12	15	18	4.5
33.0	Manufacture of primary metals: Includes assets used in the smelting and refining of ferrous and nonferrous metals from ore, pig, or scrap, the rolling, drawing, and alloying of ferrous and nonferrous metals; the manufacture of castings, forgings, and other basic products of ferrous and nonferrous metals; and the manufacture of nails, spikes, structural shapes, tubing, and wire and cable.				
33.1	Ferrous metals	14.5	18	21.5	8.0
33.11	Ferrous metals—special tools: Includes assets defined as special tools such as dies, jigs, molds, patterns, fixtures, gauges, and drawings concerning such special tools used in the activities as defined in Class 33.1, Ferrous metals. Special tools are specifically designed for the production or processing of particular products or parts and have no significant utilitarian value and cannot be adapted to further or different use after changes or improvements are made in the model design of the particular part produced by the special tools. Does not include general purpose small tools, such as wrenches and drills, both hand and power-driven, and other general pur-				

ADR Classes and Guidelines *(continued)*

Asset guide-line class	Description of assets included	Asset depreciation range (in years)			Annual asset guide-line repair allow-ance percent-age
		Lower limit	Asset guide-line period	Upper limit	
	pose equipment such as conveyors, transfer equipment, and materials handling devices.				
	Rolls, mandrels and refractories are not included in Class 33.11 but are included in Class 33.1	5	6.5	8	4.0
33.2	Nonferrous metals	11	14	17	4.5
33.21	Nonferrous metals—special tools:				
	Includes assets defined as special tools such as dies, jigs, molds, patterns, fixtures, gauges, and drawings concerning such special tools used in the activities as defined in Class 33.2. Nonferrous metals. Special tools are specifically designed for the production or processing of particular products or parts and have no significant utilitarian value and cannot be adapted to further or different use after changes or improvements are made in the model design of the particular part produced by the special tools. Does not include general purpose small tools such as wrenches and drills, both hand and power-driven, and other general purpose equipment such as conveyors, transfer equipment, and materials handling devices.				
	Rolls, mandrels and refractories are not included in Class 33.21 but are included in Class 33.2	5	6.5	8	4
34.0	Manufacture of fabricated metal products:				
	Includes assets used in the production of metal cans, tin-ware, nonelectric heating apparatus, fabricated structural metal products, metal stampings and other ferrous and nonferrous metal and wire products not elsewhere classified	9.5	12	14.5	6.0
34.01	Manufacture of fabricated metal products—special tools:				
	Includes assets defined as special tools such as dies, jigs, molds, patterns, fixtures, gauges, and returnable containers and drawings concerning such special tools used in the activities as defined in Class 34.0. Special tools are specifically designed for the production or processing of particular machine components, products or parts, and have no significant utilitarian value and cannot be adapted to further or different use after changes or improvements are made in the model design of the particular part produced by the special tools. Does not include general purpose small tools such as wrenches and drills, both hand and power-driven, and other general purpose equipment such as conveyors, transfer equipment, and materials handling devices	2.5	3.0	3.5	3.5

¶233

ADR Classes and Guidelines (continued)

Asset guide-line class	Description of assets included	Asset depreciation range (in years)			Annual asset guide-line repair allow-ance percent-age
		Lower limit	Asset guide-line period	Upper limit	
35.0	Manufacture of machinery, except electrical and transportation equipment:				
35.1	Manufacture of metalworking machinery: Includes assets used in the production of metal cutting and forming machines, special dies, tools, jigs and fixtures, and machine tool accessories	9.5	12	14.5	5.5
35.11	Manufacture of metalworking machinery—special tools: Includes assets defined as special tools, such as jigs, dies, fixtures, molds, patterns, gauges, and specialty transfer and shipping devices, used in activities as defined in Class 35.1. Special tools are specifically designed for the production or processing of particular machine components and have no significant utilitarian value and cannot be adapted to further or different use after changes or improvements are made in the model design of the particular part produced by the special tools. Does not include general purpose small tools such as wrenches and drills, both hand and power-driven, and other general purpose equipment such as conveyors, transfer equipment, and materials handling devices	5	6	7	12.5
35.2	Manufacture of other machines: Includes assets used in the production of such machinery as engines and turbines; farm machinery, construction, and mining machinery; general and special industrial machines including office machines and non-electronic computing equipment; miscellaneous machines except electrical equipment and transportation equipment	9.5	12	14.5	5.5
35.21	Manufacture of other machines—special tools: Includes assets defined as special tools, such as jigs, dies, fixtures, molds, patterns, gauges, and specialty transfer and shipping devices, used in activities as defined in Class 35.2. Special tools are specifically designed for the production or processing of particular machine components and have no significant utilitarian value and cannot be adapted to further or different use after changes or improvements are made in the model design or the particular part produced by the special tools. Does not include general purpose small tools such as wrenches and drills, both hand and power-driven, and other general purpose equipment such as conveyors, transfer equipment, and materials handling devices	5	6.5	8	12.5

ADR Classes and Guidelines (continued)

Asset guide-line class	Description of assets included	Asset depreciation range (in years)			Annual asset guide-line repair allow-ance percent-age
		Lower limit	Asset guide-line period	Upper limit	
36.0	Manufacture of electrical machinery, equipment, and supplies: Includes assets used in the production of machinery, apparatus, and supplies for the generation, storage, transmission, transformation, and utilization of electrical energy.				
36.1	Manufacture of electrical equipment: Includes assets used in the production of such machinery as electric test and distributing equipment, electrical industrial apparatus, household appliances, electric lighting and wiring equipment; electronic components and accessories, phonograph records, storage batteries and ignition systems	9.5	12	14.5	5.5
36.11	Manufacture of electrical equipment—special tools: Includes assets defined as special tools such as jigs, dies, molds, patterns, fixtures, gauges, returnable containers, and specialty transfer devices used in activities as defined in Class 36.1. Special tools are specifically designed for the production or processing of particular machine components, products or parts, and have no significant utilitarian value and cannot be adapted to further or different use after changes or improvements are made in the model design of the particular part produced by the special tools. Does not include general purpose small tools such as wrenches and drills, both hand and power-driven and other general purpose equipment such as conveyors, transfer equipment, and materials handling devices	4	5	6	
36.2	Manufacture of electronic products: Includes assets used in the production of electronic detection, guidance, control, radiation, computation, test and navigation equipment and the components thereof. Does not include the assets of manufacturers engaged only in the purchase and assembly of components	6.5	8	9.5	7.5
37.0	Manufacture of transportation equipment: Includes assets used in the production of such machinery as vehicles and equipment for the transportation of passengers and cargo.				
37.1	Manufacture of motor vehicles:				
37.11	Motor vehicle manufacturing assets: Includes assets used in the manufacture and assembly of finished automobiles, trucks, trailers, motor homes, and buses. Does not include assets used in mining,				

¶233

ADR Classes and Guidelines *(continued)*

Asset guide-line class	Description of assets included	Asset depreciation range (in years)			Annual asset guide-line repair allow-ance percent-age
		Lower limit	Asset guide-line period	Upper limit	
	printing and publishing, production of primary metals, electricity, or steam, or the manufacture of glass, industrial chemicals, batteries, or rubber products, which are classified elsewhere. Includes assets used in manufacturing activities elsewhere classified other than those excluded above,[4] where such activities are incidental to and an integral part of the manufacture and assembly of finished motor vehicles such as the manufacture of parts and subassemblies of fabricated metal products, electrical equipment, textiles, plastics, leather, and foundry and forging operations	9.5	12	14.5	9.5
	Activities will be considered incidental to the manufacture and assembly of finished motor vehicles only if 75 percent or more of the value of the products produced under one roof are used for the manufacture and assembly of finished motor vehicles. Parts which are produced as a normal replacement stock complement in connection with the manufacture and assembly of finished motor vehicles are considered used for the manufacture and assembly of finished motor vehicles. Does not include assets used in the manufacture of component parts if these assets are used by taxpayers not engaged in the assembly of finished motor vehicles.				
37.12	Motor vehicle manufacturing subsidiary assets:	2.5	3	3.5	12.5
	Includes assets defined as special tools, such as jigs, dies, fixtures, molds, patterns, gauges, and specialty transfer and shipping devices, owned by manufacturers of finished motor vehicles and used in qualified activities as defined in Class 37.11. Special tools are specifically designed for the production or processing of particular motor vehicle components and have no significant utilitarian value and cannot be adapted to further or different use after changes or improvements are made in the model design of the particular part produced by the special tools. Does not include general purpose small tools such as wrenches and drills, both hand and power-driven, and other general purpose equipment such as conveyors, transfer equipment, and materials handling devices.				
37.2	Manufacture of aerospace products:				
	Includes assets used in the production of aircraft, spacecraft, rockets, missiles and their components parts	6.5	8	9.5	7.5
37.3	Ship and boat building:				

¶233

ADR Classes and Guidelines *(continued)*

Asset guide-line class	Description of assets included	Asset depreciation range (in years)			Annual asset guide-line repair allow-ance percent-age
		Lower limit	Asset guide-line period	Upper limit	
37.31	Ship and boat building machinery and equipment: Includes assets used in the manufacture and repair of ships, boats, caissons, drilling rigs and special fabrications not included in asset guideline class 37.32. Specifically includes all manufacturing and repairing machinery and equipment, including machinery and equipment used in the operation of assets included in asset guideline 37.32. Excludes buildings and their structural components	9.5	12	14.5	8.5
37.32	Ship and boat building dry docks and land improvements: Includes assets used in the manufacture and repair of ships, boats, caissons, drilling rigs and special fabrications not included in asset guideline class 37.31. Specifically includes floating and fixed dry docks, ship basins, graving docks, shipways, piers and all other land improvements such as water, sewer, and electric systems. Excludes buildings and their structural components	13	16	19	2.5
37.33	Ships and boat building—special tools: Includes assets defined as special tools such as dies, jigs, molds, patterns, fixtures, gauges, and drawings concerning such special tools used in the activities as defined in Classes 37.31 and 37.32. Special tools are specifically designed for the production or processing of particular machine components, products or parts, and have no significant utilitarian value and cannot be adopted to further or different use after changes or improvements are made in the model design of the particular part produced by the special tools, Does not include general purpose small tools such as wrenches and drills, both hand and power-driven, and other general purpose equipment such as conveyors, transfer equipment, and materials handling devices	5	6.5	8	0.5
37.4	Manufacture of railroad transportation equipment:				
37.41	Manufacture of locomotives: Includes assets used in building or rebuilding railroad locomotives (including mining and industrial locomotives). Does not include assets of railroad transportation companies or assets of companies which manufacture components of locomotives but do not manufacture finished locomotives	9	11.5	14	7.5
37.42	Manufacture of railroad cars: Includes assets used in building or rebuilding railroad freight or passenger cars (including rail transit cars). Does not include assets of railroad transportation com-				

ADR Classes and Guidelines (continued)

Asset guide-line class	Description of assets included	Asset depreciation range (in years)			Annual asset guide-line repair allow-ance percent-age
		Lower limit	Asset guide-line period	Upper limit	
	panies or assets of companies which manufacture compo-nents of railroad cars but do not manufacture finished railroad cars	9.5	12	14.5	5.5
38.0	Manufacture of professional, scientific, and controlling instru-ments; photographic and optical goods; watches and clocks: Includes assets used in the manufacture of mechanical measuring, engineering, laboratory and scientific research instruments, optical instruments and lenses; surgical, medical and dental instruments, equipment and supplies; ophthalmic goods, photographic equipment and supplies; and watches and clocks	9.5	12	14.5	5.5
39.0	Manufacture of products not elsewhere classified: Includes assets used in the production of jewelry; musical instruments; toys and sporting goods; pens, pencils, office and art supplies. Also includes assets used in production of motor picture and television films and tapes; as waste reduction plants; and in the ginning of cotton	9.5	12	14.5	5.5
40.0	Railroad transportation: Includes the assets identified below and which are used in the commercial and contract carrying of passengers and freight by rail. Excludes any nondepreciable assets in-cluded in Interstate Commerce Commission accounts enumerated for this class. Excludes the transportation as-sets included in Class 00.2 above.				
40.1	Railroad machinery and equipment .... Includes assets classified in the following Interstate Commerce Commission accounts: Road accounts: (16) Station and office buildings (freight han-dling machinery and equipment only) (25) TOFC/COFC terminals (freight handling machinery and equipment only) (26) Communication systems (27) Signals and interlockers (37) Roadway machines (44) Shop machinery Equipment accounts: (52) Locomotives (53) Freight train cars (54) Passenger train cars (55) Highway revenue equipment (57) Work equipment (58) Miscellaneous equipment	11	14	17	10.5

ADR Classes and Guidelines *(continued)*

Asset guide-line class	Description of assets included	Asset depreciation range (in years)			Annual asset guide-line repair allow-ance percent-age
		Lower limit	Asset guide-line period	Upper limit	
40.2	Railroad structures and similar improvements .... Includes assets classified in the following Interstate Commerce Commission road accounts:	24	30	36	5.0
	(6) Bridges, trestles, and culverts				
	(7) Elevated structure				
	(13) Fences, snowsheds, and signs				
	(16) Station and office buildings (stations and other operating structures only)				
	(17) Roadway buildings				
	(18) Water stations				
	(19) Fuel stations				
	(20) Shops and enginehouses				
	(25) TOFC/COFC terminals (operating structures only)				
	(31) Power transmission systems				
	(35) Miscellaneous structures				
	(39) Public improvements construction				
40.3	Railroad wharves and docks	16	20	24	5.5
	(23) Wharves and docks				
	(24) Coal and ore wharves				
40.5	Railroad power plant and equipment:				
	Electric generating equipment:				
40.51	Hydraulic	40	50	60	1.5
40.52	Nuclear	16	20	24	3.0
40.53	Steam	22.5	28	33.5	2.5
40.54	Steam, compressed air, and other power plant equipment	22.5	28	33.5	7.5
41.0	Motor transport-passengers: Includes assets used in the urban and interurban commercial and contract carrying of passengers by road, except the transportation assets included in Class 00.2 above	6.5	8	9.5	11.5
42.0	Motor transport-freight: Includes assets used in the commercial and contract carrying of freight by road, except the transportation assets included in Class 00.2 above	6.5	8	9.5	11.0
44.0	Water transportation: Includes assets used in the commercial and contract carrying of freight and passengers by water except the transportation assets included in Class 00.2 above	16	20	24	8.0
45.0	Air transport: Includes assets used in the commercial and contract carrying of passengers and freight by air	5	6	7	14.0

¶233

ADR Classes and Guidelines *(continued)*

Asset guideline class	Description of assets included	Asset depreciation range (in years)			Annual asset guideline repair allowance percentage
		Lower limit	Asset guideline period	Upper limit	
46.0	Pipeline transportation: Includes assets used in the private, commercial, and contract carrying of petroleum, gas, and other products by means of pipe conveyors. The trunk lines related storage facilities of integrated petroleum and natural gas producers are included in this class	17.5	22	26.5	3.0
48.0	Communication: Includes assets used in the furnishing of point-to-point communication services by wire or radio, whether intended to be received aurally or visually; and radio broadcasting and television.				
48.1	Telephone: Includes the assets identified below and which are used in the provision of commercial and contract telephonic services:				
48.11	Central office buildings: Assets intended to house central office equipment as defined in Federal Communications Commission Part 31 Account No. 212 whether section 1245 or section 1250 property	36	45	54	1.5
48.12	Central office equipment: Includes central office switching and related equipment as defined in Federal Communications Commission Part 31 Account No. 221	16	20	24	6.0
48.13	Station equipment: Includes such station apparatus and connections as teletypewriters, telephones, booths, private exchanges and comparable equipment as defined in Federal Communications Commission Part 31 Account Nos. 231, 232, and 234	8	10	12	10.0
48.14	Distribution plant: Includes such assets as pole lines, cable, aerial wire and underground conduits as are classified in underground conduits, and comparable equipment as defined in Federal Communications Commission Part 31 Account Nos. 241, 242.1, 242.2, 242.3, 242.4, 243, and 244	28	35	42	2.0
48.2	Radio and television broadcasting	5	6	7	10.0
48.3	Telegraph, ocean cable, and satellite communications: Includes communications-related assets used to provide domestic and international radio-telegraph, wire-telegraph, ocean-cable, and satellite communications services.				

¶233

ADR Classes and Guidelines (continued)

Asset guide-line class	Description of assets included	Asset depreciation range (in years)			Annual asset guide-line repair allow-ance percent-age
		Lower limit	Asset guide-line period	Upper limit	
48.31	Electric power generating and distribution systems.......... Includes assets used in the provision of electric power by generation, modulation, rectification, channelization, control, and distribution. Does not include these assets when they are installed on customers' premises.	15.0	19.0	23.0	—
48.32	High frequency radio and microwave systems .... Includes assets such as transmitters and receivers, antenna supporting structures, antennas, transmission lines from equipment to antenna, transmitter cooling systems, and control and · amplification equipment. Does not include cable and long-line systems.	10.5	13.0	15.5	—
48.33	Cable and long-line systems .... Includes assets such as transmission lines, pole lines, ocean cables, buried cable and conduit, repeaters, repeater stations, and other related assets. Does not include high frequency radio or microwave systems.	21.0	26.5	32.0	—
48.34	Central office control equipment............................... Includes assets for general control, switching, and monitoring of communications signals including electromechanical switching and channeling apparatus, multiplexing equipment, patching and monitoring facilities, in-house cabling, teleprinter equipment, and associated site improvements.	13.0	16.5	20.0	—
48.35	Computerized switching, channeling, and associated control Equipment .... Includes central office switching computers, interfacing computers, other associated specialized control equipment, and site improvements.	8.5	10.5	12.5	—
48.36	Satellite ground segment property .... Includes assets such as fixed earth station equipment, antennas, satellite communications equipment, and interface equipment. Does not include general purpose equipment or equipment used in satellite space segment property.	8.0	10.0	12.0	—
48.37	Satellite space segment property .... Includes satellites and equipment used for telemetry, tracking, control, and monitoring.	6.5	8.0	9.5	—
48.38	Equipment installed on customer's premises................. Includes assets installed on customer's premises, such as computers, terminal equipment, power generation and distribution systems, private switching center, teleprinters, facsimile equipment, and other associated and related equipment.	8.0	10.0	12.0	—

¶233

ADR Classes and Guidelines *(continued)*

Asset guide-line class	Description of assets included	Asset depreciation range (in years)			Annual asset guide-line repair allow-ance percent-age
		Lower limit	Asset guide-line period	Upper limit	
48.39	Support and service equipment................................	11.0	13.5	16.0	
	Includes assets used to support but not engage in communications. Includes store, warehouse, shop, tools, and test and laboratory assets.				
48.4	Cable television:				
	Includes communications—related assets used to provide cable television (community antenna television) services. Does not include assets used to provide subscribers with two-way communications services.				
48.41	Headend	9	11	13	5
	Includes assets such as towers, antennas, preamplifiers, converters, modulation equipment, and program non-duplication systems. Does not include headend buildings and program origination assets.				
48.42	Subscriber connection and distribution systems	8	10	12	5
	Includes assets such as trunk and feeder cable, connecting hardware, amplifiers, power equipment, passive devices, directional taps, pedestals, pressure taps, drop cables, matching transformers, multiple set connecter equipment, and converters.				
48.43	Program origination	7	9	11	9
	Includes assets such as cameras, film chains, video tape recorders, lighting, and remote location equipment excluding vehicles. Does not include buildings and their structural components.				
48.44	Service and test	7	8.5	10	2.5
	Includes assets such as oscilloscopes, field strength meters, spectrum analyzers, and cable testing equipment, but does not include vehicles.				
48.45	Microwave systems	7.5	9.5	11.5	2
	Includes assets such as towers, antennas, transmitting and receiving equipment and broad band microwave assets if used in the provision of cable television services. Does not include assets used in the provision of common carrier services.				
49.0	Electric, gas and sanitary services:				
49.11	Electric utility hydraulic production plant:				
	Includes assets used in the hydraulic power production of electricity for sale, related land improvements, dams, flumes, canals, and waterways	40	50	60	1.5
49.12	Electric utility nuclear production plant:				
	Includes assets used in the nuclear power production of electricity for sale and related land improvements ...	16	20	24	3.0

¶233

ADR Classes and Guidelines *(continued)*

Asset guide-line class	Description of assets included	Asset depreciation range (in years)			Annual asset guide-line repair allow-ance percent-age
		Lower limit	Asset guide-line period	Upper limit	
49.121	Nuclear fuel assemblies: Includes initial core and replacement core nuclear fuel assemblies (i.e. the composite of fabricated nuclear fuel and container) when used in a boiling water, pressurized water, or high temperature gas reactor used in the production of electricity. Does not include nuclear fuel assemblies used in breeder reactors	4.0	5.0	6.0	
49.13	Electric utility steam production plant: Includes assets used in the steam power production of electricity for sale, combustion turbines operated in a combined cycle with a conventional steam unit, and related land improvements	22.5	28	33.5	2.5
49.14	Electric utility transmission and distribution plant: Includes assets used in the transmission and distribution of electricity for sale and related land improvements .	24	30	36	2.0
49.15	Electric utility combustion turbine production plant: Includes assets used in the production of electricity for sale by the use of such prime movers as jet engines, combustion turbines, diesel engines, gasoline engines and other internal combustion engines, their associated power turbines and/or generators, and related land improvements. Does not include combustion turbines operated in a combined cycle with a conventional steam unit	16	20	24	4.0
49.2	Gas utilities: Includes assets used in the production, transmission, and distribution of natural and manufactured gas for sale, including related land improvements and identified as:				
49.21	Distribution facilities: Including gas water heaters and gas conversion equipment installed by utility on customers' premises on a rental basis	28	35	42	2.0
49.22	Gas making facilities:				
49.221	Manufactured gas production plant: Includes assets used in the manufacture of gas having chemical and/or physical properties which do not permit complete interchangeability with domestic natural gas	24	30	36	2.0
49.222	Substitute natural gas (SNG) production plant (naphtha or lighter hydrocarbon feedstocks): Includes assets used in the catalytic conversion of feedstocks of naphtha or lighter hydrocarbons to a gaseous				

ADR Classes and Guidelines (*continued*)

Asset guide-line class	Description of assets included	Asset depreciation range (in years)			Annual asset guide-line repair allow-ance percent-age
		Lower limit	Asset guide-line period	Upper limit	
	fuel which is completely interchangeable with domestic natural gas	11	14	17	4.5
49.23	Natural gas production plant	11	14	17	4.5
49.24	Trunk pipelines and related storage facilities	17.5	22	26.5	3.0
49.3	Water utilities:				
	Includes assets used in the gathering, treatment, and commercial distribution of water	40	50	60	1.5
49.4	Central steam production and distribution:				
	Includes assets used in the production and distribution of steam for sale	22.5	28	33.5	2.5
49.5	Industrial steam and electric generation and distribution systems:				
	Includes assets used in the production and distribution of electricity with rated total capacity in excess of 500 kilowatts and/or assets used in the production and distribution of steam with rated total capacity in excess of 12,500 pounds per hour, for use by the taxpayer in his industrial manufacturing process or plant activity and not ordinarily available for sale to others.				
	Assets used to generate or distribute electricity or steam of the type described above of lesser rated capacity are not included, but are included in the appropriate manufacturing equipment classes elsewhere specified	22.5	28	33.5	2.5
50.0	Wholesale and retail trade:				
	Includes assets used in carrying out the activities of purchasing, assembling, sorting, grading, and selling of goods at both the wholesale and retail level. Also includes assets used in such activities as the operation of restaurants, cafes, coin-operated dispensing machines, and in brokerage of scrap metal	8	10	12	6.5
50.1	Wholesale and retail trade service assets:				
	Includes assets such as glassware, silverware (including kitchen utensils), crockery (usually china) and linens (generally napkins, tablecloths and towels) used in qualified activities as defined in Class 50.0	2	2.5	3	—
65.0	Building services[3]				
	Provision of the services of buildings, whether for use by others or for taxpayer's own account. Assets in the classes listed below include the structural shells of buildings and				

[3]This class is established for a three-year transition period in accordance with Section 109(e)(1) of the Revenue Act of 1971 (P.L. 92-178, CB 1972-1, 443) and will be in effect for the period beginning January 1, 1971 and ending January 1, 1974 or at such earlier date as of which asset classes incorporating the assets herein described are represcribed or modified. See Sections 1.167(a)-11(b)(3)(ii), 1.167(a)-11(b)(4)(i)(a), and 1.167(a)-11(b)(5)(vi) of the regulations for special rules relating to real property.

¶233

ADR Classes and Guidelines (continued)

Asset guide-line class	Description of assets included	Asset depreciation range (in years)			Annual asset guide-line repair allow-ance percent-age
		Lower limit	Asset guide-line period	Upper limit	

	all integral parts thereof; equipment that services normal heating, plumbing, air conditioning, illumination, fire prevention, and power requirements; equipment for the movement of passengers and freight within the building; and any additions to buildings or their components, capitalized remodeling costs, and partitions both permanent and semipermanent. Structures, closely related to the equipment they house, which are section 38 property are not included. See section 1.48-1(e)(1) of the regulations. Such structures are included in asset guideline classes appropriate to the equipment to which they are related. Depreciation periods for assets used in the provision of the services of buildings and which are not specified below shall be determined according to the facts and circumstances pertinent to each asset, except in the case of farm buildings and other building structures for which a class has otherwise been designated.	
65.1	Shelter, space, and related building services for manufacturing and for machinery and equipment repair activities:	
65.11	Factories	45
65.12	Garages	45
65.13	Machine shops..	45
65.14	Loft buildings..	50
65.2	Building services for the conduct of wholesale and retail trade, includes stores and similar structures	50
65.3	Building services for residential purposes:	
65.31	Apartments	40
65.32	Dwellings	45
65.4	Building services relating to the provision of miscellaneous services to businesses and consumers:	
65.41	Office buildings	45
65.42	Storage:	
65.421	Warehouses	60
65.422	Grain elevators..	60
65.43	Banks	50
65.44	Hotels	40
65.45	Theaters	40
70.0	Services:	
70.1	Administrative Services:	
	Includes assets used in administering normal business transactions and the maintenance of business records, their retrieval and analysis, whether these services are performed for others or for taxpayer's own account	

¶233

ADR Classes and Guidelines *(continued)*

Asset guide-line class	Description of assets included	Asset depreciation range (in years)			Annual asset guide-line repair allow-ance percent-age
		Lower limit	Asset guide-line period	Upper limit	
	and whether the assets are located in a single location or widely dispersed.				
70.11	Office furniture, fixtures, and equipment:				
	Includes furniture and fixtures which are not a structural component of a building. Includes such assets as desks, files, safes, and communications equipment (not to include communications equipment which is included in other ADR classes)	8.0	10.0	12.0	2.0
70.12	Information systems:				
	Includes computers and their peripheral equipment (does not include equipment that is an integral part of other capital equipment and which is included in other ADR classes of economic activity, i.e., computers used primarily for process or production control, switching and channeling)	5.0	6.0	7.0	7.5
	Information systems defined:				
	1) Computers: A computer is an electronically activated device capable of accepting information, applying prescribed processes to the information and supplying the results of these processes with or without human intervention. It usually consists of a central processing unit containing extensive storage, logic, arithmetic and control capabilities. Excluded from this category are adding machines, electronic desk calculators, etc.				
	2) Peripheral equipment consists of the auxiliary machines which may be placed under control of the central processing unit. Nonlimiting examples are Card readers, card punches, magnetic tape fees, high speed printers, optical character readers, tape cassettes, mass storage units, paper tape equipment, keypunches, data entry devices, teleprinters, terminals, tape drives, disc drives, disc files, disc packs, visual image projector tubes, card sorters, plotters, collators.				
	Peripheral equipment may be used on-line or off-line.				
70.13	Data handling equipment, except computers:				
	Includes typewriters, calculators, adding and accounting machines, copiers and duplicating equipment	5.0	6.0	7.0	15.0
70.2	Personal and professional services:				
	Includes assets used in the provision of personal services such as those offered by hotels and motels, laundry and dry cleaning establishments, beauty and barber shops, photographic studios and mortuaries. Includes assets used in the provision of professional services such as those offered by doctors, dentists, lawyers. accountants, ar-				

¶233

ADR Classes and Guidelines *(continued)*

Asset guide-line class	Description of assets included	Asset depreciation range (in years)			Annual asset guide-line repair allow-ance percent-age
		Lower limit	Asset guide-line period	Upper limit	
	classified in other ADR classes. Includes assets used in the provision of repair and maintenance services and those assets used in providing fire and burglary protection services and which are not classified in other ADR classes. Includes equipment or facilities used by cemetery organizations, news agencies, teletype wire services, plumbing contractors, frozen food lockers, research laboratories, hotels, and motels and which are not classified in other ADR classes	8	10	12	6.5
70.21	Personal and professional services—service assets: Includes assets such as glassware, silverware, crockery, and linens (generally sheets, pillowcases and bath towels) used in qualified activities as defined in Class 70.2	2	2.5	3	
79.0	Recreation: Includes assets used in the provision of entertainment services on payment of a fee or admission charge, as in the operation of bowling alleys, billiard and pool establishments, theaters, concert halls, and miniature golf courses. Does not include amusement parks and assets which consist primarily of specialized land improvements or structures, such as golf courses, sports stadia, etc., and buildings which house the assets used in entertainment services	8	10	12	6.5
80.0	Theme and amusement parks: Includes assets used in the provision of rides, attractions, and amusements in activities defined as theme and amusement parks, and includes appurtenances associated with a ride, attraction, amusement or theme setting within the park such as ticket booths, facades, shop interiors, and props, special purpose structures, and buildings other than warehouses, administration buildings, hotels, and motels. Includes all land improvements for or in support of park activities, (e.g. parking lots, sidewalks, waterways, bridges, fences, landscaping, etc.) and support functions (e.g. food and beverage retailing, souvenir vending, and other nonlodging accommodations) if owned by the park and provided exclusively for the benefit of park patrons. This guidelines class is a composite of all assets used in this industry except transportation equipment (general purpose trucks, cars, airplanes, etc. which are included in asset guideline classes with the prefix 00.2), assets used in the provision of administrative services in asset guideline classes with the prefix 70.1, and warehouses, administration buildings, hotels, and motels	10	12.5	15	12.5

[¶234]

Depreciation Reserve Computations

Each ADR vintage account must have a "depreciation reserve," and the amount of the reserve for each asset guideline class must be reported on the income tax return for the year. This consists of the accumulated depreciation allowable for the account, increased and decreased as follows:

Increased for

1. Annual ADR depreciation deductions
2. Ordinary retirements in which all or part of the gain is not recognized under Code provisions
3. The greater of the amount subtracted from salvage value, or the "value" of retired assets when using the alternative procedure for retirements to supplies or scrap (see Reg. §1.167(a)-11(d)(3)(viii)).

Decreased for

1. Depreciation allowable for assets which are retired in "extraordinary retirements"
2. Ordinary retirements in which all or part of the gain is not recognized under Code provisisions
3. Adjustments for property removed from vintage accounts because of a post-election change in depreciation under certain Code provisions.
4. Depreciation reserve gain recognized.

Note: The depreciation reserve may never be "decreased" to less than 0. This table is intended only as an outline since the specific increases and decreases to depreciation reserve contain numerous technical qualifications, and the Regs should be consulted when making the actual computations.

[¶235]

First Year's Depreciation on Individual-Item Basis

Assume that you build a hotel and incur the costs given for each item. Based on the ADR useful lives and disregarding salvage value, the table below indicates the first year's depreciation for each item under straight-line and 150%-declining-balance depreciation methods.

Item	Cost	Useful Life	S/L	First Year Depreciation 150% D/B
Building & Improvements				
Building only	$500,000	40	$12,500	$18,750
Air Conditioning	45,000	8	5,625	8,433
Elevators:				
Freight (2)	25,000	12	2,083	3,125
Passenger (4)	40,000	12	3,333	5,000
Boiler & Oil Burner	30,000	8	3,750	5,625
Lighting System:				
Fixtures	15,000	8	1,875	2,813
Wiring	20,000	12	1,666	2,500
Plumbing:				
Bath tubs, etc.	12,500	12	1,042	1,563
Faucets, valves, etc.	7,500	12	625	938
Pipes:				
Cold water	12,500	12	1,042	1,563
Hot water	15,000	12	1,250	1,875
Roof—copper	25,000	12	2,083	3,125
Switchboards	5,000	8	625	938
Water tank—metal	10,000	8	1,250	1,875
Fire Alarm & Prevention Equipment	15,000	8	1,875	2,813
Total Building & Improvements	$777,500			
Furniture, Fixtures & Equipment				
Refrigeration System	11,000	8	1,375	2,063
Kitchen Equipment	20,000	8	2,500	3,750
Laundry Equipment	15,000	8	1,875	2,813
House Cleaning Equipment	10,000	8	1,250	1,875
Shades & Screens	10,000	8	1,250	1,875
Blankets & Spreads	6,000	8	750	1,125
Carpets & Rugs	12,000	8	1,500	2,250
Curtains, Draperies & Scarfs	6,000	8	750	1,125
Springs, Mattresses & Pillows	6,000	8	750	1,125
Furniture:				
Dining & Guest Rooms	24,000	8	3,000	4,500
Lobby	4,000	8	500	750
Total Furniture, Fixtures & Equipment	$124,000			
TOTAL	$901,500		$56,124	$84,187

[¶236]

Comparison of First-Year Depreciation
Under Composite and Component Bases

Since the tax law limits you to 150%-declining-balance depreciation on new and straight-line on used commercial properties, you may be able to save tax dollars by figuring depreciation on a component basis rather than by using a composite rate. You might, for example, have separate accounts as follows:

	Useful Life	Cost
Building......................	40 years	$120,000
Wiring.......................	12 years	20,000
Plumbing...................	12 years	12,000
Roof	12 years	8,000
Elevator	12 years	10,000
Paving......................	8 years	5,000
Air conditioning	8 years	20,000
Ceilings	8 years	9,000
Floor........................	8 years	10,000
		$214,000

This would result in first-year depreciation aggregating $19,000 using 150%-declining-balance.

If you were to use composite depreciation, the life of the composite building would be 26.88 years. Using this and 150%-declining-balance depreciation would give first-year depreciation of only $11,941.

If you had used the 200%-declining-balance method with the composite method, your first-year depreciation would still have been only $15,923.

[¶237]

Investment Tax Credit

The investment tax credit is a special credit against the taxes due for the taxable year. The credit is granted where property subject to depreciation under Code §167 is placed into service by a taxpayer after 8/15/71. (Special rules apply to property constructed or put into service before that date.) The credit applies to all new property, and used property, if "purchased." The effect of the "purchase" requirement is to exclude used property acquired in tax-free transactions, for the most part, such as tax-free exchanges, inheritance, or from another member of a controlled group.

The credit applies at a flat rate of 7% of the "qualified investment" in the property. This means the cost or other basis if the property has a useful life of 7 years or more, but this is reduced as the following table indicates:

Life of Property	Credit as % of Full Basis of Property
7 years or more	7%
5 to 7 years	4-2/3%
3 to 5 years	2-1/3%

The credit is taken directly against the tax liability for the year without reference to the credit. However, it is limited to 50% of the tax liability as follows:

The credit for property put into service in the year, plus any unused credit from prior years, is limited to:

(1) the tax liability for the year or $25,000, whichever is less, or

(2) 50% of the tax liability for the year (without reference to the credit) in excess of $25,000. Amounts of credit limited by this rule may be carried forward to later years.

Persons who claim the credit must file Form 3468 with their income tax returns for the year claimed.

This is only a brief outline of the workings of the investment tax credit. In an actual case, the Regs should be consulted.

SECTION C
ESTATE PLANNING AIDS

[¶301]

Life Expectancy Tables

Life Expectancy at Birth in the United States

(Years)

YEAR	White			All Other			Total		
	Male	Female	Total	Male	Female	Total	Male	Female	Total
1900...........	46.6	48.7	47.6	32.5	33.5	33.0	46.3	48.3	47.3
1910...........	48.6	52.0	50.3	33.8	37.5	35.6	48.4	51.8	50.0
1920...........	54.4	55.6	54.9	45.5	45.2	45.3	53.6	54.6	54.1
1930....	59.7	63.5	61.4	47.3	49.2	48.1	58.1	61.6	59.7
1940...........	62.1	66.6	64.2	51.5	54.9	53.1	60.8	65.2	62.9
1950...........	66.5	72.2	69.1	59.1	62.9	60.8	65.6	71.1	68.2
1960...........	67.4	74.1	70.6	61.1	66.3	63.6	66.6	73.1	69.7
1961...........	67.8	74.5	71.0	61.9	67.0	64.4	67.0	73.6	70.2
1962...........	67.6	74.4	70.9	61.5	66.8	64.1	66.8	73.4	70.0
1963...........	67.5	74.4	70.8	60.9	66.5	63.6	66.6	73.4	69.9
1964...........	67.7	74.6	71.0	61.1	67.2	64.1	66.9	73.7	70.2
1965...........	67.6	74.7	71.0	61.1	67.4	64.1	66.8	73.7	70.2
1966...........	67.6	74.7	71.0	60.7	67.4	64.0	66.7	73.8	70.1
1967...........	67.8	75.1	71.3	61.1	68.2	64.6	67.0	74.2	70.5
1968...........	67.5	74.9	71.1	60.1	67.5	63.7	66.6	74.0	70.2
1969...........	67.8	75.1	71.3	60.5	68.4	64.3	66.8	74.3	70.4
1970...........	68.1	75.4	71.7	60.5	68.9	64.6	67.1	74.6	70.8
1971...........	68.3	75.7	71.9	61.3	69.4	65.2	67.4	74.9	71.1

Life Expectancy at Various Ages in the United States

(Years)

AGE	White		All Other		Total	
	Male	Female	Male	Female	Male	Female
0.............................	68.3	75.7	61.3	69.4	67.4	74.9
20.............................	50.5	57.4	44.5	52.1	49.8	56.8
40.............................	32.1	38.3	28.3	34.3	31.7	37.8
45.............................	27.7	33.7	24.7	30.2	27.4	33.3
50.............................	23.6	29.2	21.3	26.3	23.3	28.9
55.............................	19.7	25.0	18.1	22.8	19.5	24.8
60.............................	16.2	20.9	15.1	19.2	16.1	20.7
65.............................	13.2	17.0	12.8	16.2	13.2	16.9
70.............................	10.5	13.4	10.9	13.9	10.6	13.5

[¶302]

Mortality Tables

Age	American Experience (1843-1858)		Commissioners 1941 Standard Ordinary (1930-1940)		Commissioners 1958 Standard Ordinary (1950-1954)		Annuity Table for 1949–Male (1939-1949)		United States Total Population (1959-1961)	
	Deaths Per 1,000	Expectation of Life (Years)	Deaths Per 1,000	Expectation of Life (Years)	Deaths Per 1,000	Expectation of Life (Years)	Deaths Per 1,000	Expectation of Life (Years)	Deaths Per 1,000	Expectation of Life (Years)
0	154.70	41.45	22.58	62.33	7.08	68.30	4.04	73.18	25.93	69.89
1	63.49	47.94	5.77	62.76	1.76	67.78	1.58	72.48	1.70	70.75
2	35.50	50.16	4.14	62.12	1.52	66.90	.89	71.59	1.04	69.87
3	23.91	50.98	3.38	61.37	1.46	66.00	.72	70.65	.80	68.94
4	17.70	51.22	2.99	60.58	1.40	65.10	.63	69.70	.67	67.99
5	13.60	51.13	2.76	59.76	1.35	64.19	.57	68.75	.59	67.04
6	11.37	50.83	2.61	58.92	1.30	63.27	.53	67.78	.52	66.08
7	9.75	50.41	2.47	58.08	1.26	62.35	.50	66.82	.47	65.11
8	8.63	49.90	2.31	57.22	1.23	61.43	.49	65.85	.43	64.14
9	7.90	49.33	2.12	56.35	1.21	60.51	.48	64.89	.39	63.17
10	7.49	48.72	1.97	55.47	1.21	59.58	.48	63.92	.37	62.19
11	7.52	48.08	1.91	54.58	1.23	58.65	.49	62.95	.37	61.22
12	7.54	47.45	1.92	53.68	1.26	57.72	.50	61.98	.40	60.24
13	7.57	46.80	1.98	52.78	1.32	56.80	.51	61.01	.48	59.26
14	7.60	46.16	2.07	51.89	1.39	55.87	.52	60.04	.59	58.29
15	7.63	45.50	2.15	50.99	1.46	54.95	.54	59.07	.71	57.33
16	7.66	44.85	2.19	50.10	1.54	54.03	.55	58.10	.82	56.37
17	7.69	44.19	2.25	49.21	1.62	53.11	.57	57.13	.93	55.41
18	7.73	43.53	2.30	48.32	1.69	52.19	.58	56.17	1.02	54.46
19	7.77	42.87	2.37	47.43	1.74	51.28	.60	55.20	1.08	53.52
20	7.80	42.20	2.43	46.54	1.79	50.37	.62	54.23	1.15	52.58
21	7.86	41.53	2.51	45.66	1.83	49.46	.65	53.27	1.22	51.64
22	7.91	40.85	2.59	44.77	1.86	48.55	.67	52.30	1.27	50.70
23	7.96	40.17	2.68	43.88	1.89	47.64	.70	51.33	1.28	49.76
24	8.01	39.49	2.77	43.00	1.91	46.73	.73	50.37	1.27	48.83
25	8.06	38.81	2.88	42.12	1.93	45.82	.77	49.41	1.26	47.89
26	8.13	38.12	2.99	41.24	1.96	44.90	.81	48.44	1.25	46.95
27	8.20	37.43	3.11	40.36	1.99	43.99	.85	47.48	1.26	46.00
28	8.26	36.73	3.25	39.49	2.03	43.08	.90	46.52	1.30	45.06
29	8.34	36.03	3.40	38.61	2.08	42.16	.95	45.56	1.36	44.12
30	8.43	35.33	3.56	37.74	2.13	41.25	1.00	44.61	1.43	43.18
31	8.51	34.63	3.73	36.88	2.19	40.34	1.07	43.65	1.51	42.24
32	8.61	33.92	3.92	36.01	2.25	39.43	1.14	42.70	1.60	41.30
33	8.72	33.21	4.12	35.15	2.32	38.51	1.21	41.75	1.70	40.37
34	8.83	32.50	4.35	34.29	2.40	37.60	1.30	40.80	1.81	39.44
35	8.95	31.78	4.59	33.44	2.51	36.69	1.39	39.85	1.94	38.51
36	9.09	31.07	4.86	32.59	2.64	35.78	1.49	38.90	2.09	37.58
37	9.23	30.35	5.15	31.75	2.80	34.88	1.61	37.96	2.28	36.66
38	9.41	29.62	5.46	30.91	3.01	33.97	1.73	37.02	2.49	35.74
39	9.59	28.90	5.81	30.08	3.25	33.07	1.87	36.08	2.73	34.83
40	9.79	28.18	6.18	29.25	3.53	32.18	2.03	35.15	3.00	33.92

Mortality Tables *(continued)*

Age	American Experience (1843-1858)		Commissioners 1941 Standard Ordinary (1930-1940)		Commissioners 1958 Standard Ordinary (1950-1954)		Annuity Table for 1949–Male (1939-1949)		United States Total Population (1959-1961)	
	Deaths Per 1,000	Expectation of Life (Years)	Deaths Per 1,000	Expectation of Life (Years)	Deaths Per 1,000	Expectation of Life (Years)	Deaths Per 1,000	Expectation of Life (Years)	Deaths Per 1,000	Expectation of Life (Years)
41	10.01	27.45	6.59	28.43	3.84	31.29	2.22	34.22	3.30	33.02
42	10.25	26.72	7.03	27.62	4.17	30.41	2.48	33.30	3.62	32.13
43	10.52	26.00	7.51	26.81	4.53	29.54	2.80	32.38	3.97	31.25
44	10.83	25.27	8.04	26.01	4.92	28.67	3.19	31.47	4.35	30.37
45	11.16	24.54	8.61	25.21	5.35	27.81	3.63	30.57	4.76	29.50
46	11.56	23.81	9.23	24.43	5.83	26.95	4.12	29.68	5.21	28.64
47	12.00	23.08	9.91	23.65	6.36	26.11	4.66	28.80	5.73	27.79
48	12.51	22.36	10.64	22.88	6.95	25.27	5.25	27.93	6.33	26.94
49	13.11	21.63	11.45	22.12	7.60	24.45	5.88	27.07	7.00	26.11
50	13.78	20.91	12.32	21.37	8.32	23.63	6.56	26.23	7.74	25.29
51	14.54	20.20	13.27	20.64	9.11	22.82	7.28	25.40	8.52	24.49
52	15.39	19.49	14.30	19.91	9.96	22.03	8.04	24.58	9.29	23.69
53	16.33	18.79	15.43	19.19	10.89	21.25	8.84	23.78	10.05	22.91
54	17.40	18.09	16.65	18.48	11.90	20.47	9.68	22.99	10.82	22.14
55	18.57	17.40	17.98	17.78	13.00	19.71	10.56	22.20	11.61	21.37
56	19.89	16.72	19.43	17.10	14.21	18.97	11.49	21.44	12.49	20.62
57	21.34	16.05	21.00	16.43	15.54	18.23	12.46	20.68	13.52	19.87
58	22.94	15.39	22.71	15.77	17.00	17.51	13.48	19.93	14.73	19.14
59	24.72	14.74	24.57	15.13	18.59	16.81	14.54	19.20	16.11	18.42
60	26.69	14.10	26.59	14.50	20.34	16.12	15.66	18.48	17.61	17.71
61	28.88	13.47	28.78	13.88	22.24	15.44	16.87	17.76	19.17	17.02
62	31.29	12.86	31.18	13.27	24.31	14.78	18.20	17.06	20.82	16.34
63	33.94	12.26	33.76	12.69	26.57	14.14	19.67	16.37	22.52	15.68
64	36.87	11.67	36.58	12.11	29.04	13.51	21.28	15.68	24.31	15.03
65	40.13	11.10	39.64	11.55	31.75	12.90	23.07	15.01	26.22	14.39
66	43.71	10.54	42.96	11.01	34.74	12.31	25.03	14.36	28.28	13.76
67	47.65	10.00	46.56	10.48	38.04	11.73	27.19	13.71	30.53	13.15
68	52.00	9.47	50.46	9.97	41.68	11.17	29.58	13.08	33.01	12.55
69	56.76	8.97	54.70	9.47	45.61	10.64	32.20	12.46	35.73	11.96
70	61.99	8.48	59.30	8.99	49.79	10.12	35.09	11.86	38.66	11.38
71	67.67	8.00	64.27	8.52	54.15	9.63	38.27	11.28	41.82	10.82
72	73.73	7.55	69.66	8.08	58.65	9.15	41.77	10.71	45.30	10.27
73	80.18	7.11	75.50	7.64	63.26	8.69	45.62	10.15	49.15	9.74
74	87.03	6.68	81.81	7.23	68.12	8.24	49.85	9.61	53.42	9.21
75	94.37	6.27	88.64	6.82	73.37	7.81	54.50	9.09	57.99	8.71
76	102.31	5.88	96.02	6.44	79.18	7.39	59.61	8.58	62.96	8.21
77	111.06	5.49	103.99	6.07	85.70	6.98	65.22	8.10	68.67	7.73
78	120.83	5.11	112.59	5.72	93.06	6.59	71.37	7.63	75.35	7.26
79	131.73	4.74	121.86	5.38	101.19	6.21	78.11	7.17	83.02	6.81
80	144.47	4.39	131.85	5.06	109.98	5.85	85.50	6.74	92.08	6.39

¶302

Mortality Tables *(continued)*

Age	American Experience (1843-1858)		Commissioners 1941 Standard Ordinary (1930-1940)		Commissioners 1958 Standard Ordinary (1950-1954)		Annuity Table for 1949–Male 1939-1949		United States Total Population (1959-1961)	
	Deaths Per 1,000	Expectation of Life (Years)	Deaths Per 1,000	Expectation of Life (Years)	Deaths Per 1,000	Expectation of Life (Years)	Deaths Per 1,000	Expectation of Life (Years)	Deaths Per 1,000	Expectation of Life (Years)
81	158.60	4.05	142.60	4.75	119.35	5.51	93.59	6.32	102.19	5.98
82	174.30	3.71	154.16	4.46	129.17	5.19	102.44	5.92	112.44	5.61
83	191.56	3.39	166.57	4.18	139.38	4.89	112.11	5.54	121.95	5.25
84	211.36	3.08	179.88	3.91	150.01	4.60	122.67	5.18	130.67	4.91
85	235.55	2.77	194.13	3.66	161.14	4.32	134.18	4.84	143.80	4.58
86	265.68	2.47	209.37	3.42	172.82	4.06	146.71	4.51	158.16	4.26
87	303.02	2.18	225.63	3.19	185.13	3.80	160.33	4.20	173.55	3.97
88	346.69	1.91	243.00	2.98	198.25	3.55	175.12	3.90	190.32	3.70
89	395.86	1.66	261.44	2.77	212.46	3.31	191.15	3.62	208.35	3.45
90	454.55	1.42	280.99	2.58	228.14	3.06	208.49	3.36	227.09	3.22
91	532.47	1.19	301.73	2.39	245.77	2.82	227.19	3.12	245.98	3.02
92	634.26	.98	323.64	2.21	265.93	2.58	247.33	2.88	264.77	2.85
93	734.18	.80	346.66	2.03	289.30	2.33	268.96	2.67	282.84	2.69
94	857.14	.64	371.00	1.84	316.66	2.07	292.12	2.47	299.52	2.55
95	1,000.00	.50	396.21	1.63	351.24	1.80	316.83	2.28	314.16	2.43
96			447.19	1.37	400.56	1.51	343.12	2.10	329.15	2.32
97			548.26	1.08	488.42	1.18	370.97	1.94	344.50	2.21
98			724.67	.78	668.15	.83	400.35	1.79	360.18	2.10
99			1,000.00	.50	1,000.00	.50	431.20	1.65	376.16	2.01
100							463.41	1.52	392.42	1.91
101							496.87	1.40	408.91	1.83
102							531.39	1.29	425.62	1.75
103							566.76	1.20	442.50	1.67
104							602.71	1.10	459.51	1.60
105							638.96	1.02	476.62	1.53
106							675.14	.94	493.78	1.46
107							710.90	.86	510.95	1.40
108							745.82	.75	528.10	1.35
109							1,000.00	.50	545.19	1.29

[¶303]
Federal Gift Taxes

The table below sets forth the gift tax rates as they appear in the Code. These rates apply to the net taxable gift, after exclusions, after exemptions, and after splitting the gift with the donor's spouse or the marital deduction if the gift is made to the donor's spouse.

Note that the tax rates are for the total gifts, made in prior years and the current year. The current year's taxable gifts are added to the total of the prior years' taxable gifts and the gift tax for the grand total is determined. Then, from this grand total, the total gift taxes paid in prior years is subtracted. The remaining amount is the gift tax liability for the current year.

Gift	Tax
Not over $5,000	2¼% of the taxable gifts.
Over $5,000 but not over $10,000	$112.50, plus 5¼% of excess over $5,000.
Over $10,000 but not over $20,000	$375, plus 8¼% of excess over $10,000.
Over $20,000 but not over $30,000	$1,200, plus 10½% of excess over $20,000.
Over $30,000 but not over $40,000	$2,250, plus 13½% of excess over $30,000.
Over $40,000 but not over $50,000	$3,600, plus 16½% of excess over $40,000.
Over $50,000 but not over $60,000	$5,250, plus 18¾% of excess over $50,000.
Over $60,000 but not over $100,000	$7,125, plus 21% of excess over $60,000.
Over $100,000 but not over $250,000	$15,525, plus 22½% of excess over $100,000.
Over $250,000 but not over $500,000	$49,275, plus 24% of excess over $250,000.
Over $500,000 but not over $750,000	$109,275, plus 26¼% of excess over $500,000.
Over $750,000 but not over $1,000,000	$174,900, plus 27¾% of excess over $750,000.
Over $1,000,000 but not over $1,250,000	$244,275, plus 29¼% of excess over $1,000,000.
Over $1,250,000 but not over $1,500,000	$317,400, plus 31½% of excess over $1,250,000.
Over $1,500,000 but not over $2,000,000	$396,150, plus 33¾% of excess over $1,500,000.
Over $2,000,000 but not over $2,500,000	$564,900, plus 36¾% of excess over $2,000,000.
Over $2,500,000 but not over $3,000,000	$748,650, plus 39¾% of excess over $2,500,000.
Over $3,000,000 but not over $3,500,000	$947,400, plus 42% of excess over $3,000,000.
Over $3,500,000 but not over $4,000,000	$1,157,400, plus 44¼% of excess over $3,500,000.
Over $4,000,000 but not over $5,000,000	$1,378,650, plus 47¼% of excess over $4,000,000.
Over $5,000,000 but not over $6,000,000	$1,851,150, plus 50¼% of excess over $5,000,000.
Over $6,000,000 but not over $7,000,000	$2,353,650, plus 52½% of excess over $6,000,000.
Over $7,000,000 but not over $8,000,000	$2,878,650, plus 54¾% of excess over $7,000,000.
Over $8,000,000 but not over $10,000,000	$3,426,150, plus 57% of excess over $8,000,000.
Over $10,000,000	$4,566,150, plus 57¾% of excess over $10,000,000.

Note: For an example of how gifts (even taxable gifts) can save taxes, see Table on page 148.

[¶304]   How Gift Tax Varies According to Marital Status of Donor

Example of use of this table:

How much gift tax is due on a gift to a wife of $90,000 where the husband still has his full $30,000 lifetime exemption?

$90,000 gift less $3,000 annual exclusion	$87,000
Tax on $80,000 gift to spouse from table	$375
Tax on ½ the excess of $7,000 at 8¼%	289
Total gift tax	$664

Note: The $3,000 annual exclusion ($6,000 for gift in which spouse joins) must be deducted before using this table. Table utilizes $30,000 ($60,000) lifetime exemption.

Amount of Gift Exceeds	Tax on Ordinary Gifts	Rate on Excess Amount	Tax on Gift to Spouse	Tax on Gift in Which Spouse Joins	Rate on ½ Excess Amount	Saving When Spouse Joins
$ 30,000	$	2¼%	$	$	%	$
35,000	112	5¼				112
40,000	375	8¼				375
50,000	1,200	10½				1,200
60,000	2,250	13½			2¼	2,250
65,000	2,925	13½	56	112	2¼	2,813
70,000	3,600	16½	112	225	5¼	3,375
80,000	5,250	18¾	375	750	8¼	4,500
90,000	7,125	21	787	1,575	8¼	5,550
100,000	9,225	21	1,200	2,400	10½	6,825
120,000	13,425	21	2,250	4,500	13½	8,925
130,000	15,525	22½	2,925	5,850	13½	9,675
140,000	17,775	22½	3,600	7,200	16½	10,575
150,000	20,025	22½	4,425	8,850	16½	11,175
160,000	22,275	22½	5,250	10,500	18¾	11,775
180,000	26,775	22½	7,125	14,250	21	12,525
200,000	31,275	22½	9,225	18,450	21	12,825
250,000	42,525	22½	14,475	28,950	21	13,575
260,000	44,775	22½	15,525	31,050	22½	13,725
280,000	49,275	24	17,775	35,550	22½	13,725
500,000	102,075	24	42,525	85,050	22½	17,025
530,000	109,275	26¼	45,900	91.800	22½	17,475
560,000	117,150	26¼	49,275	98,550	24	18,600
750,000	167,025	26¼	72,075	144,150	24	22,875
780,000	174,900	27¾	75,675	151,350	24	23,550
1,000,000	235,950	27¾	102,075	204,150	24	31,800
1,030,000	244,275	29¼	105,675	211,350	24	32,925
1,060,000	253,050	29¼	109,275	218,550	26¼	34,500
1,280,000	317,400	31½	138,150	276,300	26¼	41,100
1,500,000	386,700	*	167,025	334,050	*	52,650

*Gift taxes on amounts in excess of $1.5 million range between 31½% and 57¾% for ordinary gifts and between 26¼% and 57¾% for marital deduction gifts.

¶304

Value of Life Estates, Annuities, Remainder Interests, and Revisions

Note: The tables following (Tables A(1), A(2), and B) are to be used only to value gifts made on or after January 1, 1971, and with regard to estates of persons dying on or after January 1, 1971. Do not use these tables if (1) the interest being valued involves more than one life (special tables must be used, or the Commissioner will supply the interest factor if he is given the facts) or (2) the person whose life is involved is in such poor health that his actual life expectancy is less than normal (actual anticipated life expectancy may be used).

Insurance company annuities are valued at replacement cost. But noncommercial annuities (e.g., $3,000 a year for life or a period of years under a will or trust), life estates, term interest and remainder interests are valued under special Treasury Tables which are set forth below. Table A(1) is used where male life expectancies are involved; Table A(2) is used where female life expectancies are involved; Table B is for interests involving a fixed number of years.

The value of a life estate is the value of the entire property less the present value of the remainder interest at the end of the life estate. *The value of a term for years* is the value of the entire property less the present value of the remainder interest at the end of the term. *The value of a remainder interest* is the value of the property times the appropriate figure under the remainder interest column in Table A(1) or Table A(2) or Table B. If the remainder interest follows a life estate, use the Table A(1) or A(2) column opposite the age of the life tenant. If the remainder interest follows a term for years, use the Table B column opposite the number of years.

The value of an annuity for life is the amount of annual payment times the figure in the "Annuity" column, opposite the appropriate age in Table A(1) or A(2). If payments are at other than annual periods make this adjustment:

If periods are	Multiply result by
Weekly	1.0291
Monthly	1.0272
Quarterly	1.0222
Semiannually	1.0148

If the first annuity payment is to be made at once, the value of the annuity is the amount of that payment *plus* the value of a similar annuity in which the first payment isn't to be made until the end of the first period.

The value of an annuity for years is the amount of annual payment times the figure in the "Annuity" column, opposite the appropriate number of years in Table B. If payments are made at other than annual periods *and* are at the end of the period, make this adjustment:

If periods are	Multiply result by
Weekly	1.0303
Monthly	1.0322
Quarterly	1.0372
Semiannually	1.0448

Value of Life Estates, Annuities,
Remainder Interests, and Revisions (continued)

Examples of the Use of These Tables

1. *Life Estate*—The donor male, age 31, who is entitled to receive the income from property worth $50,000 during his life, makes a gift of that interest. The value of the gift is $43,309; that is, $50,000 minus $6,691. The latter amount is obtained by multiplying $50,000 by .13883. The figure .13883 is from Table A(1) under the "Remainder" column at age 31.

2. *Term of Years*—A father set up a 10-year trust for his son. The trust property is worth $100,000. The value of the gift is $44,161; that is, $100,000 minus $55,839. The latter amount is obtained by multiplying $100,000 by .558395 from Table B under "Remainder" column for 10 years.

3. *Remainder*—The donor makes a gift of property worth $50,000 which he is entitled to receive on the death of his brother, who had been given the income from the property for life. The brother is 35 at the time of the gift. The value of the gift is $8,372; that is, $50,000 times .16745. The figure .16745 is from Table A(1) under the "Remainder" column at age 35.

4. *Annuity*—Under her husband's will, a widow is to receive an annuity of $10,000 a year payable annually for her life. At the time of her husband's death she is 61 years old. The value of the annuity is $103,005; that is, $10,000 times 10.3005. The figure 10.3005 is from Table A(2) under the "Annuity" column at age 61. (If the annuity had been payable at the end of each month, the value would be $105,807. That is the value of $103,005 as computed above times 1.0272, the adjustment factor for monthly payments at the end of each month.)

Table A (1)

Present Worth, at 6%, of an Annuity, a Life Estate, and a Remainder Interest—Single Life Male

1 Age	2 Annuity	3 Life Estate	4 Remainder	1 Age	2 Annuity	3 Life Estate	4 Remainder
0	15.6175	.93705	.06295	45	12.3013	.73808	.26192
1	16.0362	.96217	.03783	46	12.1158	.72695	.27305
2	16.0283	.96170	.03830	47	11.9253	.71552	.28448
3	16.0089	.96053	.03947	48	11.7308	.70885	.29615
4	15.9841	.95905	.04095	49	11.5330	.69193	.30802
5	15.9553	.95732	.04268	50	11.3329	.67997	.32003
6	15.9233	.95540	.04460	51	11.1308	.66785	.33215
7	15.8885	.95331	.04669	52	10.9267	.65560	.34440
8	15.8508	.95105	.04895	53	10.7200	.64320	.35680
9	15.8101	.94861	.05139	54	10.5100	.63060	.36940
10	15.7663	.94593	.05402	55	10.2960	.61776	.38224
11	15.7194	.94316	.05684	56	10.0777	.60466	.39534
12	15.6698	.94019	.05981	57	9.8552	.59131	.40869
13	15.6180	.93708	.06292	58	9.6297	.57778	.42222
14	15.5651	.93391	.06609	59	9.4028	.56417	.43583
15	15.5115	.93069	.06931	60	9.1753	.55052	.44948
16	15.4576	.92746	.07254	61	8.9478	.53687	.46313
17	15.4031	.92419	.07581	62	8.7202	.52321	.47679
18	15.3481	.92089	.07911	63	8.4924	.50954	.49046
19	15.2918	.91751	.08249	64	8.2642	.49585	.50415
20	15.2339	.91403	.08597	65	8.0353	.48212	.51788
21	15.1744	.91046	.08954	66	7.8060	.46836	.53164
22	15.1130	.90678	.09322	67	7.5763	.45458	.54542
23	15.0487	.90292	.09708	68	7.3462	.44077	.55923
24	14.9807	.89884	.10116	69	7.1149	.42689	.57311
25	14.9075	.89445	.10555	70	6.8823	.41294	.58706
26	14.8287	.88972	.11028	71	6.6481	.39889	.60111
27	14.7442	.88465	.11535	72	6.4123	.38474	.61526
28	14.6542	.87925	.12075	73	6.1752	.37051	.62949
29	14.5588	.87353	.12647	74	5.9373	.35624	.64376
30	14.4584	.86750	.13250	75	5.6990	.34194	.65806
31	14.3528	.86117	.13883	76	5.4602	.32761	.67239
32	14.2418	.85451	.14549	77	5.2211	.31327	.68673
33	14.1254	.84752	.15248	78	4.9825	.29895	.70105
34	14.0034	.84020	.15980	79	4.7469	.28481	.71519
35	13.8758	.83255	.16745	80	4.5164	.27098	.72902
36	13.7425	.82455	.17545	81	4.2955	.25773	.74227
37	13.6036	.81622	.18378	82	4.0879	.24527	.75473
38	13.4591	.80755	.19245	83	3.8924	.23354	.76646
39	13.3090	.79854	.20146	84	3.7029	.22217	.77783
40	13.1538	.78923	.21077	85	3.5117	.21070	.78930
41	12.9934	.77960	.22040	86	3.3259	.19955	.80045
42	12.8279	.76967	.23033	87	3.1450	.18870	.81130
43	12.6574	.75944	.24056	88	2.9703	.17822	.82178
44	12.4819	.74891	.25109	89	2.8052	.16831	.83169

Table A (1) *(continued)*

Present Worth, at 6%, of an Annuity, a Life Estate,
and a Remainder Interest—Single Life Male

1 Age	2 Annuity	3 Life Estate	4 Remainder	1 Age	2 Annuity	3 Life Estate	4 Remainder
90	2.6536	.15922	.84078	100	1.6812	.10087	.89913
91	2.5162	.15097	.84903	101	1.6101	.09661	.90339
92	2.3917	.14350	.85650	102	1.5416	.09250	.90750
93	2.2801	.13681	.86319	103	1.4744	.08846	.91154
94	2.1802	.13081	.86919	104	1.4065	.08439	.91561
95	2.0891	.12535	.87465	105	1.3334	.08000	.92000
96	1.9997	.11998	.88002	106	1.2452	.07471	.92529
97	1.9145	.11487	.88513	107	1.1196	.06718	.93282
98	1.8331	.10999	.89001	108	.9043	.05426	.94574
99	1.7554	.10532	.89468	109	.4717	.02830	.97170

Table A (2)

Present Worth, at 6%, of an Annuity, a Life Estate,
and a Remainder Interest—Single Life Female

1 Age	2 Annuity	3 Life Estate	4 Remainder	1 Age	2 Annuity	3 Life Estate	4 Remainder
0	15.8972	.95383	.04617	25	15.3959	.92375	.07625
1	16.2284	.97370	.02630	26	15.3322	.91993	.08007
2	16.2287	.97372	.02628	27	15.2652	.91591	.08409
3	16.2180	.97308	.02692	28	15.1946	.91168	.08832
4	16.2029	.97217	.02783	29	15.1208	.90725	.09275
5	16.1850	.97110	.02890	30	15.0432	.90259	.09741
6	16.1648	.96989	.03011	31	14.9622	.89773	.10227
7	16.1421	.96853	.03147	32	14.8775	.89265	.10735
8	16.1172	.96703	.03297	33	14.7888	.88733	.11267
9	16.0901	.96541	.03459	34	14.6960	.88176	.11824
10	16.0608	.96365	.03635	35	14.5989	.87593	.12407
11	16.0293	.96176	.03824	36	14.4975	.86935	.13015
12	15.9958	.95975	.04025	37	14.3915	.86349	.13651
13	15.9607	.95764	.04236	38	14.2811	.85687	.14313
14	15.9239	.95543	.04457	39	14.1663	.84998	.15002
15	15.8856	.95314	.04686	40	14.0468	.84281	.15719
16	15.8460	.95076	.04924	41	13.9227	.83536	.16464
17	15.8048	.94829	.05171	42	13.7940	.82764	.17236
18	15.7620	.94572	.05428	43	13.6604	.81962	.18038
19	15.7172	.94303	.05697	44	13.5219	.81131	.18869
20	15.6701	.94021	.05979	45	13.3781	.80269	.19731
21	15.6207	.93724	.06276	46	13.2290	.79374	.20626
22	15.5687	.93412	.06588	47	13.0746	.78448	.21552
23	15.5141	.93085	.06915	48	12.9147	.77438	.22512
24	15.4565	.92739	.07261	49	12.7496	.76498	.23502

¶305

Table A (2) *(continued)*

Present Worth, at 6%, of an Annuity, a Life Estate,
and a Remainder Interest—Single Life Female

1 Age	2 Annuity	3 Life Estate	4 Remainder	1 Age	2 Annuity	3 Life Estate	4 Remainder
50	12.5793	.75476	.24524	80	5.0195	.30117	.69883
51	12.4039	.74423	.25577	81	4.7482	.28489	.71511
52	12.2232	.73339	.26661	82	4.4892	.26935	.73065
53	12.0367	.72220	.27730	83	4.2398	.25439	.74561
54	11.8436	.71062	.28938	84	3.9927	.23956	.76044
55	11.6432	.69859	.30141	85	3.7401	.22441	.77559
56	11.4353	.68612	.31388	86	3.5016	.21010	.78990
57	11.2200	.67320	.32680	87	3.2790	.19674	.80326
58	10.9980	.65988	.34012	88	3.0719	.18431	.81569
59	10.7703	.64622	.35378	89	2.8808	.17285	.82715
60	10.5376	.63226	.36774	90	2.7068	.16241	.83759
61	10.3005	.61803	.38197	91	2.5502	.15301	.84699
62	10.0587	.60352	.39648	92	2.4116	.14470	.85530
63	9.8118	.58871	.41129	93	2.2901	.13741	.86259
64	9.5592	.57355	.42645	94	2.1839	.13103	.86897
65	9.3005	.55803	.44197	95	2.0891	.12535	.87465
66	9.0352	.54211	.45789	96	1.9997	.11998	.88002
67	8.7639	.52583	.47417	97	1.9145	.11487	.88513
68	8.4874	.50924	.49076	98	1.8331	.10999	.89001
69	8.2068	.49241	.50759	99	1.7554	.10532	.89468
70	7.9234	.47540	.52460	100	1.6812	.10087	.89913
71	7.6371	.45823	.54177	101	1.6101	.09661	.90339
72	7.3480	.44088	.55912	102	1.5416	.09250	.90750
73	7.0568	.42341	.57659	103	1.4744	.08846	.91154
74	6.7645	.40587	.59413	104	1.4065	.08439	.91561
75	6.4721	.38833	.61167	105	1.3334	.08000	.92000
76	6.1788	.37073	.62927	106	1.2452	.07471	.92529
77	5.8845	.35307	.64693	107	1.1196	.06718	.93282
78	5.5910	.33546	.66454	108	.9043	.05426	.94574
79	5.3018	.31811	.68189	109	.4717	.02830	.97170

Table B

Present Worth, at 6%, of an Annuity for a Term Certain, an Income Interest for a Term Certain, and a Remainder Interest Postponed for a Term Certain.

1 Number of Years	2 Annuity	3 Term Certain	4 Remainder	1 Number of Years	2 Annuity	3 Term Certain	4 Remainder
1	0.9434	.056604	.943396	31	13.9291	.835745	.164255
2	1.8334	.110004	.889996	32	14.0840	.845043	.154957
3	2.6730	.160381	.839619	33	14.2302	.853814	.146186
4	3.4651	.207906	.792094	34	14.3681	.862088	.137912
5	4.2124	.252742	.747258	35	14.4982	.869895	.130105
6	4.9173	.295039	.704961	36	14.6210	.877259	.122741
7	5.5824	.334943	.665057	37	14.7368	.884207	.115793
8	6.2098	.372588	.627412	38	14.8460	.890761	.109239
9	6.8017	.408102	.591893	39	14.9491	.896944	.103056
10	7.3601	.441605	.558395	40	15.0463	.902778	.097222
11	7.8869	.473212	.526788	41	15.1380	.908281	.091719
12	8.3838	.503031	.496969	42	15.2245	.913473	.086527
13	8.8527	.531161	.468839	43	15.3062	.918370	.081630
14	9.2950	.557699	.442301	44	15.3832	.922991	.077009
15	9.7122	.582735	.417265	45	15.4558	.927350	.072650
16	10.1059	.606354	.393646	46	15.5244	.931462	.068538
17	10.4773	.628636	.371364	47	15.5890	.935342	.064653
18	10.8276	.649656	.350344	48	15.6500	.939002	.060998
19	11.1581	.669487	.330513	49	15.7076	.942454	.057546
20	11.4699	.688195	.311805	50	15.7619	.945712	.054288
21	11.7641	.705845	.294155	51	15.8131	.948785	.051215
22	12.0416	.722495	.277505	52	15.8614	.951684	.048816
23	12.3034	.738203	.261797	53	15.9070	.954418	.045582
24	12.5504	.753021	.246979	54	15.9500	.956999	.043001
25	12.7834	.767001	.232999	55	15.9905	.959433	.040567
26	13.0032	.780190	.219310	56	16.0288	.961729	.038271
27	13.2105	.792632	.207368	57	16.0649	.963895	.036105
28	13.4062	.804370	.195630	58	16.0990	.965939	.034061
29	13.5907	.815443	.184557	59	16.1311	.967867	.032133
30	13.7648	.828590	.174110	60	16.1614	.969686	.030314

[¶306]

Gift Tax Weighed over Life Expectancy

In his work, "You, Your Heirs and Your Estate," George Byron Gordon gives us this table to use in weighing projected tax savings for loss of future income on the gift tax payment.

Gift Tax Rate	*Years of Expectancy*							
	5	10	15	20	25	30	35	40
				Interest Factor at 2.5%				
	1.131	1.28	1.448	1.639	1.854	2.098	2.373	2.685
8¼	9.33	10.56	11.95	13.52	15.3	17.31	19.58	22.15
10½	11.88	13.44	15.2	17.21	19.47	22.03	24.92	28.19
13½	15.27	17.28	19.55	22.13	25.03	28.32	32.04	36.25
16½	18.66	21.12	23.89	27.04	30.59	34.62	39.15	44.3
18¾	21.21	24.00	27.15	30.73	34.76	39.34	44.49	50.34
21	23.75	26.88	30.41	34.42	38.93	44.06	49.83	56.39
22½	25.45	28.8	32.58	36.88	41.72	47.21	53.39	60.41
24	27.14	30.72	34.75	39.34	44.5	50.35	56.95	64.44
26¼	29.69	33.6	38.01	43.02	48.67	55.07	62.29	70.48
27¾	31.39	35.52	40.18	45.48	51.45	58.22	65.85	74.51
29¼	32.08	37.44	42.35	47.94	54.23	61.37	69.41	78.54
31½	35.63	40.32	45.61	51.63	58.4	66.09	74.75	84.58
33¾	38.17	43.2	48.87	55.32	62.57	70.81	80.09	90.62
36¾	41.56	47.04	53.21	60.23	68.13	77.1	87.21	98.67
39¾	44.96	50.88	57.56	65.15	73.7	83.4	94.33	106.73
42	47.5	53.76	60.82	68.84	77.87	88.12	99.67	112.77

Gifts are economical to the point of equality between the weighted gift tax rate above and the effective estate tax rate on the amount of gift.

Example: A net gift of $20,000 subject to an effective gift tax rate of 16½% is contemplated. It represents the top $20,000 of a taxable estate of $250,000. The applicable estate tax rate is 26.28%. The gift would be economical for a donor having a life expectancy up to 15 years.

Note: To find the factor for a different interest rate, go to the table at page 5 and find the factor for the number of years of life expectancy. Divide the factor you find there by the factor for 2.5% at the top of this table, and multiply that number times the number you get in this table.

Example: To change from 2½% to 6%, 20 year expectancy, 24% gift tax rate. From table on page 5, at 20 years and 6% (3.2075)

Divided by factor from this table (1.639) = 1.957

Times number from this table at 24% (39.34) = 76.99

Table Illustrating Estate Tax Savings by Making Gifts

The following table shows the dramatic tax savings which are made possible by the use of gifts. Column 7 shows the percent of estate tax saved when ½ the estate is given away during life, and Column 8 shows the further tax-free gain possible if the amounts given away are used to buy life insurance on the life of the donor.

(1) Decedent needs estate (after debts and probate expenses) of	(2) To leave heirs net of	(3) Estate tax on col. 1	(4) Gift tax on lifetime gifts of half of amount in col. 1	(5) Estate tax on balance of estate after gifts and gift tax are deducted	(6) Net to heirs [col. 1 − (cols. 4 + 5)]	(7) Percent of col. 3 estate tax saved through life-time gifts	(8) Net to heirs if lifetime gifts are of 10 pay life premiums on estate owner (age 45)
$ 65,161	$ 65,000	$ 161	—	—	$ 65,161	100%	$ 81,451
81,860	80,000	1,860	—	—	81,860	100%	101,075
119,333	110,000	9,333	—	—	119,333	100%	149,166
189,571	160,000	29,571	2,896	3,340	183,335	79%	230,728
333,088	260,000	73,088	16,997	17,773	298,318	52%	381,590
480,147	360,000	120,147	33,541	34,660	411,946	43%	531,982
784,151	560,000	224,151	68,973	69,893	645,285	38%	841,322
1,611,273	1,060,000	551,273	173,754	170,859	1,266,660	38%	1,669,478
3,665,365	2,060,000	1,605,365	488,180	437,691	2,739,494	42%	3,655,835
6,527,667	3,060,000	3,467,667	1,033,010	836,904	4,657,753	46%	6,289,670

Cost of Charitable Contributions

If Taxable Income Before Deducting Contribution is:	Single Taxpayer	Cost per $100 of Contribution is: Married Filing Jointly	Married Filing Separately	Head of Household
$ 10,000	$75	$78	$72	$77
13,000	71	75	64	73
17,000	66	72	58	69
22,000	62	68	52	65
30,000	55	61	47	58
50,000	40	50	40	45
60,000	38	47	38	42
80,000	34	42	34	38
110,000	30	38	30	34
130,000	30	36	30	33
150,000	30	34	30	32
170,000	30	32	30	31
190,000	30	31	30	30
250,000	30	30	30	30

How Much Federal Estate Tax?

(A) Taxable estate equal to or more than–	(B) Taxable estate less than–	(C) Tax on amount in column (A)	(D) Rate of tax on excess over amount in column (A) Percent
. . . .	$ 5,000	. . . . . .	3
$ 5,000	10,000	$ 150	7
10,000	20,000	500	11
20,000	30,000	1,600	14
30,000	40,000	3,000	18
40,000	50,000	4,800	22
50,000	60,000	7,000	25
60,000	100,000	9,500	28
100,000	250,000	20,700	30
250,000	500,000	65,700	32
500,000	750,000	145,700	35
750,000	1,000,000	233,200	37
1,000,000	1,250,000	325,700	39
1,250,000	1,500,000	423,200	42
1,500,000	2,000,000	528,200	45
2,000,000	2,500,000	753,200	49
2,500,000	3,000,000	998,200	53
3,000,000	3,500,000	1,263,200	56
3,500,000	4,000,000	1,543,200	59
4,000,000	5,000,000	1,838,200	63
5,000,000	6,000,000	2,468,200	67
6,000,000	7,000,000	3,138,200	70
7,000,000	8,000,000	3,838,200	73
8,000,000	10,000,000	4,568,200	76
10,000,000	. . . .	6,088,200	77

State Death Taxes

The following table contains the rate and nature of taxation (indicated by "√") imposed by each of the several states and territories. In most areas, an additional tax (col. 4), sometimes referred to as the "gap" tax, is imposed to absorb the maximum credit allowed against the Federal estate tax for death taxes paid to other jurisdictions. Four states, Alabama, Arkansas, Florida, and Georgia, impose an initial estate tax (col. 5) which is based solely on the Federal estate tax provisions, and the tax, in the last analysis, will equal the Federal credit allowed.

State	Inheritance Tax	Estate Tax	Rates in %	Additional Est. Tax; Picks up Fed. Credit	Estate Tax for Federal Credit only
ALABAMA					√
ALASKA	√		1-17½	√	
ARIZONA		√	.8-16	√	
ARKANSAS					√
CALIFORNIA	√		2-24	√	
COLORADO	√		2-16	√	
CONNECTICUT	√		2-14	√	
DELAWARE	√		1-8	√	
DIST. OF COLUMBIA	√		1-15	√	
FLORIDA					√
GEORGIA					√
HAWAII	√		1½-9	√	
IDAHO	√		2-30	√	
ILLINOIS	√		2-30	√	
INDIANA	√		1-20	√	
IOWA	√		1-15	√	
KANSAS	√		1½-15	√	
KENTUCKY	√		2-16	√	
LOUISIANA	√		2-10	√	
MAINE	√		2-18	√	
MARYLAND	√		1-7½	√	
MASSACHUSETTS	√		1-15	√	
MICHIGAN	√		2-15	√	
MINNESOTA	√		1½-30	√	
MISSISSIPPI		√	1-16		
MISSOURI	√		1-30	√	
MONTANA	√		2-32	√	
NEBRASKA	√		6-18	√	
NEVADA (No Tax)					
NEW HAMPSHIRE	√		10	√	
NEW JERSEY	√		1-16	√	
NEW MEXICO	√		1-5	√	
NEW YORK		√	2-21	√	
NORTH CAROLINA	√		1-17	√	
NORTH DAKOTA		√	2-23		

State Death Taxes *(continued)*

State	Inheri-tance Tax	Estate Tax	Rates in %	Additional Est. Tax; picks up Fed. Credit	Estate Tax for Federal Credit only
OHIO	√		1-11	√	
OKLAHOMA		√	1-10	√	
OREGON	√	√	1-20		
PENNSYLVANIA	√		2-15	√	
PUERTO RÍCO	√		5-70		
RHODE ISLAND	√	√	2-15	√	
SOUTH CAROLINA		√	4-6	√	
SOUTH DAKOTA	√		1-20		
TENNESSEE	√		1-15	√	
TEXAS	√		1-20	√	
UTAH		√	3-10		
VERMONT	√		2-12	√	
VIRGINIA	√		1-15	√	
WASHINGTON	√		1-25	√	
WEST VIRGINIA	√		3-30		
WISCONSIN	√		2-40	√	
WYOMING	√		2-6	√	

[¶311]

Maximum Credit for State Death Taxes

The maximum credit for state death taxes is illustrated by the following table. The credit is 0 if the taxable estate is less than $40,000. Also note that the credit is limited to the amount actually paid as state death taxes.

(A) Taxable estate equal to or more than–	(B) Taxable estate less than–	(C) Credit on amount in column (A)	(D) Rates of credit on excess over amount in column (A) Percent
$ 40,000	$ 90,000	. . . .	0.8
90,000	140,000	$ 400	1.6
140,000	240,000	1,200	2.4
240,000	440,000	3,600	3.2
440,000	640,000	10,000	4.0
640,000	840,000	18,000	4.8
840,000	1,040,000	27,600	5.6
1,040,000	1,540,000	38,800	6.4
1,540,000	2,040,000	70,800	7.2
2,040,000	2,540,000	106,800	8.0
2,540,000	3,040,000	146,800	8.8
3,040,000	3,540,000	190,800	9.6
3,540,000	4,040,000	238,800	10.4
4,040,000	5,040,000	290,800	11.2
5,040,000	6,040,000	402,800	12.0
6,040,000	7,040,000	522,800	12.8
7,040,000	8,040,000	650,800	13.6
8,040,000	9,040,000	786,800	14.4
9,040,000	10,040,000	930,800	15.2
10,040,000	. . . .	1,082,800	16.0

Estate Tax With and Without Credit
For State Death Taxes

Taxable Estate	Net Tax (Without Credit)	Net Tax (With Maximum Credit)
$ 5,000	$ 150	$ 150
10,000	500	500
20,000	1,600	1,600
30,000	3,000	3,000
40,000	4,800	4,800
50,000	7,000	6,920
60,000	9,500	9,340
100,000	20,700	20,140
250,000	65,700	61,780
500,000	145,700	133,300
750,000	233,200	209,920
1,000,000	325,700	289,140
1,250,000	423,200	370,960
1,500,000	528,200	459,960
2,000,000	753,200	649,280
2,500,000	998,200	854,600
3,000,000	1,263,200	1,075,920
3,500,000	1,543,200	1,308,240
4,000,000	1,838,200	1,551,560
5,000,000	2,468,200	2,069,880
6,000,000	3,138,200	2,620,200
7,000,000	3,838,200	3,192,520
8,000,000	4,568,200	3,786,840
10,000,000	6,088,200	5,011,480

How Much of Estate Is Left After Taxes

If Taxable Estate Is:	Tax Is:	Amount of Estate Left Is:
$ 5,000	$ 150	$ 4,850
10,000	500	9,500
20,000	1,600	18,400
30,000	3,000	27,000
40,000	4,800	35,200
50,000	7,000	43,000
60,000	9,500	50,500
100,000	20,700	79,300
250,000	65,700	184,300
500,000	145,700	354,300
750,000	233,200	516,800
1,000,000	325,700	674,300
1,250,000	423,200	826,800
1,500,000	528,200	971,800
2,000,000	753,200	1,246,800
2,500,000	998,200	1,501,800
3,000,000	1,263,200	1,736,800
3,500,000	1,543,200	1,956,800
4,000,000	1,838,200	2,161,800
5,000,000	2,468,200	2,531,800
6,000,000	3,138,200	2,861,800
7,000,000	3,838,200	3,161,800
8,000,000	4,568,200	3,431,800
10,000,000	6,088,200	3,911,800

[¶314]
Using Treasury Bonds to Pay Estate Taxes

Many estate owners enable their executors to take advantage of a peculiar feature of certain U.S. Treasury Bonds. Most of these bonds can be purchased at a discount from their par value. Yet, assuming they are still selling below par at the time of the decedent's death, they can, nevertheless, be turned in at par for the purpose of paying federal estate tax. Even a purchase in anticipation of death will permit the estate to get the benefit of the discount. It doesn't matter how long the decedent has held the bonds, as long as they were owned by him and formed a part of his estate at the time of his death.

Since these bonds form a part of the decedent's estate, the question arises as to how they are to be valued in the estate. The 2nd Circuit has ruled (*Bankers Trust Co.*, 284 F. 2d 537, 1960; *Fried*, 445 F. 2d 979, 1970) that to the extent the bonds can be used at par to pay estate taxes, whether or not they are so applied, they are to be valued at least at par in the gross estate. This limits somewhat, but does not eliminate, the estate tax advantage inherent in such bonds when bought at a discount from par. At the same time, if the estate is to be the recipient of substantial income, valuing the bonds at par rather than at market value may produce an income tax savings. If valued at the market and then the executor redeems at par, there would be a capital gains tax to pay, based on the difference. If valued in the estate at par there would be no capital gain at redemption.

[¶315]
Treasury Bonds Which Can Be Redeemed at Par to Pay Federal Estate Taxes*

Series

4¼s, May 1975-78
3¼s, June 1978-83
4s,   Feb. 1980
3½s, Nov. 1980
3¼s, May 1985
4¼s, Aug. 1987-92
4s,   Feb. 1988-93
4⅛s, May 1989-94
3½s, Feb. 1990
3s,   Feb. 1995
3½s, Nov. 1998

Example: 3¼s, May 1985, Bid 73.20, par 100. However, some of the premium will be lost to the estate tax. If the expected rate of the estate tax is 40%, then 40% will be lost. So, the net premium would be 60% of 36.6%, or 21.96%.

---

*The market prices of these bonds can usually be found in many daily newspapers under the heading "Government Securities." To calculate the percent of estate tax premium, divide the par value by the bid price.

How Much Estate Tax Without Marital Deduction

Taxable Estate Before Exemption		Federal Tax Without Marital Deduction		Maximum Credit for State Death Tax		Federal Tax with Maximum Credit for State Tax	
		Tax on lowest amount in first column	Plus this % of excess	Credit on lowest amount in first column	Plus this % of excess	Tax on lowest amount in first column	Plus this % of excess
$ 0	$ 60,000		0				0
60,000	65,000		3%				3%
65,000	70,000	$ 150	7			$ 150	7
70,000	80,000	500	11			500	11
80,000	90,000	1,600	14			1,600	14
90,000	100,000	3,000	18			3,000	18
100,000	110,000	4,800	22		.8%	4,800	21.2
110,000	120,000	7,000	25	$ 80	.8	6,920	24.2
120,000	150,000	9,500	28	160	.8	9,340	27.2
150,000	160,000	17,900	28	400	1.6	17,500	26.4
160,000	200,000	20,700	30	560	1.6	20,140	28.4
200,000	300,000	32,700	30	1,200	2.4	31,500	27.6
300,000	310,000	62,700	30	3,600	3.2	59,100	26.8
310,000	500,000	65,700	32	3,920	3.2	61,780	28.8
500,000	560,000	126,500	32	10,000	4.0	116,500	28.0
560,000	700,000	145,700	35	12,400	4.0	133,300	31.0
700,000	810,000	194,700	35	18,000	4.8	176,700	30.2
810,000	900,000	233,200	37	23,280	4.8	209,920	32.2
900,000	1,060,000	266,500	37	27,600	5.6	238,900	31.4
1,060,000	1,100,000	325,700	39	36,560	5.6	289,140	33.4
1,100,000	1,310,000	341,300	39	38,800	6.4	302,500	32.6
1,310,000	1,560,000	423,200	42	52,240	6.4	370,960	35.6
1,560,000	1,600,000	528,200	45	68,240	6.4	459,960	38.6
1,600,000	2,060,000	546,200	45	70,800	7.2	475,400	37.8
2,060,000	2,100,000	753,200	49	103,920	7.2	649,280	41.8
2,100,000	2,560,000	772,800	49	106,800	8.0	666,000	41.0
2,560,000	2,600,000	998,200	53	143,600	8.0	854,600	45.0
2,600,000	3,060,000	1,019,400	53	146,800	8.8	872,600	44.2
3,060,000	3,100,000	1,263,200	56	187,280	8.8	1,075,920	47.2
3,100,000	3,560,000	1,285,600	56	190,800	9.6	1,094,800	46.4
3,560,000	3,600,000	1,543,200	59	234,960	9.6	1,308,240	49.4
3,600,000	4,060,000	1,566,800	59	238,800	10.4	1,328,000	48.6
4,060,000	4,100,000	1,838,200	63	286,640	10.4	1,551,560	52.6
4,100,000	5,060,000	1,863,400	63	290,800	11.2	1,572,600	51.8
5,060,000	5,100,000	2,468,200	67	398,320	11.2	2,069,880	55.8
5,100,000	6,060,000	2,495,000	67	402,800	12.0	2,092,200	55.0
6,060,000	6,100,000	3,138,200	70	518,000	12.0	2,620,200	58.0
6,100,000	7,060,000	3,166,200	70	522,800	12.8	2,643,400	57.2
7,060,000	7,100,000	3,838,200	73	645,680	12.8	3,192,520	60.2
7,100,000	8,060,000	3,867,400	73	650,800	13.6	3,216,600	59.4

How Much Estate Tax Without Marital Deduction *(continued)*

Taxable Estate Before Exemption		Federal Tax Without Marital Deduction		Maximum Credit for State Death Tax		Federal Tax with Maximum Credit for State Tax	
		Tax on low-est amount in first column	Plus this % of ex-cess	Credit on lowest this % amount in first column	Plus est amount of ex-cess	Tax on low-this % in first column	Plus of ex-cess
8,060,000	8,100,000	4,568,200	76	781,360	13.6	3,786,840	62.4
8,100,000	9,100,000	4,598,600	76	786,800	14.4	3,811,800	61.6
9,100,000	10,060,000	5,358,600	76	930,800	15.2	4,427,800	60.8
10,060,000	10,100,000	6,088,200	77	1,076,720	15.2	5,011,480	61.8
10,100,000		6,119,000	77	1,082,800	16.0	5,036,200	61.0

NOTE: The federal tax both with and without the maximum credit for state tax is given. In general, state tax will at least equal the amount of credit. To the extent that your state inheritance tax exceeds the maximum credit, the total of state and federal tax will exceed the maximum federal tax.

[¶317]   How Much Estate Tax With Marital Deduction

In order to compute the estate tax with the marital deduction, it is necessary to compute the adjusted gross estate. Where community property is not involved, adjusted gross estate is simply the gross estate less any Code §2053 or 2054 expenses. The amount of the marital deduction is limited to one-half of the adjusted gross estate or the amount of property passing to the surviving spouse, whichever is less.

Where community property is involved, the adjusted gross estate is reduced by the amount of the community property, and the §2053 and 2054 deductions are also adjusted (see Regs. §20,2056(c)-2(j), example 1).

Once the amount of the marital deduction is computed, the amount is deducted from the estate before the tax rates are applied. The effect of this is to reduce the estate tax considerably in most cases, since the tax will be computed on only about one-half the estate. Of course, when the surviving spouse later dies, these amounts will presumably be included in his or her estate and taxed then. However, the surviving spouse may give away much of the property, so long as the gifts are not in contemplation of death, and thus avoid the estate tax entirely. Also, the surviving spouse's estate will have an exemption of $60,000, so even more of the estate can be sheltered from tax.

On the negative side, the marital deduction is an exceedingly complex provision which must be carefully analyzed in order to prevent serious blunders. But if employed properly, it can result in substantial tax savings.

[¶318] # Funeral and Administration Costs

The estate planner invariably considers the impact of estate taxes in computing shrinkage and in determining the amount of cash, insurance, and liquid assets which should be available to the estate. While the impact of combined funeral and administration costs may not generally be as great, especially in larger estates, it should not be neglected. Payment of these costs are, with few exceptions, as sure as taxes, and the necessity for cash to meet them could make inroads into a thin controlling business interest or necessitate the sale of property which was intended to be held.

Included in the administration category are executors', attorneys', accountants' and appraisers' fees, court costs, and miscellaneous items such as traveling and storage expenses, publication fees, title search fees, etc.

The table which follows is based on a U.S. Treasury report on Fiduciary, Gift, and Estate Tax. Actual projections for any particular estate, of course, cannot be presumed from this table. They will vary from case to case, depending upon the nature of the estate property, its intended distribution pattern, and place of probate and location of the property, as well as other factors.

Gross Estate		% Claimed for Funeral and Administration Expenses
$ 60,000 under 70,000		5.1
70,000 under 80,000		5.0
80,000 under 90,000		4.7
90,000 under 100,000		4.6
100,000 under 120,000		4.5
120,000 under 150,000		4.4
150,000 under 200,000		4.4
200,000 under 300,000		4.5
300,000 under 500,000		4.4
500,000 under 1,000,000		4.0
1,000,000 under 2,000,000		3.9
2,000,000 under 3,000,000		3.6
3,000,000 under 5,000,000		3.9
5,000,000 under 10,000,000		3.8
10,000,000 under 20,000,000		3.4

[¶319]   **Executors' Commissions—State by State**

Most of the states statutorily set forth a schedule of fees for executors and administrators, while some merely call for reasonable fees, the reasonableness to be determined by the courts. In a number of states, fees for testamentary trustees are the same as those allowed to executors. Some states provide for a distinct statutory fee for trustees, but most provide for reasonable fees to be determined by the court, more often than not based in large measure upon trust receipts—with 5% annually being a fairly reasonable national average. In the case of both executors and testamentary trustees, additional reasonable fees may usually be charged for extraordinary services.

The table that follows relates exclusively to compensation of executors and administrators and is based on both statutory fee allowances and the usual fees charged by corporate executors where a statutory rate does not apply. For the latter, appreciation is expressed to the American College of Probate Counsel, which was kind enough to supply a compilation from which we have borrowed some of the nonstatutory fees.

Alabama:	Not more than 2½% of receipts and disbursements.
Alaska:	First $1000—7% Next $1000—5% Next $2000—4% All above $4000—2%
Arizona:	First $1000—7% Next $9000—5% All above $10,000—4%
Arkansas:	First $1000—Not more than 10% Next $4000—5% All above $5000—3%
California:	First $1000—7% Next $100,000—2% Next $9000—4% Next $350,000—1½% Next $40,000—3% All above $500,000—1%
Colorado:	First $25,000—6% Next $75,000—4% All above $100,000—3%
Connecticut:	No statutory fee schedule or minimums. The following schedule has been suggested as reasonable: First $1000—5% Next $490,000—2% Next $4000—4% Next $1,000,000—1½% Next $5000—3% All above $1,500,000—1%
Delaware:	No statutory fee schedule or minimums. Fees are determined by the Register of Wills. Some typical percentage fees allowed on total gross estates are as follows: To $50,000—6% To $300,000—4% To $100,000—5% To $500,000—3.5% To $200,000—4.5% To $1,000,000—3% To $2,000,000—2.5%
District of Columbia:	Not under 1% nor more than 10%.
Florida:	First $1000—6% Next $4000—4% All above $5000—2½%
Georgia:	2½% of receipts and disbursements.

Executors' Commissions—State by State *(continued)*

Hawaii:	On receipts, 7% of first $5000—all above $5000, 5%. On principal: First $1000—5% Next $9000—4% Next $10,000—3% All above $20,000—2%
Idaho:	First $1000—5% Next $9000—4% All above $10,000—3%
Illinois:	No statutory fee schedule or minimums. The following is an example of customary rates: First $25,000—5% Next $150,000—2½% Next $25,000—3½% Next $750,000—2% Next $50,000—3% All above $1,000,000—1½%
Indiana:	No statutory fee schedule or minimums. The following is an example of customary rates: First $10,000—5% Next $150,000—2½% Next $40,000—3½% All above $250,000—2% Next $50,000—3%
Iowa:	First $1000—not more than 6% Next $4000—4% All above $5000—2%
Kansas:	No statutory fee schedule or minimums. The following is an example of customary rates: First $10,000—5% Next $50,000—2% Next $15,000—4% All above $100,000—1% Next $25,000—3%
Kentucky:	Not more than 5% of income and 5% of personal estate.
Louisiana:	Fee is 2½% of the inventory of the estate. It may be increased by the court upon showing that usual commission is inadequate.
Maine:	Not more than 5% of personal estate, with rate being reduced as a matter of practice in larger estates.
Maryland:	First $20,000—not more than 10%. All above $20,000—not more than 4%.
Massachusetts:	No statutory fee schedule or minimums. The following is an example of customary rates: 3% of gross personal estate, plus 3% on any real estate that is sold, with rate being reduced as a matter of practice as the size of the estate increases.
Michigan:	First $1000—5% Next $4000—2½% All above $5000—2%
Minnesota:	No statutory fee schedule or minimums. The following is an example of customary rates: First $100,000—3% Next $400,000—2% Next $500,000—1½% All above $1,000,000—1%
Mississippi:	Not less than 1% nor more than 7% on amount of estate administered.
Missouri:	First $5000—5% Next $300,000—2¾% Next $20,000—4% Next $600,000—2½% Next $75,000—3% All above $1,000,000—2%
Montana:	First $1000—7% Next $9000—5% Next $10,000—4% All above $20,000—2%
Nebraska:	First $1000—5% Next $4000—2½% All above $5000—2%

¶319

Executors' Commissions—State by State *(continued)*

Nevada:
First $1000—6%
Next $4000—4%
All above $5000—2%

New Hampshire:
No statutory fee schedule or minimums. Suggested schedule by the New Hampshire Bar Association is as follows:

First $1000—7%   Next $60,000—3%
Next $4000—5%   Next $50,000—2½%
Next $10,000—4%   All above $125,000—2%

New Jersey:
On income—6%. On corpus not exceeding $100,000—5%.
On excess over $100,000—the percentage, not in excess of 5%, in discretion of the court. Usual rates—5% of first $100,000 and 3½% of excess.

New Mexico:
First $3000—10%
All above $3000—5%
For cash, U.S. savings bonds, or life insurance proceeds, the compensation is 5% on the first $5000 and 1% on everything above that figure.

New York:
First $ 25,000—4%
Next $125,000—3½%
Next $150,000—3%
All above $300,000—2%

North Carolina:
Not more than 5% of receipts and disbursements.

North Dakota:
First $1000—5%
Next $5000—3%
Next $44,000—2%
All above $50,000, within the discretion of the court, but not above 2%

Ohio:
First $1000—6%
Next $4000—4%
All above $5000—2%

Oklahoma:
First $1000—5%
Next $4000—4%
All above $5000—2½%

Oregon:
First $1000—7%
Next $9000—4%
Next $40,000—3%
All above $50,000—2%

Pennsylvania:
No statutory fee schedule or minimums. The following is an example of suggested fees:

On principal:

First $50,000—5%   Next $250,000—2½%
Next $50,000—4%   All over $500,000—2%
Next $150,000—3%

Additional commissions on gross income might be as follows:

First $25,000—5%
Next $25,000—4%
All over $50,000—3%

Rhode Island:
No statutory fee schedules or minimums. The following is an example of suggested rates:
First $500,000—3 to 3½%
Over $500,000—subject to lesser rates

South Carolina:
Not more than 2½% of receipts and disbursements.

South Dakota:
First $1000—5%
Next $4000—4%
All above $5000—2½%

¶319

Executors' Commissions—State by State *(continued)*

Tennessee: No statutory fee schedule or minimums. The following is an example of suggested reasonable rates:

First $20,000—5%   Next $200,000—3%

Next $80,000—4%   All above $300,000—2%

Texas: Not more than 5% of the value of the administered estate.

Utah: First $1000—5%   Next $40,000—2%

Next $4000—4%   Next $50,000—1½%

Next $5000—3%   All above $100,000—1%

Vermont: Statute provides $4 a day. Customary charges, in addition, are 4% of probate estate on the first $100,000, with declining rates on sums over $100,000.

Virginia: No statutory fee schedule or minimums. The following is an example of suggested reasonable fees.

On principal:

First $50,000—5%

Next $50,000—4%

Next $900,000—3%

All above $1,000,000—2%

On income:

5% on all receipts

Washington: No statutory fee schedule or minimums. The following is an example of reasonable fees.

First $5000—5%   Next $250,000—1½%

Next $45,000—3%   All above $500,000—1%

Next $200,000—2%

West Virginia: No statutory fee schedule or minimums—an example of the customary rate of compensation is 5% on receipts.

Wisconsin: 2%

Wyoming: First $1000—10%

Next $4000—5%

Next $15,000—3%

All above $20,000—2%

[¶320]   Liquidity Value of Assets

Cash will be needed to meet funeral expenses, executor's and trustee's fees, and estate taxes. Potential liabilities and the ability to convert assets into cash to meet them must be estimated as accurately as possible during life.

The cash an experienced professional executor may reasonably expect to get for estate assets may be estimated by applying the percentages indicated below.

Assets	Percentage of Market Value Which Experienced Executor May Realize in Cash
Cash in bank	100
U.S. accumulation bonds	100
Government bonds	100
Municipals	90
Listed bonds	90
Listed common stocks:	
High-grade investment	85
High-grade speculative	70
Preferred stocks	90
Unlisted bonds:	
High-grade	85
Other	60
Unlisted stocks:	
High-grade	80
Other	30
Mortgages:	
High-grade	100
Other	70
Real estate	Your estimate
Close corporation stock:	
Subject to Buy-Sell Agreement	Price stipulated
Other	100-30
Partnership interests:	
Subject to Buy-Sell Agreement	Price stipulated
Other	Liquidated value
Proprietorship interests:	
Subject to Buy-Sell Agreement	Price stipulated
Other	Liquidated value
Interests in trusts and estates, based on underlying assets	100-30
Personal effects	Your estimate
Life insurance	100
Annuities	100
Other assets	Your estimate

The market value of a business that is to be liquidated is its liquidation value with some variation, depending upon the nature of the business. The ratio of the liquidating value of assets to their book value may reasonably be figured as follows:

Assets	Percentage of Market Value
Cash	100
Accounts receivable	85-25
Inventory	100-35
Realty	100-50
Fixtures	50-10
Equipment	75-25

¶320

[¶321]

Basic Annual Premiums Per $1,000 of Permanent Life Insurance

There are two primary kinds of permanent life insurance policies, namely, Ordinary Life and Endowment.

Under an Ordinary Life policy, the benefits consist of the payment of the sum insured to the beneficiary at the death of the insured. The consideration for this policy usually takes the form of premium payments throughout the lifetime of the insured. If the premium payments are for a limited number of years, e.g., 20 years or paid up at 65, it is called a limited payment life policy (which is simply a variation of Ordinary Life).

An endowment policy provides for the payment of the sum insured in case of the death of the insured, and in addition guarantees the payment of the sum insured at the end of a stipulated period if the insured is then living. The premium for this policy may be paid throughout the whole period of the endowment, or it may be paid in a shorter number of years or in a single payment as in single premium Ordinary Life policies.

The table which follows illustrates the rates for the various types of basic policies as described. These are for nonparticipating policies (i.e., those which do not pay dividends). They are rule-of-thumb amounts only to give you a quick idea what the net cost of a particular policy at a particular age will be. To get exact premium costs and make comparisons, check the rate books.

[¶322]

Basic Annual Premiums Per $1,000 of Insurance (Nonparticipating)

Age	Ordi- nary Life Male	20 Pay. Life Male	Life Pay. to 65 Male	Life Pay. to 65 Fem.	20 Year End. Male	30 Year End. Male	Age	Ordi- nary Life Male	20 Pay. Life Male	Life Pay. to 65 Male	Life Pay. to 65 Fem.	20 Year End. Male	30 Year End. Male
5	7.59	15.22	8.00	7.57	42.92	24.90	38	21.07	31.73	26.00	24.17	44.78	28.67
10	8.45	16.90	9.01	8.52	42.97	24.95	39	21.91	32.52	27.30	25.42	44.99	29.09
15	9.74	18.79	10.65	9.83	43.02	25.00	40	22.78	33.35	28.75	26.57	45.24	29.55
16	10.03	19.21	11.00	10.18	43.04	25.05	41	23.68	34.21	30.26	27.99	45.52	30.04
17	10.33	19.65	11.36	10.55	43.08	25.17	42	24.61	35.10	31.87	29.50	45.83	30.57
18	10.65	20.10	11.73	10.93	43.12	25.32	43	25.58	36.02	33.65	31.14	46.16	31.13
19	10.97	20.56	12.11	11.30	43.16	25.48	44	26.60	36.97	35.67	32.99	46.53	31.76
20	11.31	21.02	12.52	11.68	43.20	25.60	45	27.68	37.94	37.94	35.10	46.94	32.46
21	11.65	21.47	12.94	12.07	43.23	25.69	46	28.81	38.93	40.35	37.40	47.38	33.24
22	12.00	21.93	13.37	12.48	43.25	25.77	47	29.98	39.95	42.85	39.83	47.86	34.08
23	12.37	22.39	13.82	12.92	43.27	25.84	48	31.21	40.99	45.58	42.51	48.38	34.98
24	12.75	22.86	14.31	13.37	43.29	25.91	49	32.52	42.08	48.74	45.57	48.93	35.95
25	13.16	23.36	14.83	13.83	43.33	25.98	50	33.90	43.20	52.57	49.18	49.51	36.98
26	13.60	23.88	15.39	14.32	43.38	26.04	51	35.36	44.35	56.69	53.03	50.09	—
27	14.05	24.42	15.97	14.85	43.43	26.09	52	36.88	45.51	60.99	57.05	50.67	—
28	14.53	24.97	16.59	15.43	43.48	26.15	53	38.48	46.73	66.07	61.89	51.31	—
29	15.04	25.55	17.26	16.06	43.55	26.23	54	40.17	48.02	72.66	68.13	52.04	—
30	15.58	26.15	18.00	16.74	43.64	26.36	55	41.98	49.43	81.51	76.45	52.93	—
31	16.15	26.77	18.80	17.46	43.74	26.54	56	43.86	50.92	—	—	53.93	—
32	16.76	27.42	19.66	18.23	43.85	26.76	57	45.81	52.48	—	—	55.01	—
33	17.39	28.09	20.58	19.07	43.98	27.02	58	47.87	54.13	—	—	56.23	—
34	18.06	28.78	21.56	19.96	44.12	27.30	59	50.11	55.93	—	—	57.65	—
35	18.76	29.49	22.60	20.85	44.27	27.61	60	52.56	57.92	—	—	59.35	—
36	19.50	30.22	23.68	21.90	44.43	27.94	65	67.80	70.33	—	—	71.45	—
37	20.27	30.96	24.80	22.99	44.60	28.29	70	91.42	—	—	—	—	—

[¶323]

Cash Value and Net Cost of Nonparticipating
Policies After 10 and 20 Years
(Per $1,000—CSO 3%)

Ordinary Life

	Age 30 $15.60		Age 35 $18.75		Age 40 $22.80		Age 45 $27.70	
Annual Premiums								
	10 Yrs	20 Yrs	10 Yrs	20 Yrs	10 Yrs	20 Yrs	10 Yrs	20 Yrs
Total Premiums	$156.00	$312.00	$187.50	$375.00	$228.00	$456.00	$277.00	$554.00
Cash Value	124.00	303.00	147.00	345.00	173.00	390.00	201.00	437.00
Net Cost or Gain*	32.00	9.00	40.50	30.00	55.00	66.00	76.00	117.00
Average Cost or Gain*	3.20	.45	4.05	1.50	5.50	3.30	7.60	5.85

Life Paid-Up at 65

	Age 30 $18.00		Age 35 $22.60		Age 40 $28.77		Age 45 $38.00	
Annual Premiums								
	10 Yrs	20 Yrs	10 Yrs	20 Yrs	10 Yrs	20 Yrs	10 Yrs	20 Yrs
Total Premiums	$180.00	$360.00	$226.00	$452.00	$287.70	$575.40	$380.00	$760.00
Cash Value	145.00	355.00	181.00	430.00	228.00	538.00	300.00	716.00
Net Cost or Gain*	35.00	5.00	45.00	22.00	59.70	37.40	80.00	44.00
Average Cost or Gain*	3.50	.25	4.50	1.10	5.97	1.87	8.00	2.20

20–Pay Life

	Age 30 $26.20		Age 35 $29.50		Age 40 $33.40		Age 45 $38.00	
Annual Premiums								
	10 Yrs	20 Yrs	10 Yrs	20 Yrs	10 Yrs	20 Yrs	10 Yrs	20 Yrs
Total Premiums	$262.00	$524.00	$295.00	$590.00	$334.00	$668.00	$380.00	$760.00
Cash Value	223.00	549.00	249.00	605.00	274.00	661.00	299.00	716.00
Net Cost or Gain*	39.00	25.00*	46.00	15.00*	60.00	7.00	81.00	44.00
Average Cost or Gain*	3.90	1.25*	4.60	.75*	6.00	.35	8.10	2.20

20–Year Endowment

	Age 30 $43.65		Age 35 $44.30		Age 40 $45.25		Age 45 $47.00	
Annual Premiums								
	10 Yrs	20 Yrs	10 Yrs	20 Yrs	10 Yrs	20 Yrs	10 Yrs	20 Yrs
Total Premiums	$436.50	$873.00	$443.00	$886.00	$452.50	$905.00	$470.00	$940.00
Cash Value	404.00	1000.00	408.00	1000.00	401.00	1000.00	399.00	1000.00
Net Cost or Gain*	32.50	127.00*	40.00	114.00*	51.50	95.00*	71.00	60.00*
Average Cost or Gain*	3.25	6.35*	4.00	5.70*	5.15	4.75*	7.10	3.00*

*Gain—Cash value exceeds total premiums.

¶323

[¶324]   Basic Annual Premiums Per $1,000 of Term Life Insurance

The benefits under a term policy consist of payment of the sum insured to the beneficiary at the death of the insured, provided death takes place within a stipulated number of years called the term of the policy. At the end of that period the policy becomes null and void. Premium payments are made throughout the term period.

	5 YEAR TERM RENEWABLE TO AGE 65, CONVERTIBLE Minimum $5,000		10 YEAR NON-RENEWABLE TERM Minimum $5,000		TERM TO AGE 65 Minimum $5,000	
	Male Lives	Female Lives	Male Lives	Female Lives	Male Lives	Female Lives
Age	Flat Rate	Flat Rate	Flat Rate	Flat Rate	Flat Rate	Flat Rate
20	$4.35	$4.32	$4.37	$4.34	$7.55	$6.94
21	4.36	4.33	4.38	4.35	7.73	7.10
22	4.37	4.34	4.40	4.36	7.93	7.26
23	4.38	4.35	4.43	4.37	8.13	7.43
24	4.40	4.36	4.46	4.38	8.34	7.61
25	4.43	4.37	4.51	4.40	8.56	7.79
26	4.47	4.38	4.57	4.43	8.79	7.98
27	4.53	4.40	4.66	4.46	9.02	8.17
28	4.61	4.43	4.77	4.51	9.26	8.37
29	4.71	4.47	4.90	4.57	9.52	8.58
30	4.83	4.53	5.05	4.66	9.80	8.81
31	4.99	4.61	5.23	4.77	10.10	9.06
32	5.19	4.71	5.45	4.90	10.42	9.32
33	5.41	4.83	5.69	5.05	10.76	9.60
34	5.65	4.99	5.96	5.23	11.12	9.89
35	5.92	5.19	6.26	5.45	11.49	10.19
36	6.20	5.41	6.57	5.69	11.88	10.51
37	6.50	5.65	6.90	5.96	12.28	10.83
38	6.82	5.92	7.27	6.26	12.71	11.18
39	7.19	6.20	7.68	6.57	13.16	11.54
40	7.60	6.50	8.15	6.90	13.64	11.92
41	8.05	6.82	8.67	7.27	14.14	12.32
42	8.52	7.19	9.23	7.68	14.67	12.73
43	9.05	7.60	9.85	8.15	15.22	13.16
44	9.65	8.05	10.54	8.67	15.82	13.62
45	10.35	8.52	11.31	9.23	16.47	14.12
46	11.16	9.05	12.17	9.85	17.18	14.68
47	12.06	9.65	13.10	10.54	17.96	15.29
48	13.04	10.35	14.10	11.31	18.77	15.92
49	14.08	11.16	15.20	12.17	19.60	16.56
50	15.16	12.06	16.40	13.10	20.42	17.16
51	16.21	13.04	17.66	14.10	21.24	17.74
52	17.24	14.08	18.98	15.20	22.06	18.30
53	18.36	15.16	20.40	16.40	22.90	18.86
54	19.69	16.21	22.00	17.66	23.75	19.40

Basic Annual Premiums Per $1,000 of Term Life Insurance (continued)

Age	5 YEAR TERM RENEWABLE TO AGE 65, CONVERTIBLE Minimum $5,000		10 YEAR NON-RENEWABLE TERM Minimum $5,000		TERM TO AGE 65 Minimum $5,000	
	Male Lives Flat Rate	Female Lives Flat Rate	Male Lives Flat Rate	Female Lives Flat Rate	Male Lives Flat Rate	Female Lives Flat Rate
55	21.35	17.24	23.83	18.98	24.60	19.94
56	23.30	18.36				
57	25.54	19.69				
58	27.95	21.35				
59	30.39	23.30				
60	32.75	25.54				
61	34.18	27.95				
62	35.50	30.39				
63	36.85	32.75				
64	38.27	34.18				

Decreasing Term

Male Age	Female Age	10 YEAR Flat Rate	15 YEAR Flat Rate	20 YEAR Flat Rate	25 YEAR Flat Rate	30 YEAR Flat Rate
	20	$5.37	$5.39	$5.42	$5.45	$5.49
	21	5.38	5.40	5.43	5.46	5.50
	22	5.39	5.41	5.44	5.47	5.51
20	23	5.40	5.42	5.45	5.48	5.53
21	24	5.41	5.43	5.46	5.50	5.58
22	25	5.42	5.44	5.48	5.53	5.64
23	26	5.43	5.46	5.50	5.57	5.72
24	27	5.44	5.48	5.52	5.62	5.82
25	28	5.45	5.50	5.56	5.68	5.93
26	29	5.47	5.53	5.60	5.75	6.06
27	30	5.49	5.55	5.64	5.83	6.20
28	31	5.52	5.58	5.70	5.92	6.36
29	32	5.56	5.63	5.77	6.04	6.54
30	33	5.62	5.70	5.86	6.19	6.76
31	34	5.70	5.79	5.97	6.37	7.01
32	35	5.79	5.90	6.10	6.57	7.28
33	36	5.90	6.03	6.25	6.80	7.58
34	37	6.02	6.17	6.42	7.06	7.92
35	38	6.15	6.34	6.63	7.36	8.30
36	39	6.28	6.51	6.86	7.68	8.71

Male Age	Female Age	10 YEAR Flat Rate	15 YEAR Flat Rate	20 YEAR Flat Rate	25 YEAR Flat Rate	30 YEAR Flat Rate
37	40	$6.41	$6.69	$7.11	$8.02	$9.14
38	41	6.55	6.90	7.40	8.39	9.62
39	42	6.74	7.15	7.73	8.83	10.16
40	43	6.98	7.45	8.13	9.33	10.78
41	44	7.28	7.81	8.59	9.91	
42	45	7.63	8.23	9.11	10.55	
43	46	8.02	8.69	9.68	11.25	
44	47	8.46	9.19	10.30	12.01	
45	48	8.94	9.73	10.96	12.82	
46	49	9.46	10.31	11.65		
47	50	10.02	10.93	12.38		
48	51	10.62	11.59	13.16		
49	52	11.27	12.30	14.00		
50	53	11.97	13.07	14.92		
51	54	12.71	13.88			
52	55	13.49	14.72			
53	56	14.32	15.62			
54	57	15.20	16.61			
55	58	16.16	17.71			
56	59	17.17				
57	60	18.23				
58		19.35				
59		20.57				
60		21.89				

¶324

[¶325]   **Family Income Rider Cost**

The job of covering family security requirements without committing all of one's savings to insurance investments can be greatly facilitated by the use of a family income policy. This type of policy was developed to provide a minimum adequate program of family life insurance protection at minimum cost. It is made up of a base of permanent protection plus term insurance attached as a rider to the permanent policy. The rider is reducing term insurance attached to a basic policy having a fixed face amount. The beneficiary gets a fixed amount of income until a predetermined year, in the event of the insured's death. This is particularly useful to meet the family income needs during a period when the children are growing or to bridge the gap between the husband's death and when his widow becomes eligible for Social Security.

The family income rider is commonly offered at a rate of $10 or $20 per month for each $1,000 of permanent insurance. The following table shows the basic annual premium for this coverage, which would be added to the cost of the permanent insurance.

		10 YEAR	15 YEAR	20 YEAR	25 YEAR	10 YEAR	15 YEAR	20 YEAR	25 YEAR
		\$10 Monthly Income				\$20 Monthly Income			
MALE AGE	FEMALE AGE	FLAT RATE	FLAT RATE	FLAT RATE	FLAT RATE	FLAT RATE	FLAT RATE	FLAT RATE	FLAT RATE
	20	$ 1.81	$ 2.56	$ 3.30	$ 4.07	$ 3.98	$ 5.63	$ 7.25	$ 8.94
	21	1.81	2.57	3.33	4.11	3.98	5.65	7.32	9.03
	22	1.81	2.58	3.36	4.17	3.98	5.67	7.38	9.16
20	23	1.82	2.60	3.39	4.24	4.00	5.71	7.45	9.32
21	24	1.82	2.61	3.42	4.31	4.00	5.74	7.52	9.47
22	25	1.82	2.62	3.45	4.38	4.00	5.76	7.58	9.63
23	26	1.83	2.63	3.48	4.46	4.02	5.78	7.65	9.80
24	27	1.84	2.65	3.54	4.57	4.04	5.82	7.78	10.04
25	28	1.86	2.69	3.62	4.73	4.09	5.91	7.96	10.39
26	29	1.89	2.75	3.73	4.93	4.15	6.04	8.20	10.83
27	30	1.93	2.83	3.87	5.16	4.24	6.22	8.50	11.34
28	31	1.98	2.92	4.02	5.42	4.35	6.42	8.83	11.91
29	32	2.04	3.02	4.21	5.73	4.48	6.64	9.25	12.59
30	33	2.11	3.15	4.43	6.09	4.64	6.92	9.74	13.38
31	34	2.18	3.29	4.68	6.49	4.79	7.23	10.28	14.26
32	35	2.26	3.44	4.96	6.93	4.97	7.56	10.90	15.23
33	36	2.35	3.61	5.27	7.42	5.16	7.93	11.58	16.31
34	37	2.45	3.81	5.62	7.97	5.38	8.37	12.35	17.51
35	38	2.58	4.04	6.01	8.57	5.67	8.88	13.21	18.83
36	39	2.73	4.31	6.45	9.24	6.00	9.47	14.17	20.31
37	40	2.90	4.60	6.93	9.99	6.37	10.11	15.23	21.95
38	41	3.08	4.93	7.44	10.78	6.77	10.83	16.35	23.06
39	42	3.28	5.28	8.00	11.60	7.21	11.60	17.58	25.49
40	43	3.49	5.64	8.59	12.45	7.67	12.39	18.88	27.36
41	44	3.68	5.99	9.18	13.32	8.09	13.16	20.17	29.27
42	45	3.86	6.33	9.76	14.21	8.48	13.91	21.45	31.23
43	46	4.07	6.71	10.40	15.14	8.94	14.75	22.86	33.27
44	47	4.33	7.16	11.16	16.09	9.52	15.73	24.53	35.36
45	48	4.69	7.75	12.08	17.08	10.31	17.03	26.55	37.54
46	49	5.16	8.49	13.17		11.34	18.66	28.94	

¶325

Family Income Rider Cost *(continued)*

		10 YEAR	15 YEAR	20 YEAR	25 YEAR	10 YEAR	15 YEAR	20 YEAR	25 YEAR
		\$10 Monthly Income				\$20 Monthly Income			
MALE AGE	FEMALE AGE	FLAT RATE	FLAT RATE	FLAT RATE	FLAT RATE	FLAT RATE	FLAT RATE	FLAT RATE	FLAT RATE
47	50	5.73	9.34	14.39		12.59	20.53	31.62	
48	51	6.35	10.28	15.74		13.95	22.59	34.59	
49	52	7.01	11.27	17.23		15.41	24.77	37.86	
50	53	7.67	12.29	18.84		16.86	27.01	41.40	
51	54	8.32	13.33			18.28	29.29		
52	55	8.97	14.41			19.71	31.67		
53	56	9.65	15.54			21.21	34.15		
54	57	10.39	16.71			22.83	36.72		
55	58	11.21	17.92			24.64	39.38		
56	59	12.11				26.61			
57	60	13.07				28.72			
58		14.10				30.99			
59		15.19				33.38			
60		16.34				35.91			

[¶326] Family Maintenance Rider

The family maintenance rider differs from the family income rider in that the insurance company will pay \$10 per month per \$1,000 of insurance for a designated period *commencing at the date of death* of the insured for the period specified.

The following table gives typical premium costs for the family maintenance rider.

Age	15 Year Rider Male	20 Year Rider Male	Age	15 Year Rider Male	20 Year Rider Male
16	\$ __	\$ __	35	\$ 7.86	11.68
20	4.25	5.61	36	8.44	12.55
21	4.31	5.70	37	9.07	13.49
22	4.35	5.76	38	9.76	14.48
23	4.39	5.85	39	10.51	15.58
24	4.48	5.97	40	11.32	16.79
25	4.60	6.18	41	12.20	18.08
26	4.76	6.48	42	13.11	19.46
27	4.97	6.86	43	14.10	20.92
28	5.21	7.28	44	15.19	22.53
29	5.48	7.76	45	16.40	24.29
30	5.79	8.29	46	17.72	
31	6.13	8.86	47	19.13	
32	6.49	9.46	48	20.65	
33	6.89	10.14	49	22.30	
34	7.35	10.87	50	24.10	

Dividend Options

Life insurance companies that issue participating policies pay dividends which reflect the difference between the premium charged for a given class of policies and the actual cost based on the company's experience. Under nonparticipating plans, policies are written at a net premium. Dividends are not paid—but the premiums are lower than the gross premiums in comparable participating policies.

The following is a list of how these dividends are paid or used.

(1) Cash: The policyholder can receive his dividend in cash. The most frequent use of this option is for paid-up policies. Another situation where this option is attractive is if insured is disabled and premiums are being waived. In most other cases, the advantages of the other options should be considered.

(2) Reduce Premiums: Insured can apply dividends as part payment of premiums. This dividend is used when insured needs this money to help meet premium obligations, or if a low net expenditure for insurance is desired. It is used on minimum deposit plans when reducing instead of level coverage is wanted.

(3) Accumulating at Interest: This election permits the insurance company to retain dividends on deposit and have it build up at a guaranteed rate of interest. If the company's earnings are less than the rate it guarantees, the policyholder will still be credited with the specified interest rate; should the earnings be greater, he will receive the higher interest.

Dividend accumulations are often used when insured wants to increase the guaranteed retirement income provided under policy's retirement options.

The interest earned on dividend accumulations is taxable. Insurance companies are required to report this interest to the government and send you a notice to the effect that the interest has been reported (on Form 1099). Many insureds choose to apply their dividends to buy dividend additions (see below) and thereby avoid any tax since then the dividends are immediately applied to the purchase of additional insurance and never accumulate interest.

Death proceeds including accumulations are received by the beneficiary income tax free. But accumulated dividends (including interest) are classified for estate tax purposes as property owned by the insured at the time of his death. Whether these amounts are included or excluded from his estate tax would depend upon the same rules used in determining whether general property interests were includible in the gross estate. There would be noticeable shrinkage of a sizeable estate where dividends were left to accumulate.

(4) Paid-up Additions: Dividends are applied to buy additional paid-up insurance. The increased protection acquired via dividend additions requires no medical examination and serves as a valuable tool when health impairment otherwise prevents other additional insurance.

(5) The "Fifth" Dividend Option: This is a relatively new choice which provides for the purchase of one-year term insurance usually equal to the yearly increase in cash value. The balance of the dividend is left on deposit to accumulate for future cash value purchases.

What the option accomplishes is to increase the face value of the policy by the amount of the cash value. The net result here is to eliminate the policy owner from becoming a co-insurer on the policy.

[¶328]

Dividend Checklist

Following is a rundown of some additional pointers and planning considerations about dividends.

(1) Some companies pay a *terminal dividend* which is an additional one-shot dividend upon surrender of policy, death, or both. Often this dividend does not come into effect until a late policy year, such as the tenth year, and may gradually increase each year until, say, the twentieth year.

(2) There often is a policy provision for the payment of dividends for the fraction of the policy year in which the insured dies. This may be the full year's dividend or a proportionate amount.

(3) Fractional dividends are usually not paid upon surrender of the policy.

(4) Companies have an *automatic option* when no dividend option is selected by the insured. This varies among companies—it may be cash, accumulation at interest, or paid-up additions.

(5) The dividend due at end of first policy year is usually contingent upon payment of second year premium. Dividends for later years may not be contingent upon payment.

(6) Paid-up additions are usually not available on term insurance.

(7) Some companies include a provision automatically applying accumulated dividends to any past due premium. Similarly the automatic premium loan provision could apply to cash values of paid-up additions.

(8) The valuable guaranteed annuity options at retirement or death often apply to accumulated dividends and cash values or proceeds of paid-up additions.

(9) Many companies provide a *paid-up option*. When accumulated dividends (or the cash value of additions) plus the cash value of the policy equal the net single premium for a paid-up policy of the same amount at the attained age of the insured, the insured can request that policy be endorsed as fully paid up.

(10) An *endowment option* is offered by many companies. Under one type, when accumulated dividends plus the cash value equal the face amount, the policy can, upon request, be paid as an endowment. Another type lets dividends be used each year to shorten the endowment period by adding them directly to policy reserve.

[¶329]

Group Life Insurance Rates

Monthly Rates per $1,000 Insurance—One Year Term

Attained Age	Monthly Premium	Attained Age	Monthly Premium	Atttained Age	Monthly Premium
15	$ .20	42	$ .65	69	$5.02
16	.20	43	.69	70	5.44
17	.21	44	.74	71	5.89
18	.21	45	.79	72	6.39
19	.22	46	.85	73	6.92
20	.22	47	.91	74	7.50
21	.23	48	.98	75	8.13
22	.24	49	1.05	76	8.80
23	.25	50	1.13	77	9.54
24	.26	51	1.22	78	10.32
25	.27	52	1.31	79	11.17
26	.28	53	1.42	80	12.09
27	.29	54	1.53	81	13.07
28	.30	55	1.65	82	14.13
29	.31	56	1.78	83	15.27
30	.33	57	1.93	84	16.49
31	.34	58	2.08	85	17.80
32	.36	59	2.25	86	19.20
33	.38	60	2.44	87	20.69
34	.40	61	2.64	88	22.28
35	.42	62	2.86	89	23.97
36	.45	63	3.10	90	25.76
37	.47	64	3.36	91	27.66
38	.50	65	3.64	92	29.67
39	.54	66	3.94	93	31.78
40	.57	67	4.27	94	34.01
41	.61	68	4.63	95	36.32

The above premiums are applied against the amounts of insurance at each age to obtain the basic premium. The initial gross premium will be the sum of this basic premium and a policy charge. The policy charge when the volume of insurance exceeds $75,000 is $11.25 monthly per policy. The policy charge when the volume of insurance is less than $75,000 is $.15 monthly per $1,000 of insurance.

Example: A corporate employer has two executives, age 45, who are insured for $2,500 each. The remaining 20 employees are insured for $1,000 each and range in age as follows:

Two are 23; three are 25; five are 30; five are 32; five are 35; three are 38; and two are 40.

Age	No. of Employees	*Premium per $1,000	Monthly Premium	Insurance
23	2	$.40	$ .80	$ 2,000
25	3	.42	1.26	3,000
30	5	.48	2.40	5.000
32	5	.51	2.55	5,000
35	5	.57	2.85	5,000
38	3	.65	1.95	3,000
40	2	.72	1.44	2,000
45	2	.94	4.70	5,000
Total	27		$17.95	$30,000

*Includes 15¢ loading.

¶329

Proceeds at Guaranteed Interest—Payable
in Equal Monthly Installments

The following table shows the percentage of the proceeds of an insurance policy which would be paid to the beneficiary each month at guaranteed rates of interest. To find the amount of the monthly payments, multiply the number you get from the table for the proper interest rate and term by the amount of the proceeds left with the company.

Number of Years Payable	3%	3½%	4%	4½%	5%	5½%	6%
10	.009632	.009860	.010091	.010325	.010563	.010803	.011047
11	.008882	.009112	.009346	.009583	.009824	.010068	.010316
12	.008257	.008490	.008727	.008967	.009211	.009459	.009710
13	.007730	.007965	.008204	.008447	.008695	.008946	.009202
14	.007279	.007516	.007758	.008004	.008255	.008510	.008769
15	.006889	.007128	.007373	.007622	.007875	.008134	.008397
16	.006548	.006790	.007037	.007289	.007546	.007808	.008074
17	.006248	.006493	.006742	.006997	.007257	.007522	.007792
18	.005982	.006229	.006481	.006738	.007002	.007270	.007544
19	.005745	.005994	.006248	.006509	.006775	.007047	.007325
20	.005532	.005783	.006040	.006303	.006573	.006848	.007129
21	.005340	.005593	.005853	.006119	.006391	.006670	.006954
22	.005166	.005421	.005683	.005952	.006227	.006509	.006797
23	.005008	.005265	.005529	.005801	.006079	.006364	.006656
24	.004863	.005123	.005389	.005663	.005945	.006233	.006528
25	.004730	.004992	.005261	.005538	.005822	.006113	.006411
26	.004608	.004872	.005144	.005423	.005710	.006004	.006306
27	.004496	.004762	.005036	.005318	.005607	.005905	.006209
28	.004392	.004660	.004936	.005221	.005513	.005813	.006121
29	.004295	.004565	.004844	.005131	.005426	.005730	.006040
30	.004205	.004478	.004759	.005048	.005346	.005652	.005966
31	.004122	.004396	.004680	.004972	.005272	.005581	.005898
32	.004044	.004320	.004606	.004901	.005204	.005516	.005835
33	.003971	.004250	.004537	.004834	.005140	.005455	.005777
34	.003903	.004183	.004473	.004773	.005081	.005398	.005723
35	.003839	.004121	.004413	.004715	.005026	.005346	.005674
36	.003779	.004063	.004357	.004662	.004975	.005298	.005628
37	.003722	.004008	.004305	.004612	.004927	.005252	.005585
38	.003669	.003957	.004256	.004564	.004883	.005210	.005546
39	.003618	.003909	.004209	.004521	.004841	.005171	.005509
40	.003571	.003863	.004166	.004479	.004802	.005135	.005475
41	.003526	.003820	.004125	.004440	.004766	.005100	.005443
42	.003483	.003779	.004086	.004404	.004732	.005069	.005414
43	.003443	.003741	.004050	.004370	.004700	.005039	.005386
44	.003405	.003704	.004015	.004337	.004670	.005011	.005361
45	.003368	.003670	.003983	.004307	.004641	.004985	.005336
46	.003334	.003637	.003952	.004278	.004615	.004960	.005314
47	.003301	.003606	.003923	.004251	.004590	.004937	.005293
48	.003270	.003577	.003896	.004226	.004566	.004916	.005274
49	.003240	.003549	.003869	.004201	.004544	.004895	.005255
50	.003212	.003522	.003845	.004179	.004523	.004877	.005238

[¶331]

How Long Proceeds Will Last if Paid Out
Monthly Until Principal and Interest Are Exhausted

% of proceeds paid each year	Dollars per Mo. per $1,000 of proceeds	Fund Will Last–When Guaranteed Rate of Interest Is:					
		2%		2½%		3%	
		Yrs.	Mos.	Yrs.	Mos.	Yrs.	Mos.
5.0%	$4.17	25	5	27	6	30	2
5.4	4.50	23	0	24	8	26	9
5.5	4.58	22	6	24	1	26	0
6.0	5.00	20	2	21	5	22	11
6.5	5.42	18	3	19	3	20	5
6.6	5.50	17	11	18	11	20	0
7.0	5.83	16	9	17	7	18	6
7.2	6.00	16	2	16	11	17	10
7.5	6.25	15	5	16	1	16	11
7.8	6.50	14	9	15	4	16	1
8.0	6.67	14	4	14	11	15	7
8.4	7.00	13	6	14	0	14	7
8.5	7.08	13	4	13	10	14	5
9.0	7.50	12	6	12	11	13	5
9.5	7.92	11	9	12	2	12	7
10.0	8.33	11	1	11	5	11	10
10.2	8.50	10	10	11	2	11	6
10.5	8.75	10	6	10	10	11	2
10.8	9.00	10	2	10	6	10	9
11.0	9.17	10	0	10	3	10	6
11.4	9.50	9	7	9	10	10	1
12.0	10.00	9	1	9	3	9	6
12.5	10.42	8	8	8	10	9	1
13.0	10.83	8	4	8	6	8	8
13.5	11.25	8	0	8	2	8	4
14.0	11.67	7	8	7	10	8	0
15.0	12.50	7	1	7	3	7	5
16.0	13.33	6	7	6	9	6	10
17.0	14.17	6	2	6	4	6	5
18.0	15.00	5	10	5	11	6	0
19.0	15.83	5	6	5	7	5	8
20.0	16.67	5	3	5	3	5	4
21.0	17.50	4	11	5	0	5	1
22.0	18.33	4	9	4	9	4	10
23.0	19.17	4	6	4	7	4	7
24.0	20.00	4	4	4	4	4	5
25.0	20.83	4	1	4	2	4	3

¶331

[¶332]

Monthly Life Income Per $1000 of Proceeds
at Various Ages

(2½% Interest)

| Age | | Life Income Only | 5 Years Certain and Life | 10 Years Certain and Life | 15 Years Certain and Life | 20 Years Certain and Life | Installment Refund |
Male	Female						
25	30	$3.08	$3.08	$3.08	$3.07	$3.05	$3.01
30	35	3.27	3.27	3.26	3.24	3.22	3.17
31	36	3.31	3.31	3.30	3.28	3.25	3.20
32	37	3.36	3.36	3.34	3.32	3.29	3.24
33	38	3.41	3.40	3.39	3.36	3.33	3.28
34	39	3.45	3.45	3.43	3.41	3.37	3.32
35	40	3.50	3.50	3.48	3.45	3.41	3.36
36	41	3.56	3.55	3.53	3.50	3.45	3.40
37	42	3.61	3.61	3.59	3.55	3.50	3.44
38	43	3.67	3.66	3.64	3.60	3.54	3.49
39	44	3.73	3.72	3.70	3.65	3.59	3.53
40	45	3.79	3.78	3.76	3.71	3.64	3.58
41	46	3.86	3.85	3.82	3.77	3.69	3.63
42	47	3.93	3.92	3.88	3.82	3.74	3.68
43	48	4.00	3.99	3.95	3.88	3.79	3.74
44	49	4.08	4.06	4.02	3.95	3.84	3.80
45	50	4.15	4.14	4.09	4.01	3.90	3.85
46	51	4.24	4.22	4.17	4.08	3.95	3.91
47	52	4.33	4.31	4.25	4.15	4.01	3.98
48	53	4.42	4.40	4.33	4.22	4.07	4.04
49	54	4.51	4.49	4.42	4.29	4.12	4.11
50	55	4.61	4.59	4.50	4.37	4.18	4.18
51	56	4.72	4.69	4.60	4.44	4.24	4.26
52	57	4.83	4.80	4.69	4.52	4.30	4.33
53	58	4.95	4.91	4.79	4.60	4.36	4.42
54	59	5.07	5.03	4.90	4.69	4.41	4.50
55	60	5.20	5.15	5.01	4.77	4.47	4.59
56	61	5.34	5.28	5.12	4.86	4.53	4.68
57	62	5.48	5.42	5.23	4.94	4.59	4.77
58	63	5.64	5.56	5.35	5.03	4.64	4.87
59	64	5.80	5.72	5.48	5.12	4.70	4.98
60	65	5.97	5.87	5.61	5.21	4.75	5.08
61	66	6.15	6.04	5.74	5.30	4.80	5.20
62	67	6.34	6.22	5.87	5.39	4.85	5.31
63	68	6.54	6.40	6.01	5.48	4.90	5.44
64	69	6.75	6.59	6.16	5.56	4.94	5.57
65	70	6.97	6.79	6.30	5.65	4.98	5.70
70	75	8.32	7.95	7.07	6.05	5.14	6.48

[¶333]

How Much Retirement Capital Does It Take
to Yield $100 per Month?

The following table assumes that, upon retirement, a fixed amount of capital is put at interest. All interest payments are made monthly, and interest is compounded monthly. None of the principal is ever used up, so that, at any time, the full amount can be withdrawn.

Interest Rate	Amount of Capital Needed for $100 per Month
3.0%	$40,000
3.5%	34,285
4.0%	30,000
4.5%	26,667
5.0%	24,000
5.5%	21,818
6.0%	20,000
6.5%	18,462
7.0%	17,143
7.5%	16,000
8.0%	15,000
8.5%	14,118
9.0%	13,333
9.5%	12,632
10.0%	12,000

[¶334]

Net Value of Life Insurance in the Estate

The value of insurance in the estate plan will vary sharply depending on how it is set up. When we provide insurance to meet the estate tax liability, that insurance itself will increase the liability which it is designed to meet, unless we find a way to arrange it so that it will not fall subject to estate tax and yet be available and sufficiently flexible so that it can be fed back into the estate to meet estate liabilities.

The following table will show you how much insurance is needed to yield a given net estate if the insurance itself is included in the estate and if it is made free of estate tax.

Net Estate After Any Marital De-deduction but before $60,000 Exemption	Insurance Needed to Pay Federal Estate Taxes	
	If Insurance Included in Estate	If Insurance Not Included in Estate
$ 100,000	$ 6,154	$ 4,800
200,000	46,714	32,700
300,000	91,912	62,700
400,000	138,971	94,500
500,000	191,846	126,500
750,000	335,737	212,200
1,000,000	505,173	303,500

Here is a table which shows you how a change in the ownership of an insurance policy can increase every net $1,000 of insurance value you carry for your family.

Size of Taxable Estate	Net Insurance Value	
	Before Change	After Change
$ 100,000	$1,000	$1,300
200,000	1,000	1,400
500,000	1,000	1,500
600,000	1,000	1,550
1,000,000	1,000	1,600
2,000,000	1,000	1,900
5,000,000	1,000	2,700

These calculations have been rounded off with a typical state tax and you will have to check your own state law to make precise calculations with your own figures.

[¶335]

Single Premium Deferred Annuity

The following table shows how much of an annuity you get for a single premium paid at an age earlier than the age when the annuity is to begin.

Monthly Life Income Per $1,000 of Single Premium

Age When Annuity is Elected

Age at Issue		M 55 or F 60		M 60 or F 65		M 65 or F 70		MALE 70	
			120		120		120		120
		Cash Refund	Mos. Certain	Cash Refund	Mos. Certain	Cash Refund	Mos. Certain	Cash Refund	Mos. Certain
Male	Female								
30	35	$8.10	$8.96	$10.06	$11.33	$12.62	$14.39	$16.04	$18.32
31	36	7.91	8.75	9.81	11.05	12.32	14.05	15.65	17.88
32	37	7.71	8.53	9.57	10.78	12.02	13.70	15.27	17.44
33	38	7.52	8.32	9.34	10.52	11.72	13.37	14.90	17.02
34	39	7.34	8.12	9.11	10.26	11.44	13.04	14.53	16.60
35	40	7.16	7.92	8.89	10.01	11.16	12.72	14.18	16.19
36	41	6.98	7.73	8.67	9.77	10.89	12.42	13.84	15.80
37	42	6.81	7.54	8.46	9.53	10.62	12.11	13.50	15.41
38	43	6.65	7.36	8.26	9.30	10.36	11.82	13.17	15.04
39	44	6.49	7.18	8.05	9.07	10.11	11.53	12.85	14.67
40	45	6.33	7.00	7.86	8.85	9.86	11.25	12.53	14.31
41	46	6.16	6.82	7.66	8.63	9.62	10.98	12.23	13.97
42	47	5.99	6.63	7.48	8.42	9.39	10.70	11.93	13.62
43	48	5.83	6.45	7.30	8.22	9.16	10.45	11.64	13.29
44	49	5.68	6.28	7.12	8.01	8.93	10.19	11.36	12.97
45	50	5.52	6.11	6.94	7.82	8.72	9.94	11.08	12.65
46	51	5.36	5.93	6.76	7.61	8.50	9.70	10.81	12.35
47	52	5.21	5.75	6.57	7.40	8.30	9.46	10.54	12.04
48	53	5.06	5.59	6.40	7.21	8.09	9.23	10.29	11.75
49	54	4.91	5.43	6.23	7.01	7.89	9.00	10.04	11.46
50	55	4.77	5.27	6.06	6.83	7.71	8.79	9.79	11.18
51	56	4.62	5.12	5.89	6.63	7.50	8.55	9.55	10.91
52	57	4.48	4.96	5.71	6.44	7.29	8.32	9.32	10.64
53	58	4.35	4.81	5.55	6.25	7.10	8.10	9.09	10.39
54	59	4.22	4.67	5.39	6.07	6.91	7.88	8.87	10.13
55	60			5.23	5.89	6.72	7.67	8.66	9.88
56	61			5.07	5.71	6.53	7.45	8.43	9.62
57	62			4.92	5.54	6.34	7.23	8.19	9.36
58	63			4.77	5.37	6.16	7.02	7.98	9.11
59	64			4.63	5.21	5.98	6.82	7.76	8.86
60	65					5.80	6.62	7.55	8.63
61						5.63	6.42	7.34	8.38
62						5.45	6.22	7.12	8.13
63						5.29	6.04	6.92	7.90
64						5.13	5.85	6.71	7.67
65								6.52	7.44

Annual Premium Deferred Annuity

[¶336]

The following table shows how much of an annuity you get beginning at a deferred date for annual premiums beginning at the ages indicated.

Monthly Life Income per $100 of Annual Premium

Age When Annuity is Elected

Age At Issue	MALE 55		MALE 60		MALE 65		MALE 70		FEMALE 55		FEMALE 60		FEMALE 65		FEMALE 70	
	Cash Re-fund	120 Mos. Certain	Cash Re-fund	120 Mos. Certain	Cash Re-fund	120 Mos. Certain	Cash Re-fund	120 Mos. Certain	Cash Re-fund	120 Mos. Certain	Cash Re-fund	120 Mos. Certain	Cash Re-fund	120 Mos. Certain	Cash Re-fund	120 Mos. Certain
40	7.31	8.09	11.52	12.97	17.18	19.59	24.88	28.41	6.72	7.31	10.50	11.62	15.48	17.44	22.14	25.26
41	6.75	7.46	10.83	12.19	16.31	18.60	23.79	27.16	6.20	6.75	9.87	10.92	14.70	16.56	21.18	24.15
42	6.19	6.85	10.15	11.43	15.46	17.63	22.72	25.95	5.69	6.19	9.25	10.23	13.94	15.70	20.23	23.07
43	5.65	6.25	9.48	10.67	14.64	16.69	21.67	24.75	5.19	5.65	8.64	9.56	13.19	14.86	19.29	22.00
44	5.11	5.66	8.82	9.94	13.82	15.76	20.65	23.58	4.70	5.11	8.04	8.89	12.45	14.03	18.38	20.96
45	4.59	5.08	8.18	9.21	13.02	14.85	19.64	22.43	4.22	4.59	7.46	8.25	11.73	13.21	17.49	19.94
46	4.06	4.49	7.52	8.47	12.19	13.90	18.59	21.23	3.73	4.06	6.85	7.58	10.99	12.37	16.55	18.87
47	3.54	3.91	6.87	7.74	11.38	12.98	17.56	20.05	3.25	3.54	6.26	6.93	10.26	11.56	15.63	17.83
48	3.03	3.35	6.25	7.04	10.59	12.08	16.56	18.91	2.78	3.03	5.69	6.30	9.55	10.75	14.74	16.81
49	2.54	2.81	5.63	6.35	9.82	11.20	15.58	17.79	2.33	2.54	5.13	5.68	8.85	9.97	13.87	15.82
50	2.06	2.28	5.03	5.67	9.08	10.35	14.62	16.70	1.89	2.06	4.59	5.08	8.18	9.21	13.02	14.85
51	1.61	1.78	4.45	5.01	8.34	9.51	13.69	15.64	1.48	1.61	4.06	4.49	7.52	8.47	12.19	13.90
52	1.16	1.29	3.88	4.37	7.63	8.70	12.79	14.60	1.07	1.16	3.54	3.91	6.87	7.74	11.38	12.98
53	.74	.81	3.32	3.75	6.93	7.90	11.90	13.59	.68	.74	3.03	3.35	6.25	7.04	10.59	12.08
54	.32	.35	2.78	3.14	6.25	7.13	11.04	12.60	.29	.32	2.54	2.81	5.63	6.35	9.82	11.20
55			2.26	2.54	5.59	6.37	10.20	11.64			2.06	2.28	5.03	5.67	9.08	10.35
56			1.77	1.99	4.94	5.64	9.38	10.71			1.61	1.78	4.46	5.02	8.35	9.52
57			1.29	1.46	4.32	4.92	8.58	9.80			1.18	1.30	3.89	4.38	7.64	8.72
58			.83	.94	3.71	4.23	7.81	8.92			.76	.84	3.34	3.77	6.95	7.93
59			.38	.43	3.11	3.55	7.05	8.05			.35	.38	2.81	3.16	6.28	7.16
60					2.53	2.89	6.31	7.21					2.28	2.57	5.62	6.41
61					2.01	2.30	5.66	6.46					1.81	2.04	5.04	5.75
62					1.49	1.70	4.97	5.67					1.34	1.51	4.42	5.04
63					.97	1.11	4.31	4.92					.87	.98	3.84	4.38
64					.44	.51	3.62	4.14					.40	.45	3.23	3.68
65							2.96	3.38							2.63	3.00

[¶337]

Single Premium Straight Life Immediate Annuity

The following table gives the single premium cost of an immediate straight life annuity. In addition to acting as a guide for such premium costs, it is valuable in computing gain or loss on an exchange of property for a private annuity. It is also useful in computing the amount of a deductible contribution where a gift is made to a charity in exchange for a life annuity.

	Male Lives				Female Lives			
Age Last Birth-day	Price of Annuity of		Annuity Purchased By $1,000		Price of Annuity of		Annuity Purchased By $1,000	
	$100 Ann.	$10 Monthly	Ann. Pay-ment	Mo. Pay-ment	$100 Ann.	$10 Monthly	Ann. Pay-ment	Mo. Pay-ment
35	$2438.00	$2983.80	$41.02	$3.35	$2616.60	$3198.12	$38.22	$3.13
36	2405.30	2944.56	41.57	3.40	2588.40	3164.28	38.63	3.16
37	2371.70	2904.24	42.16	3.44	2559.50	3129.60	39.07	3.20
38	2337.40	2863.08	42.78	3.49	2529.90	3094.08	39.53	3.23
39	2302.30 ·	2820.96	43.43	3.54	2499.50	3057.60	40.01	3.27
40	2266.40	2777.88	44.12	3.60	2468.40	3020.28	40.51	3.31
41	2229.60	2733.72	44.85	3.66	2436.60	2982.12	41.04	3.35
42	2191.90	2688.48	45.62	3.72	2404.00	2943.00	41.60	3.40
43	2153.50	2642.40	46.44	3.78	2370.60	2902.92	42.18	3.44
44	2114.50	2595.60	47.29	3.85	2336.60	2862.12	42.80	3.49
45	2075.20	2548.44	48.19	3.92	2301.70	2820.24	43.45	3.55
46	2035.50	2500.80	49.13	4.00	2266.10	2777.52	44.13	3.60
47	1995.10	2452.32	50.12	4.08	2229.70	2733.84	44.85	3.66
48	1954.40	2403.48	51.17	4.16	2192.60	2689.32	45.61	3.72
49	1913.40	2354.28	52.26	4.25	2154.70	2643.84	46.41	3.78
50	1872.40	2305.08	53.41	4.34	2116.10	2597.52	47.26	3.85
51	1831.30	2255.76	54.61	4.43	2076.70	2550.24	48.15	3.92
52	1789.90	2206.08	55.87	4.53	2036.60	2502.12	49.10	4.00
53	1748.40	2156.28	57.20	4.64	1995.70	2453.04	50.11	4.08
54	1706.80	2106.36	58.59	4.75	1954.10	2403.12	51.17	4.16
55	1665.00	2056.20	60.06	4.86	1911.80	2352.36	52.31	4.25
56	1622.90	2005.68	61.62	4.99	1868.60	2300.52	53.52	4.35
57	1580.50	1954.80	63.27	5.12	1824.40	2247.48	54.81	4.45
58	1538.10	1903.92	65.02	5.25	1779.60	2193.72	56.19	4.56
59	1495.90	1853.28	66.85	5.40	1734.40	2139.48	57.66	4.67
60	1454.30	1803.36	68.76	5.55	1689.30	2085.36	59.20	4.80
61	1413.30	1754.16	70.76	5.70	1644.10	2031.12	60.82	4.92
62	1372.80	1705.56 ·	72.84	5.86	1598.50	1976.40	62.56	5.06
63	1332.60	1657.32	75.04	6.03	1552.80	1921.56	64.40	5.20
64	1292.70	1609.44	77.36	6.21	1507.00	1866.60	66.36	5.36
65	1253.00	1561.80	79.81	6.40	1461.40	1811.88	68.43	5.52
66	1213.70	1514.64	82.39	6.60	1415.90	1757.28	70.63	5.69
67	1174.80	1467.96	85.12	6.81	1370.60	1702.92	72.96	5.87
68	1136.10	1421.52	88.02	7.03	1325.20	1648.44	75.46	6.07
69	1097.40	1375.08	91.12	7.27	1279.70	1593.84	78.14	6.27

Single Premium Straight Life Immediate Annuity *(continued)*

	Male Lives				Female Lives			
	Price of Annuity of		*Annuity Purchased By $1,000*		*Price of Annuity of*		*Annuity Purchased By $1,000*	
Age Last Birth- day	*$100 Ann.*	*$10 Monthly*	*Ann. Pay- ment*	*Mo. Pay- ment*	*$100 Ann.*	*$10 Monthly*	*Ann. Pay- ment*	*Mo. Pay- ment*
70	1058.50	1328.40	94.47	7.53	1234.20	1539.24	81.02	6.50
71	1018.50	1280.40	98.18	7.81	1187.80	1483.56	84.19	6.74
72	977.70	1231.44	102.28	8.12	1140.70	1427.04	87.67	7.01
73	937.10	1182.72	106.71	8.46	1093.80	1370.76	91.42	7.30
74	898.10	1135.92	111.35	8.80	1048.30	1316.16	95.39	7.60
75	861.80	1092.36	116.04	9.15	1005.20	1264.44	99.48	7.91
76	828.80	1052.76	120.66	9.50	964.90	1216.08	103.64	8.22
77	798.40	1016.28	125.25	9.84	926.70	1170.24	107.91	8.55
78	769.60	981.72	129.94	10.19	890.00	1126.20	112.36	8.88
79	741.50	948.00	134.86	10.55	854.40	1083.48	117.04	9.23
80	713.30	914.16	140.19	10.94	819.40	1041.48	122.04	9.60
81	685.50	880.80	145.88	11.35	785.30	1000.56	127.34	9.99
82	658.70	848.64	151.81	11.78	752.40	961.08	132.91	10.40
83	632.10	816.72	158.20	12.24	720.20	922.44	138.85	10.84
84	604.70	783.84	165.37	12.76	688.20	884.04	145.31	11.31
85	575.60	748.92	173.73	13.35	655.90	845.28	152.46	11.83

¶337

[¶338]

How Much Annuity Income Is Taxed?

Only part of each annuity payment is taxable. You determine the tax-free portion as follows: You take your annual payments and multiply them by your life expectancy. This will give you the expected return on the contract. You then take your total investment and divide it by your expected return to arrive at your annual exclusion ratio.

The table which follows does all this for you. Just locate the age and type of policy and the percentage given is the tax-free portion of the annuity.

Note, however, that the figures given in this table are approximate and should be used only as an indication of how much tax-free income a policy is likely to produce. This table should not be used to determine the actual tax-free portion. (See government tables following this table.)

[¶339]

Immediate Single Life Annuities
($100 of monthly income)

Approximate Exclusion Ratio

Age	Male			Female		
	No. Ref.	10 Yr. Cert.	Instal. Ref.	No. Ref.	10 Yr. Cert.	Instal. Ref.
50	66.3%	65.4%	61.9%	62.6%	61.7%	58.4%
51	67.1	65.9	62.8	63.5	62.6	59.3
52	67.7	66.5	62.9	64.2	63.3	60.0
53	68.6	67.0	63.2	64.8	63.6	60.1
54	69.5	68.0	63.5	65.5	64.3	60.3
55	70.1	68.3	64.4	66.2	65.0	61.1
56	70.8	69.1	64.5	67.0	65.4	61.9
57	71.5	69.4	64.7	67.4	65.9	61.9
58	72.2	70.3	66.4	68.2	65.9	62.1
59	73.0	70.7	66.7	69.0	66.3	63.0
60	73.8	71.1	66.2	69.5	67.2	63.0
61	74.7	71.6	68.2	70.0	67.8	63.0
62	75.2	72.3	68.2	70.5	68.0	64.5
63	76.1	72.8	67.8	71.0	68.7	64.6
64	76.6	73.0	69.6	71.5	68.7	64.7
65	77.2	73.3	69.7	72.0	68.9	65.6
66	77.8	73.5	69.1	72.6	69.2	65.7
67	78.4	73.9	71.1	72.7	69.6	65.5
68	79.0	74.3	70.5	73.3	69.8	66.6
69	79.8	74.0	70.9	73.4	69.8	66.5
70	80.0	74.1	72.8	73.6	69.8	67.3
71	80.3	74.3	72.9	73.9	69.9	67.3
72	81.4	74.5	72.8	74.2	70.1	67.5
73	81.9	75.0	72.1	74.6	70.4	68.5
74	81.7	73.9	70.8	75.1	70.1	68.8
75	82.4	73.8	71.2	75.0	70.1	68.7

[¶340]
Expected Return Per $1 Annual Payment for Single Life Annuity
(Government Table I)

Example of use of this table:

Find exempt portion of annuity of $100 per month for single male annuitant. Annuitant is 65. Contract cost $14,000.

$$\frac{\text{Cost of contract (\$14,000)}}{\text{Annual payments (\$1200)} \times \text{multiple from table (15)}} = \frac{14}{18} \times \$100 = \$77.78$$

Ages		Expected Return Per $1 Annual Payment	Ages		Expected Return Per $1 Annual Payment	Ages		Expected Return Per $1 Annual Payment
Male	Female		Male	Female		Male	Female	
16	21	55.8	41	46	33.0	66	71	14.4
17	22	54.9	42	47	32.1	67	72	13.8
18	23	53.9	43	48	31.2	68	73	13.2
19	24	53.0	44	49	30.4	69	74	12.6
20	25	52.1	45	50	29.6	70	75	12.1
21	26	51.1	46	51	28.7	71	76	11.6
22	27	50.2	47	52	27.9	72	77	11.0
23	28	49.3	48	53	27.1	73	78	10.5
24	29	48.3	49	54	26.3	74	79	10.1
25	30	47.4	50	55	25.5	75	80	9.6
26	31	46.5	51	56	24.7	76	81	9.1
27	32	45.6	52	57	24.0	77	82	8.7
28	33	44.6	53	58	23.2	78	83	8.3
29	34	43.7	54	59	22.4	79	84	7.8
30	35	42.8	55	60	21.7	80	85	7.5
31	36	41.9	56	61	21.0	81	86	7.1
32	37	41.0	57	62	20.3	82	87	6.7
33	38	40.0	58	63	19.6	83	88	6.3
34	39	39.1	59	64	18.9	84	89	6.0
35	40	38.2	60	65	18.2	85	90	5.7
36	41	37.3	61	66	17.5	86	91	5.4
37	42	36.5	62	67	16.9	87	92	5.1
38	43	35.6	63	68	16.2	88	93	4.8
39	44	34.7	64	69	15.6	89	94	4.5
40	45	33.8	65	70	15.0	90	95	4.2

If annuity payments are other than monthly or if first annuity payment is earlier than regular period for payment thereafter, the figures in the table must be adjusted. Add or subtract as follows:

If the number of whole months from the annuity starting date to the first payment date is	0-1	2	3	4	5	6	7	8	9	10	11	12
And payments under the contract are to be made:												
Annually	+.5	+.4	+.3	+.2	+.1	0	0	−.1	−.2	−.3	−.4	−.5
Semiannually	+.2	+.1	0	0	−.1	−.2						
Quarterly	+.1	0	−.1									

¶340

[¶341]

Expected Return on Joint and Survivor Annuity—
Wife Younger—Uniform Payment
(Government Table II)

Where wife is same age or older than husband use table on next page.

Example of use of this table:

Find exempt portion of annuity of $100 per month for a married couple. Husband is 67; wife is 62. Contract cost $21,000.

$$\frac{\text{Cost of contract (\$21,000)}}{\text{Annual payment (\$1200)} \times \text{multiple from table (23)}} = \frac{210}{276} \times 100 = \$76.09$$

Note: See p. 276 for use of Table IIA where annuity pays lesser amount to specified survivor. For adjustment for early or other than monthly payment, see Table I, p. 273.

Age of Husband	1 yr.	2 yrs.	3 yrs.	4 yrs.	5 yrs.	6 yrs.	7 yrs.	8 yrs.
					Wife Younger by			
45	39.9	40.5	41.1	41.7	42.3	—	—	—
46	38.9	39.5	40.1	40.7	41.4	42.0	—	—
47	38.0	38.6	39.2	39.8	40.4	41.1	41.8	—
48	37.1	37.7	38.3	38.9	39.5	40.2	40.8	41.5
49	36.2	36.8	37.3	38.0	38.6	39.2	39.9	40.6
50	35.3	35.8	36.4	37.0	37.7	38.3	39.0	39.6
51	34.4	34.9	35.5	36.1	36.7	37.4	38.0	38.7
52	33.5	34.0	34.6	35.2	35.8	36.5	37.1	37.8
53	32.6	33.1	33.7	34.3	34.9	35.6	36.2	36.9
54	31.7	32.2	32.8	33.4	34.0	34.7	35.3	36.0
55	30.8	31.4	31.9	32.5	33.1	33.8	34.4	35.1
56	29.9	30.5	31.1	31.6	32.2	32.9	33.5	34.2
57	29.1	29.6	30.2	30.8	31.4	32.0	32.6	33.3
58	28.2	28.8	29.3	29.9	30.5	31.1	31.7	32.4
59	27.4	27.9	28.5	29.0	29.6	30.2	30.9	31.5
60	26.5	27.1	27.6	28.2	28.8	29.4	30.0	30.6
61	25.7	26.2	26.8	27.3	27.9	28.5	29.1	29.8
62	24.9	25.4	25.9	26.5	27.1	27.7	28.3	28.9
63	24.1	24.6	25.1	25.7	26.2	26.8	27.4	28.1
64	23.3	23.8	24.3	24.9	25.4	26.0	26.6	27.2
65	22.5	23.0	23.5	24.1	24.6	25.2	25.8	26.4
66	21.7	22.2	22.7	23.3	23.8	24.4	25.0	25.6
67	21.0	21.4	21.9	22.5	23.0	23.6	24.1	24.7
68	20.2	20.7	21.2	21.7	22.2	22.8	23.4	23.9
69	19.5	19.9	20.4	20.9	21.5	22.0	22.6	23.2
70	18.7	19.2	19.7	20.2	20.7	21.2	21.8	22.4
71	18.0	18.5	19.0	19.5	20.0	20.5	21.0	21.6
72	17.3	17.8	18.2	18.7	19.2	19.8	20.3	20.9
73	16.7	17.1	17.5	18.0	18.5	19.0	19.6	20.1
74	16.0	16.4	16.9	17.3	17.8	18.3	18.8	19.4
75	15.3	15.7	16.2	16.6	17.1	17.6	18.1	18.7
76	14.7	15.1	15.5	16.0	16.4	16.9	17.4	18.0
77	14.1	14.5	14.9	15.3	15.8	16.3	16.7	17.3
78	13.5	13.8	14.3	14.7	15.1	15.6	16.1	16.6
79	12.9	13.2	13.6	14.1	14.5	15.0	15.4	15.9
80	12.3	12.7	13.0	13.5	13.9	14.3	14.8	15.3

[¶342]

Expected Return on Joint and Survivor Annuity—
Wife Older—Uniform Payment
(Government Table II)

Example of use of this table:

Find exempt portion of annuity of $100 per month for a married couple, husband is 62 and wife is 67. Contract cost $21,000.

$$\frac{\text{Cost of contract (\$21,000)}}{\text{Annual payment (\$1200)} \times \text{multiple from table (22.2)}} = \frac{210}{266.4} \times 100 = \$78.83$$

Note: See p. 277 where annuity pays lesser amount to specific survivor. For adjustment for early or other than monthly payment see Table I, p. 273.

Age of Husband	Wife Same Age	Wife Older by							
		1 yr.	2 yrs.	3 yrs.	4 yrs.	5 yrs.	6 yrs.	7 yrs.	8 yrs.
45	39.3	38.8	38.2	37.7	37.2	36.8	36.3	35.9	35.5
46	38.4	37.8	37.3	36.8	36.3	35.9	35.4	35.0	34.6
47	37.5	36.9	36.4	35.9	35.4	35.0	34.5	34.1	33.7
48	36.5	36.0	35.5	35.0	34.5	34.0	33.6	33.2	32.8
49	35.6	35.1	34.6	34.1	33.6	33.1	32.7	32.3	31.9
50	34.7	34.2	33.7	33.2	32.7	32.3	31.8	31.4	31.0
51	33.8	33.3	32.8	32.3	31.8	31.4	30.9	30.5	30.1
52	32.9	32.4	31.9	31.4	30.9	30.5	30.1	29.7	29.3
53	32.0	31.5	31.0	30.5	30.1	29.6	29.2	28.8	28.4
54	31.2	30.6	30.1	29.7	29.2	28.8	28.3	27.9	27.6
55	30.3	29.8	29.3	28.8	28.3	27.9	27.5	27.1	26.7
56	29.4	28.9	28.4	27.9	27.5	27.1	26.7	26.3	25.9
57	28.6	28.1	27.6	27.1	26.7	26.2	25.8	25.4	25.1
58	27.7	27.2	26.7	26.3	25.8	25.4	25.0	24.6	24.3
59	26.9	26.4	25.9	25.4	25.0	24.6	24.2	23.8	23.5
60	26.0	25.5	25.1	24.6	24.2	23.8	23.4	23.0	22.7
61	25.2	24.7	24.3	23.8	23.4	23.0	22.6	22.2	21.9
62	24.4	23.9	23.5	23.0	22.6	22.2	21.8	21.5	21.1
63	23.6	23.1	22.7	22.2	21.8	21.4	21.1	20.7	20.4
64	22.8	22.3	21.9	21.5	21.1	20.7	20.3	20.0	19.6
65	22.0	21.6	21.1	20.7	20.3	19.9	19.6	19.2	18.9
66	21.3	20.8	20.4	20.0	19.6	19.2	18.8	18.5	18.2
67	20.5	20.1	19.6	19.2	18.8	18.5	18.1	17.8	17.5
68	19.8	19.3	18.9	18.5	18.1	17.8	17.4	17.1	16.8
69	19.0	18.6	18.2	17.8	17.4	17.1	16.7	16.4	16.1
70	18.3	17.9	17.5	17.1	16.7	16.4	16.1	15.8	15.5
71	17.6	17.2	16.8	16.4	16.1	15.7	15.4	15.1	14.8
72	16.9	16.5	16.1	15.8	15.4	15.1	14.8	14.5	14.2
73	16.2	15.8	15.5	15.1	14.8	14.4	14.1	13.8	13.6
74	15.6	15.2	14.8	14.5	14.1	13.8	13.5	13.2	13.0
75	14.9	14.5	14.2	13.8	13.5	13.2	12.9	12.6	12.4
76	14.3	13.9	13.6	13.2	12.9	12.6	12.3	12.1	11.8
77	13.7	13.3	13.0	12.6	12.3	12.1	11.8	11.5	11.3
78	13.1	12.7	12.4	12.1	11.8	11.5	11.2	11.0	10.7
79	12.5	12.2	11.8	11.5	11.2	11.0	10.7	10.5	10.2
80	11.9	11.6	11.3	11.0	10.7	10.4	10.2	10.0	9.7

[¶343]

Expected Return on Joint and Survivor Annuity—
Wife Younger—Different Amount After First Death
(Government Table IIA)

Where wife is same age or older than husband use table on next page.

Example of use of this table:

Find exclusion ratio of joint and survivor contract of $100 per month so long as both husband and wife live and $50 per month after death of one. Husband is 70; wife is 67. Cost of contract is $13,500.

Multiple from Table II on p. 274 for husband age 70, and for wife at 3 years younger	19.7
Multiple from Table below	9.3
Difference	10.4
Portion of expected return (reduced payment) 10.4 × $600	$ 6,240
Portion of expected return (full payment) 9.3 × $1,200	$11,160
Expected Return	$17,400

Exclusion Ratio   13,500 (cost of contract)
Exclusion Ratio 17,400 (expected return)

Note: See Table I, p. 273 for early or other than monthly payments.

Age of Husband	1 yr.	2 yrs.	3 yrs.	4 yrs.	5 yrs.	6 yrs.	7 yrs.	8 yrs.
				Wife Younger by				
45	24.4	24.7	25.0	25.2	25.5	—	—	—
46	23.6	23.9	24.2	24.4	24.7	24.9	—	—
47	22.9	23.1	23.4	23.7	23.9	24.2	24.4	—
48	22.1	22.4	22.7	22.9	23.2	23.4	23.6	23.8
49	21.4	21.6	21.9	22.2	22.4	22.6	22.9	23.1
50	20.6	20.9	21.2	21.4	21.7	21.9	22.1	22.3
51	19.9	20.2	20.5	20.7	20.9	21.2	21.4	21.6
52	19.2	19.5	19.8	20.0	20.2	20.4	20.7	20.9
53	18.5	18.8	19.1	19.3	19.5	19.7	19.9	20.1
54	17.9	18.1	18.4	18.6	18.8	19.0	19.2	19.4
55	17.2	17.5	17.7	17.9	18.1	18.4	18.6	18.7
56	16.6	16.8	17.0	17.3	17.5	17.7	17.9	18.1
57	15.9	16.2	16.4	16.6	16.8	17.0	17.2	17.4
58	15.3	15.5	15.8	16.0	16.2	16.4	16.6	16.7
59	14.7	14.9	15.1	15.3	15.5	15.7	15.9	16.1
60	14.1	14.3	14.5	14.7	14.9	15.1	15.3	15.5
61	13.5	13.7	13.9	14.1	14.3	14.5	14.7	14.9
62	12.9	13.2	13.4	13.6	13.7	13.9	14.1	14.3
63	12.4	12.6	12.8	13.0	13.2	13.3	13.5	13.7
64	11.8	12.0	12.2	12.4	12.6	12.8	12.9	13.1
65	11.3	11.5	11.7	11.9	12.1	12.2	12.4	12.5
66	10.8	11.0	11.2	11.4	11.5	11.7	11.9	12.0
67	10.3	10.5	10.7	10.9	11.0	11.2	11.3	11.5
68	9.8	10.0	10.2	10.4	10.5	10.7	10.8	11.0
69	9.4	9.6	9.7	9.9	10.0	10.2	10.3	10.5

Expected Return on Joint and Survivor Annuity—Wife Younger—Different Amount After First Death (Government Table IIA) *(continued)*

Age of Husband	Wife Younger by							
	1 yr.	2 yrs.	3 yrs.	4 yrs.	5 yrs.	6 yrs.	7 yrs.	8 yrs.
70	8.9	9.1	9.3	9.4	9.6	9.7	9.8	10.0
71	8.5	8.7	8.8	9.0	9.1	9.3	9.4	9.5
72	8.1	8.2	8.4	8.5	8.7	8.8	8.9	9.1
73	7.7	7.8	8.0	8.1	8.2	8.4	8.5	8.6
74	7.3	7.4	7.6	7.7	7.8	8.0	8.1	8.2
75	6.9	7.0	7.2	7.3	7.4	7.6	7.7	7.8
76	6.5	6.7	6.8	6.9	7.1	7.2	7.3	7.4
77	6.2	6.3	6.4	6.6	6.7	6.8	6.9	7.0
78	5.9	6.0	6.1	6.2	6.3	6.4	6.5	6.6
79	5.5	5.7	5.8	5.9	6.0	6.1	6.2	6.3
80	5.2	5.3	5.5	5.6	5.7	5.8	5.9	6.0

[¶344] Expected Return on Joint and Survivor Annuity—Wife Older—Different Amount After First Death (Government Table IIA)

Example of use of this table:

Find the exclusion of joint and survivor contract which pays $100 per month while both husband and wife are living and $50 per month after the death of one. Husband is 67 and wife is 70. Cost of the contract is $13,500.

Multiple from Table II on p. 275 for husband age 67, and for wife 3 years older	19.2
Multiple from Table below	9.5
Difference	9.7
Expected return for $100 payments 9.5 × $1,200	$11,400
Expected return for $50 payments 9.7 × $600	5,820
Total Expected Return	$17,220

Exclusion

$$\frac{\text{Costs of Contract } (\$13,500)}{\text{Expected Return } (\$17,220)} \times 100 = \$78.33.$$

Note: See Table I, p. 273 for early or other than monthly payments.

Expected Return on Joint and Survivor Annuity—
Wife Older—Different Amount after First Death
(Government Table IIA) *(continued)*

Age of Husband	Wife Same Age	Wife Older by							
		1 yr.	*2 yrs.*	*3 yrs.*	*4 yrs.*	*5 yrs.*	*6 yrs.*	*7 yrs.*	*8 yrs.*
45	24.1	23.8	23.4	23.1	22.7	22.4	22.0	21.6	21.2
46	23.3	23.0	22.7	22.3	22.0	21.6	21.2	20.9	20.5
47	22.6	22.2	21.9	21.6	21.2	20.9	20.5	20.1	19.8
48	21.8	21.5	21.2	20.9	20.5	20.2	19.8	19.4	19.1
49	21.1	20.8	20.5	20.1	19.8	19.5	19.1	18.8	18.4
50	20.4	20.1	19.8	19.4	19.1	18.8	18.4	18.1	17.7
51	19.7	19.4	19.1	18.8	18.4	18.1	17.8	17.4	17.0
52	19.0	18.7	18.4	18.1	17.8	17.4	17.1	16.8	16.4
53	18.3	18.0	17.7	17.4	17.1	16.8	16.4	16.1	15.8
54	17.6	17.3	17.0	16.8	16.4	16.1	15.8	15.5	15.1
55	16.9	16.7	16.4	16.1	15.8	15.5	15.2	14.9	14.5
56	16.3	16.0	15.8	15.5	15.2	14.9	14.6	14.3	13.9
57	15.7	15.4	15.1	14.9	14.6	14.3	14.0	13.7	13.4
58	15.1	14.8	14.5	14.3	14.0	13.7	13.4	13.1	12.8
59	14.4	14.2	13.9	13.7	13.4	13.1	12.8	12.6	12.3
60	13.9	13.6	13.4	13.1	12.8	12.6	12.3	12.0	11.7
61	13.3	13.0	12.8	12.6	12.3	12.0	11.8	11.5	11.2
62	12.7	12.5	12.3	12.0	11.8	11.5	11.2	11.0	10.7
63	12.2	11.9	11.7	11.5	11.2	11.0	10.7	10.5	10.2
64	11.6	11.4	11.2	11.0	10.7	10.5	10.2	10.0	9.7
65	11.1	10.9	10.7	10.5	10.2	10.0	9.8	9.5	9.3
66	10.6	10.4	10.2	10.0	9.8	9.5	9.3	9.1	8.8
67	10.1	9.9	9.7	9.5	9.3	9.1	8.9	8.6	8.4
68	9.7	9.5	9.3	9.1	8.9	8.6	8.4	8.2	8.0
69	9.2	9.0	8.8	8.6	8.4	8.2	8.0	7.8	7.6
70	8.8	8.6	8.4	8.2	8.0	7.8	7.6	7.4	7.2
71	8.3	8.1	8.0	7.8	7.6	7.4	7.2	7.0	6.8
72	7.9	7.7	7.6	7.4	7.2	7.0	6.8	6.6	6.4
73	7.5	7.3	7.2	7.0	6.8	6.7	6.5	6.3	6.1
74	7.1	7.0	6.8	6.6	6.5	6.3	6.1	6.0	5.8
75	6.8	6.6	6.4	6.3	6.1	6.0	5.8	5.6	5.5
76	6.4	6.3	6.1	6.0	5.8	5.6	5.5	5.3	5.2
77	6.1	5.9	5.8	5.6	5.5	5.3	5.2	5.0	4.9
78	5.7	5.6	5.5	5.3	5.2	5.0	4.9	4.7	4.6
79	5.4	5.3	5.2	5.0	4.9	4.7	4.6	4.5	4.3
80	5.1	5.0	4:9	4.7	4.6	4.5	4.3	4.2	4.1

¶344

[¶345]

Cost of Contract With Refund or Payment Certain Feature
(Government Table III)

Example of use of this table:

Find cost of contract of $100 per month to husband, age 65. Purchase price is $21,053 and refund of purchase price guaranteed.

Purchase price...		$21,053
Annual payment...	$ 1,200	
Years guaranteed ($21,053 refund—$1,200 annual payment).........................	17.5	
Rounded to nearest year...	18	
% in table at age 65 for 18 years..	30%	
Value of refund: 30% of	$21,053	6,316
Cost of Contract		$14,737

Note: See Table I for early or other than monthly payment.

| Age | | Years Guaranteed | | | | | | | | | | | |
Male	Female	5	8	10	12	15	18	20	22	25	28	30	35
40	45	1%	2%	3%	3%	4%	6%	7%	8%	9%	11%	13%	17%
41	46	1	2	3	3	5	6	7	8	10	12	14	18
42	47	1	2	3	4	5	6	8	9	11	13	15	19
43	48	1	2	3	4	5	7	8	9	12	14	16	21
44	49	1	3	3	4	6	7	9	10	12	15	17	22
45	50	2	3	4	5	6	8	9	11	13	16	18	23
46	51	2	3	4	5	7	9	10	12	14	17	19	25
47	52	2	3	4	5	7	9	11	12	15	18	20	26
48	53	2	3	5	6	8	10	12	13	16	19	22	28
49	54	2	4	5	6	8	11	12	14	17	21	23	29
50	55	2	4	5	7	9	11	13	15	18	22	24	31
51	56	3	4	6	7	9	12	14	16	20	23	26	32
52	57	3	5	6	8	10	13	15	17	21	25	27	34
53	58	3	5	7	8	11	14	16	19	22	26	29	38
54	59	3	5	7	9	12	15	17	20	24	28	31	38
55	60	3	6	8	9	13	16	18	21	25	29	32	39
56	61	4	6	8	10	13	17	20	22	27	31	34	41
57	62	4	7	9	11	14	18	21	24	28	33	36	43
58	63	4	7	9	12	15	19	22	25	30	34	37	45
59	64	5	8	10	12	16	21	24	27	31	36	39	47
60	65	5	8	11	13	18	22	25	28	33	38	41	48
61	66	5	9	12	14	19	23	27	30	35	40	43	50
62	67	6	10	12	15	20	25	28	32	37	42	45	52
63	68	6	10	13	16	21	26	30	33	39	44	47	54
64	69	7	11	14	17	23	28	32	35	41	46	49	55
65	70	7	12	15	19	24	30	33	37	42	47	50	57
66	71	8	13	16	20	26	31	35	39	44	49	52	59
67	72	8	14	17	21	27	33	37	41	46	51	54	61
68	73	9	14	18	23	29	35	39	43	48	53	56	62
69	74	9	16	20	24	30	37	41	45	50	55	58	64
70	75	10	17	21	26	32	39	43	47	52	57	60	65
71	76	11	18	22	27	34	41	45	49	54	59	61	67
72	77	12	19	24	29	36	43	47	51	56	60	63	68
73	78	12	20	25	30	38	45	49	53	58	62	65	70
74	79	13	22	27	32	40	47	51	55	60	64	66	71
75	80	14	23	29	34	42	49	53	57	62	66	68	72

Medicare Benefits, Coverage and Conditions

Service	Some Items Covered	What Patient Buys
BASIC PLAN (Part A) Inpatient hospital care	Semi-private room Private room (if medically required) Board Hospital nursing services Resident physicians and dentists and interns in training Operating room Wheelchair, crutches, prosthetic devices Drugs and supplies Blood in excess of first 3 pints Physical therapy Laboratory, diagnostic tests	$68 of costs for first 60 days, $17 a day per next 30 days in each spell of illness. In addition, patient has lifetime reserve of 90 days during which he pays $34 a day. After, patient pays all costs.
Post-hospital extended care in approved nursing home or convalescent part of a hospital	Semi-private room Board Facility's nursing services Resident physicians and interns in training and other diagnostic and therapeutic services of affiliated hospital Wheelchair, crutches, prosthetic devices Drugs and supplies Physical, occupational and speech therapy	$8.50 a day after first 20 days and all costs after 100 days in each spell of illness
Post-hospital home health services	Resident physicians and interns of affiliated hospital Part-time nursing Physical, occupational or speech therapy Part-time service of home health aide Medical supplies other than drugs Wheelchair, crutches, prosthetic devices Visits to institution to use equipment that cannot be taken to patient's home	Costs after 100 visits to home in the 365 days following discharge from hospital, nursing home or other post-hospital extended-care institution
SUPPLEMENTARY PLAN (Part B) Medical, surgical and health services at home, office, clinic or institution	Physicians and surgeons including osteopaths, pathologists, radiologists, anesthesiologists Operations on jaw by doctor of dentistry or doctor of oral surgery X-ray Radium and radioactive isotope therapy Surgical dressing Splints, casts, braces	First $50 of costs plus 20% of remainder in any one calendar year

Medicare Benefits, Coverage and Conditions *(continued)*

Service	Some Items Covered	What Patient Buys
	Artificial arms, legs, eyes Iron lungs, oxygen tents, hospital beds Wheelchair and crutches Ambulance, if needed	
Home health services	Same as those covered under basic plan	
Outpatient hospital diagnostic services	X-ray Basal metabolism tests Urinalysis Gastro-Intestinal series Cardiograms Electroencephalograms Pulmonary function tests	
Treatment of mental disorders	Medical treatment Therapy Psychoanalysis	First $50 of cost and 50% of remainder in any one calendar year
Inpatient hospital care	Physicians, surgeons, radiologists, anesthesiologists, pathologists First 3 pints of blood Private nurses	Private room covered if medically necessary, otherwise patient may take private room and pay the difference Government payments for care in psychiatric hospital limited to 190 days during patient's lifetime Cost of blood in excess of 3 pints not covered if arrangements made by patient to replace
Post-hospital extended care in approved nursing home or convalescent part of a hospital	Physicians and surgeons Private nurses Custodial care	Patient must have spent at least 3 days in hospital and be admitted to extended-care institution within 14 days of discharge from hospital. Readmission for same illness without further hospitalization. Patient may take private room and pay difference
Outpatient hospital diagnostic services	Services performed in physicians's office	If tests continue past one period of 20 consecutive days, patient pays another $20 and 20% of cost for tests during next 20-day period $20 payments are credited toward $50 patient pays under supplementary plan
Post-hospital home health services	Physicians and surgeons Private nurses Transportation to hospital or other institution to use special equipment Food service arrangements Housekeepers	Patient must be in home health plan within 14 days after discharge from hospital or extended-care institution

(Left margin label spanning table: BASIC PLAN)

Medicare Benefits, Coverage and Conditions *(continued)*

SUPPLEMENTARY PLAN

Service	Some Items Covered	What Patient Buys
Medical, surgical and health services at home, office, clinic or institution	Routine physical check-ups Eyeglass examinations and eyeglasses Hearing aid examinations and hearing aids Immunizations Orthopedic shoes and arches Ordinary dental work Cosmetic surgery except when medically required Drugs that can be self-administered	Ambulance cost is covered only if ambulance is medically necessary and if trip is made to or from nearest locality with institution having appropriate facilities
Home health services		Government payments limited to 100 visits to home during any one calendar year, patient need not have been hospitalized previously
Treatment for mental disorders	Treatment by psychologists, lay analysts and others without medical degrees	Government payments limited to $250 in any one calendar year